Student Art and Lecture Notebook

for

Principles of

Biochemistry

With a Human Focus

Reginald H. Garrett and Charles M. Grisham

University of Virginia

HARCOURT COLLEGE PUBLISHERS

Fort Worth | Philadelphia | San Diego | New York | Orlando | Austin

San Antonio | Toronto | Montreal | London | Sydney | Tokyo

ISBN 0-03-034596-0

012 202 7654321

Preface

This Student Art and Lecture Notebook accompanies the first edition of *Principles of Biochemistry with a Human Focus* and is designed to increase productivity in the biochemistry lecture. This notebook contains copies of the art most often used by professors in the classroom. Students can effectively take notes right on the art being referenced. This prevents valuable class time from being wasted redrawing figures and avoids the possibility of errors being introduced in sketched reproductions. Each figure has ample space for marking and there is a blank page across from each piece of art for additional notes.

TABLE OF CONTENTS

Figure 1.8 Molecular organization of the cell.

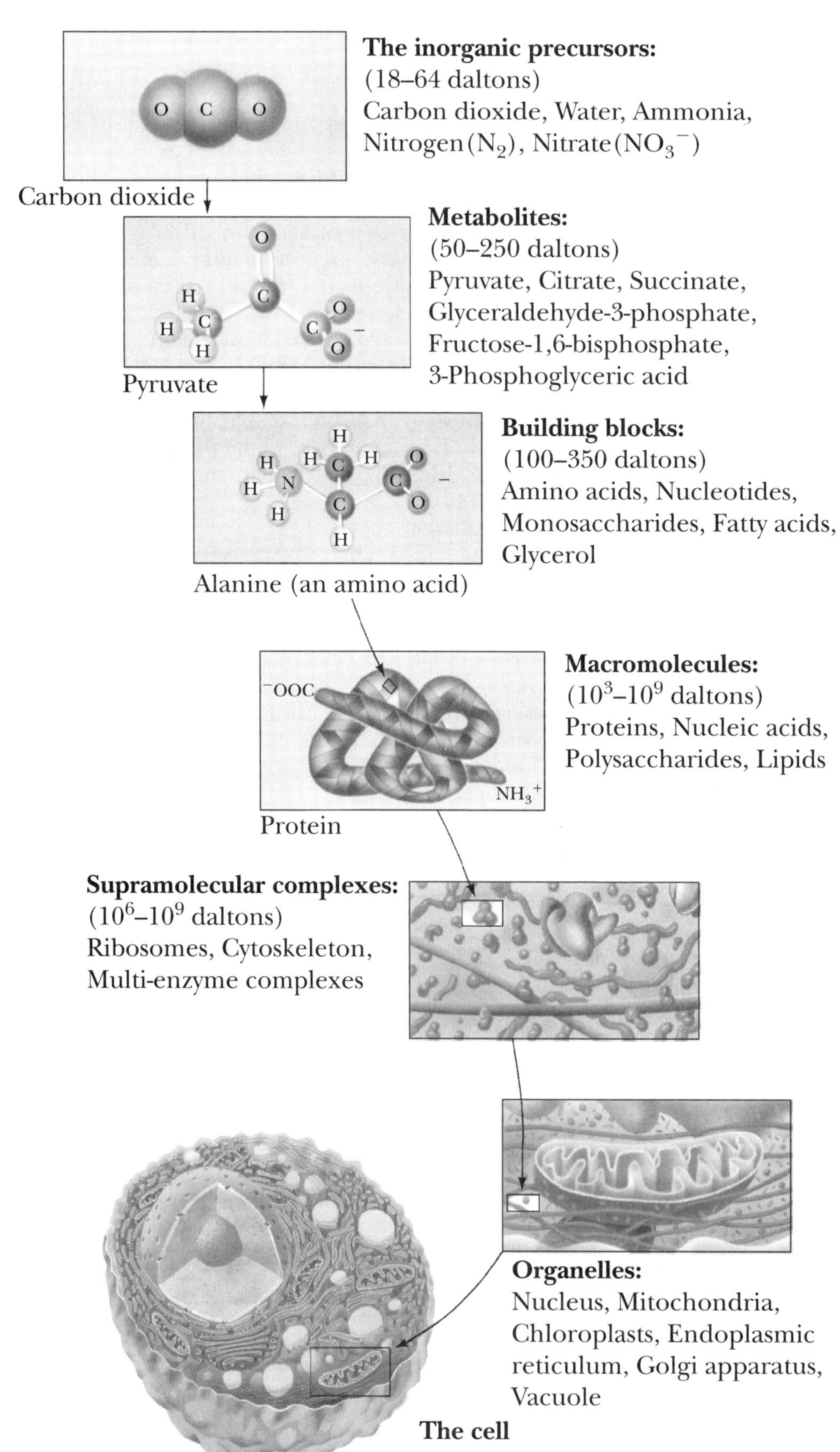

Figure 1.20   A bacterial cell.

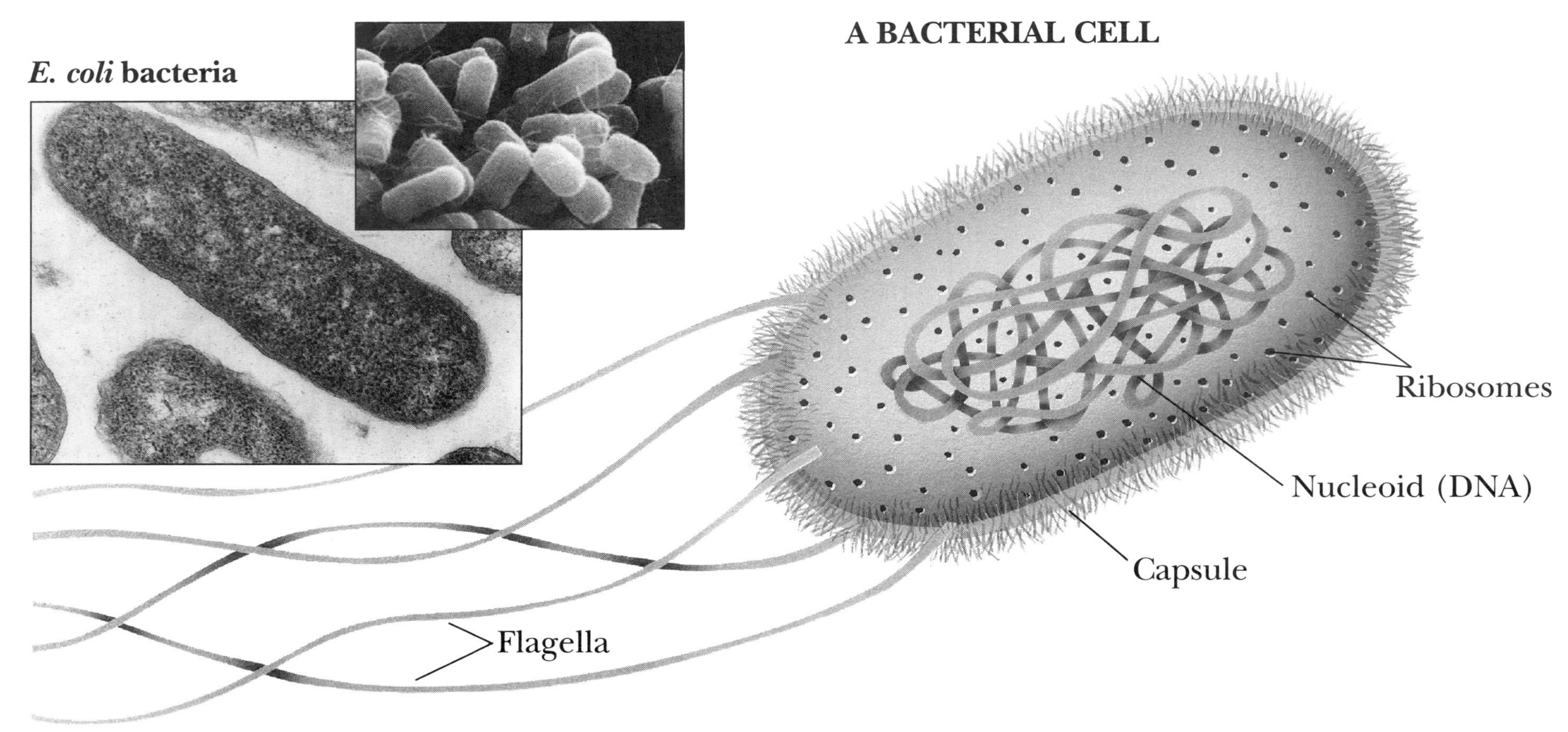

Figure 1.23 Virus life cycle.

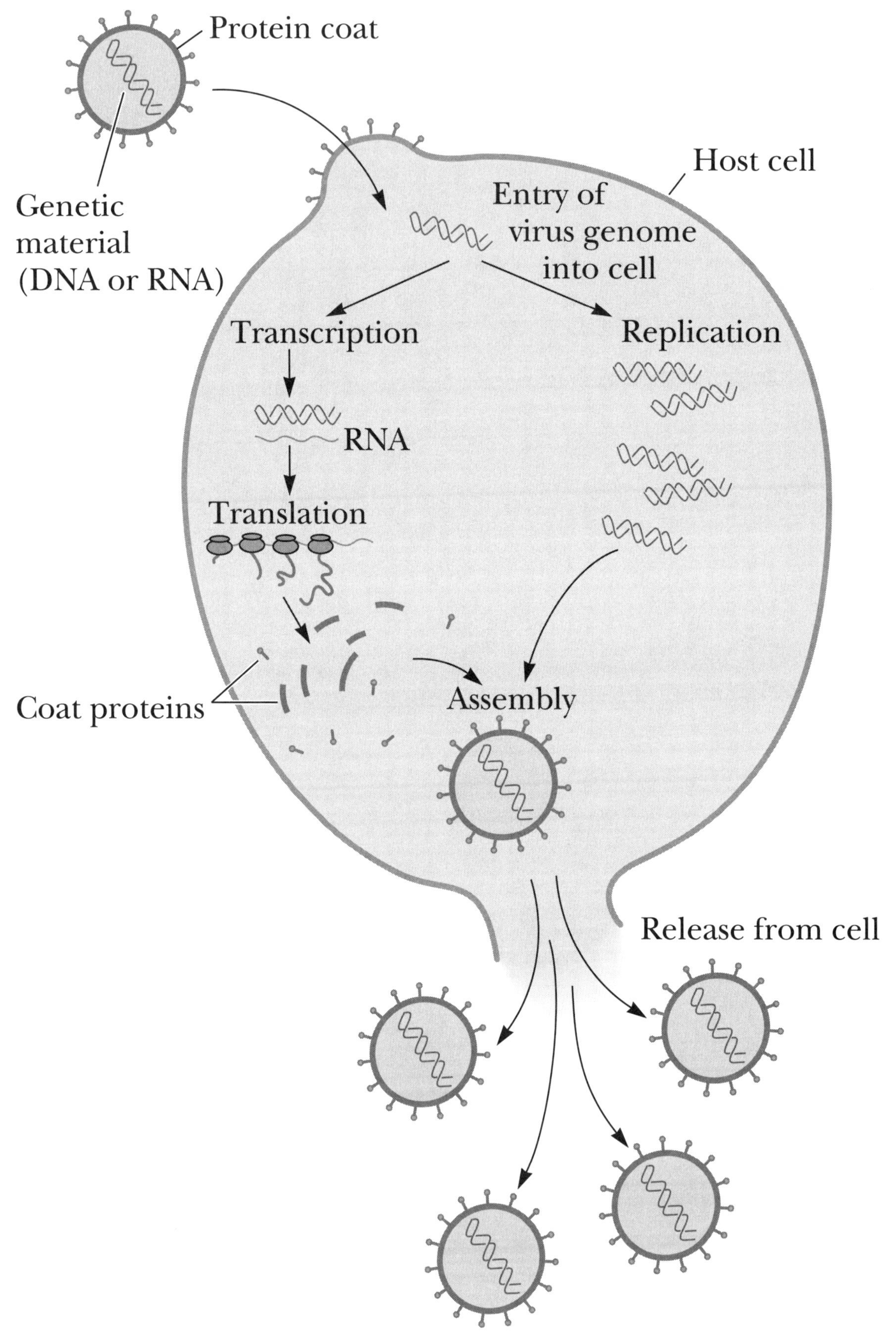

Figure 4.3a   20 Common amino acids.

(a)   Nonpolar (hydrophobic)

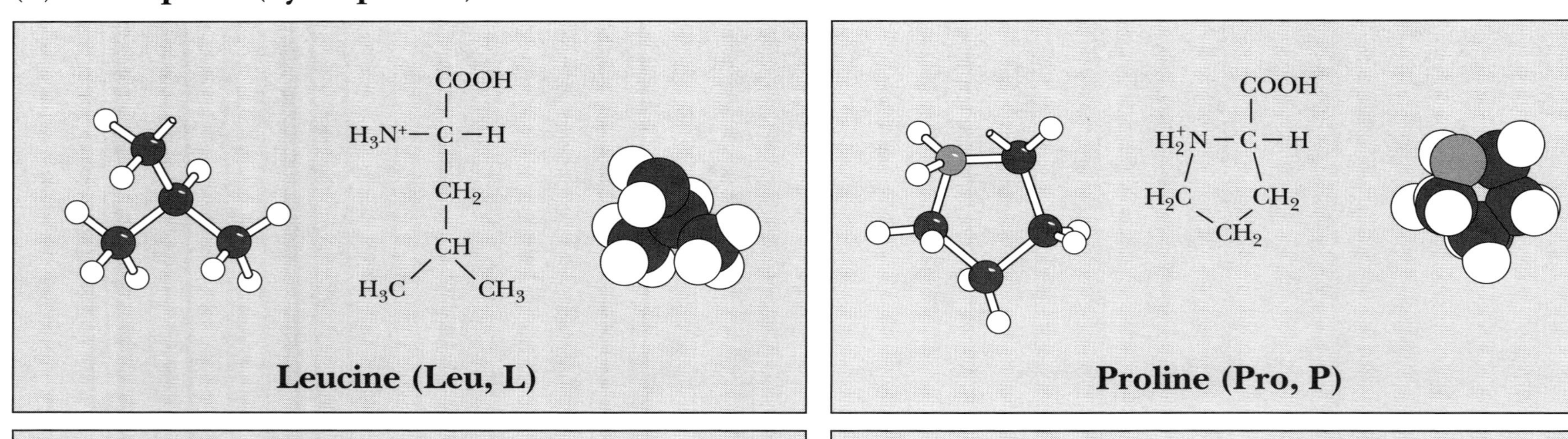

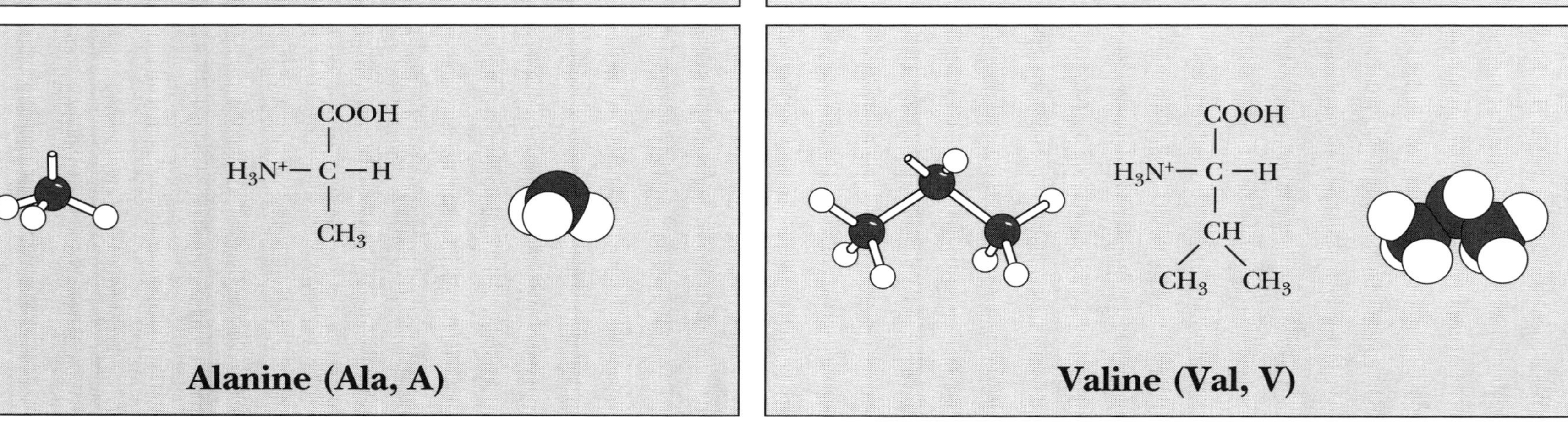

6

© Harcourt, Inc.

Figure 4.3b   20 Common amino acids.

(b)   Polar, uncharged

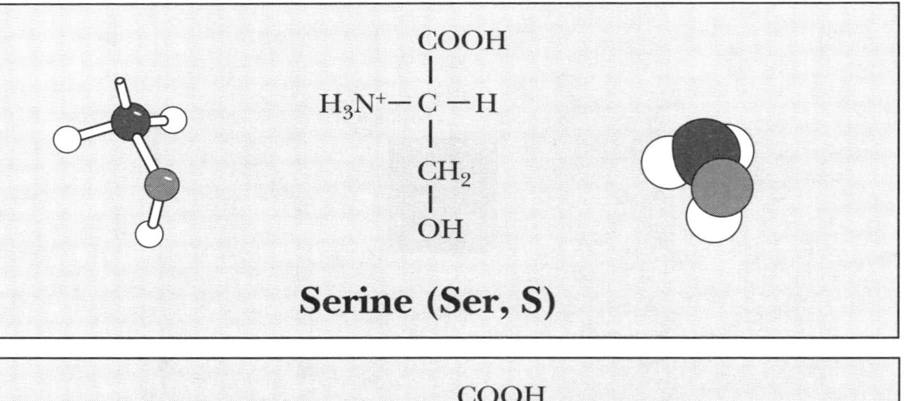

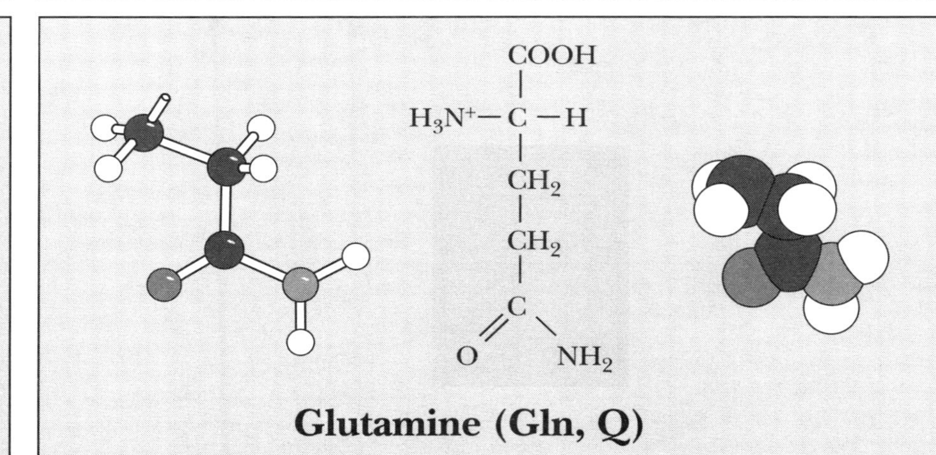

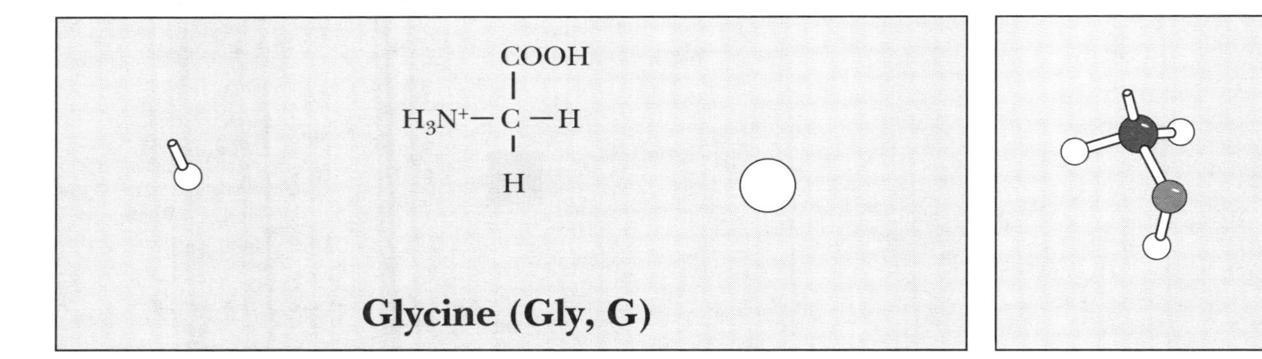

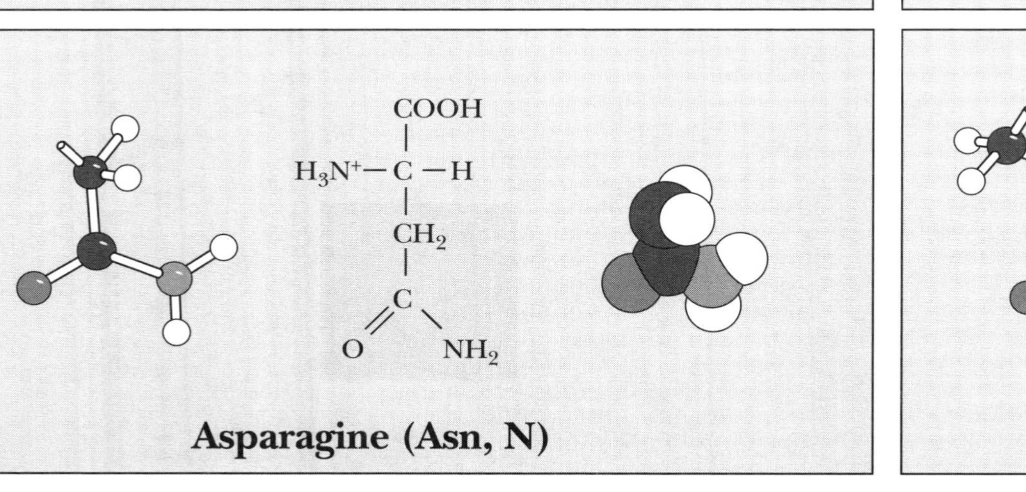

Figure 4.3c   20 Common amino acids.

(c)   Acidic

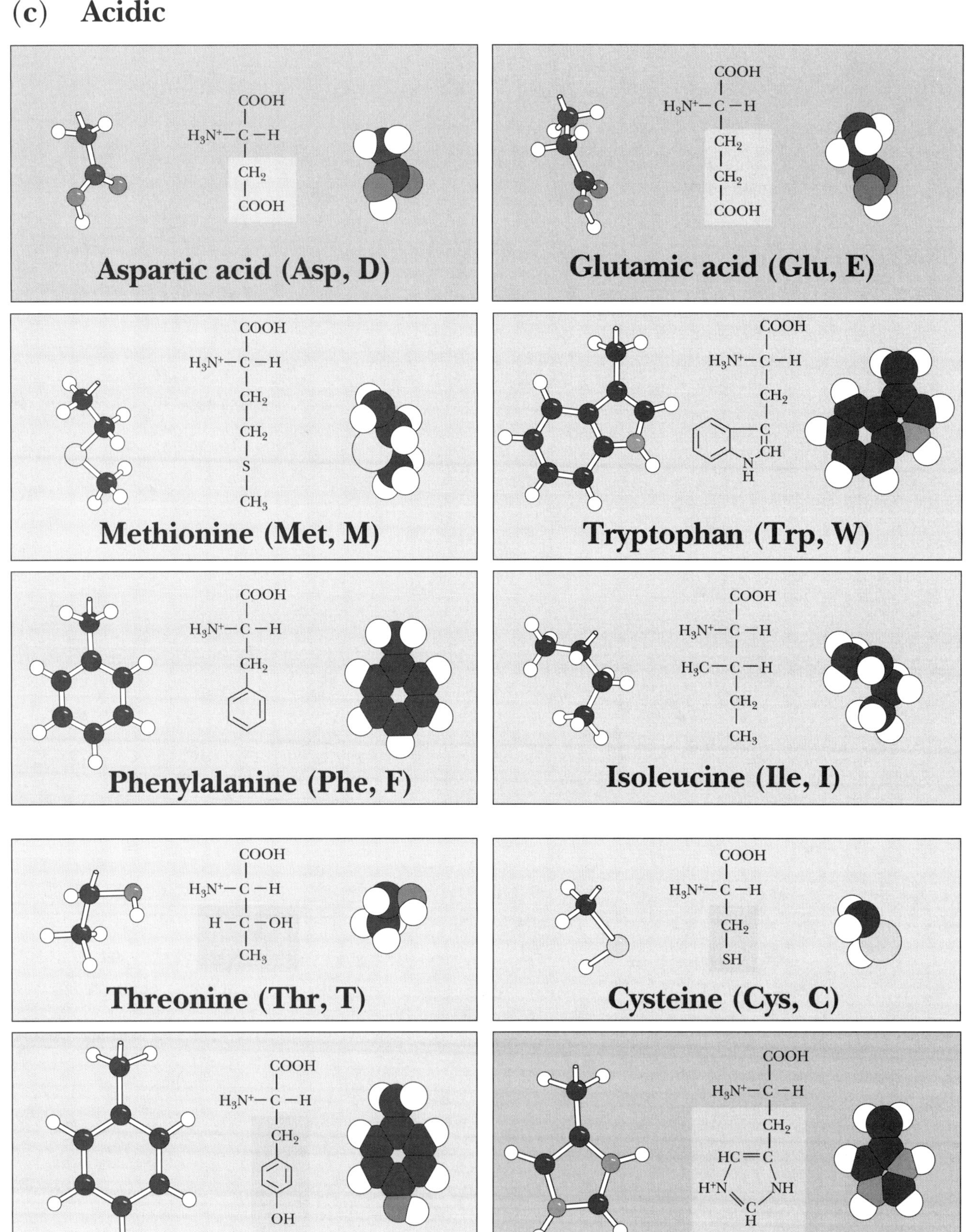

Aspartic acid (Asp, D)
Glutamic acid (Glu, E)
Methionine (Met, M)
Tryptophan (Trp, W)
Phenylalanine (Phe, F)
Isoleucine (Ile, I)
Threonine (Thr, T)
Cysteine (Cys, C)
Tyrosine (Tyr, Y)
Histidine (His, H)

Figure 4.3d   20 Common amino acids.

(d)   **Basic**

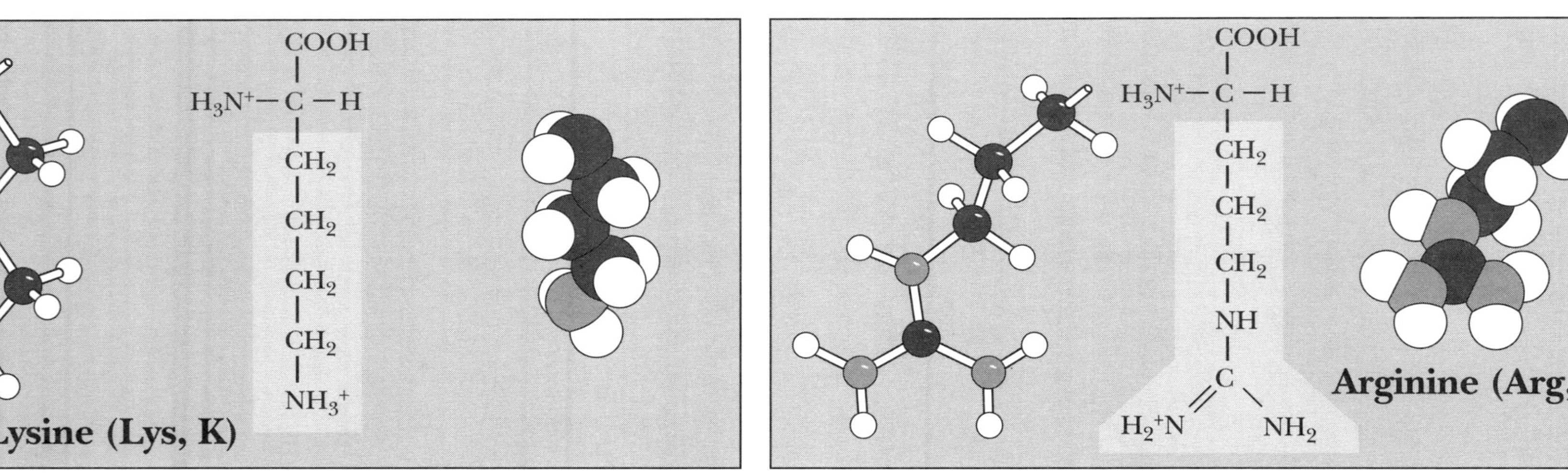

Figure 4.7 Titrations of glutamic acid and lysine.

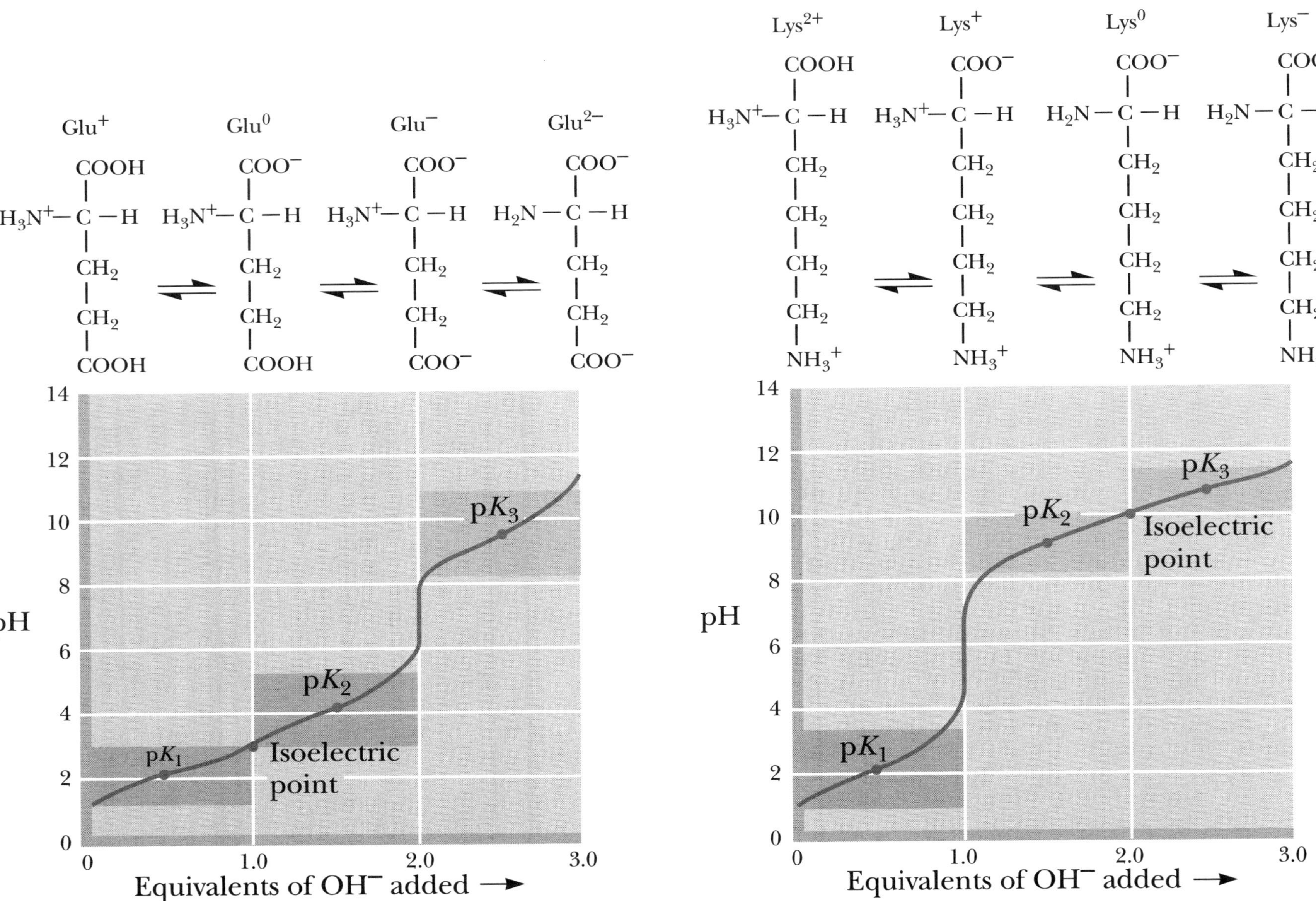

Figure 4.8 Reactions of amino acid side-chain functional groups.

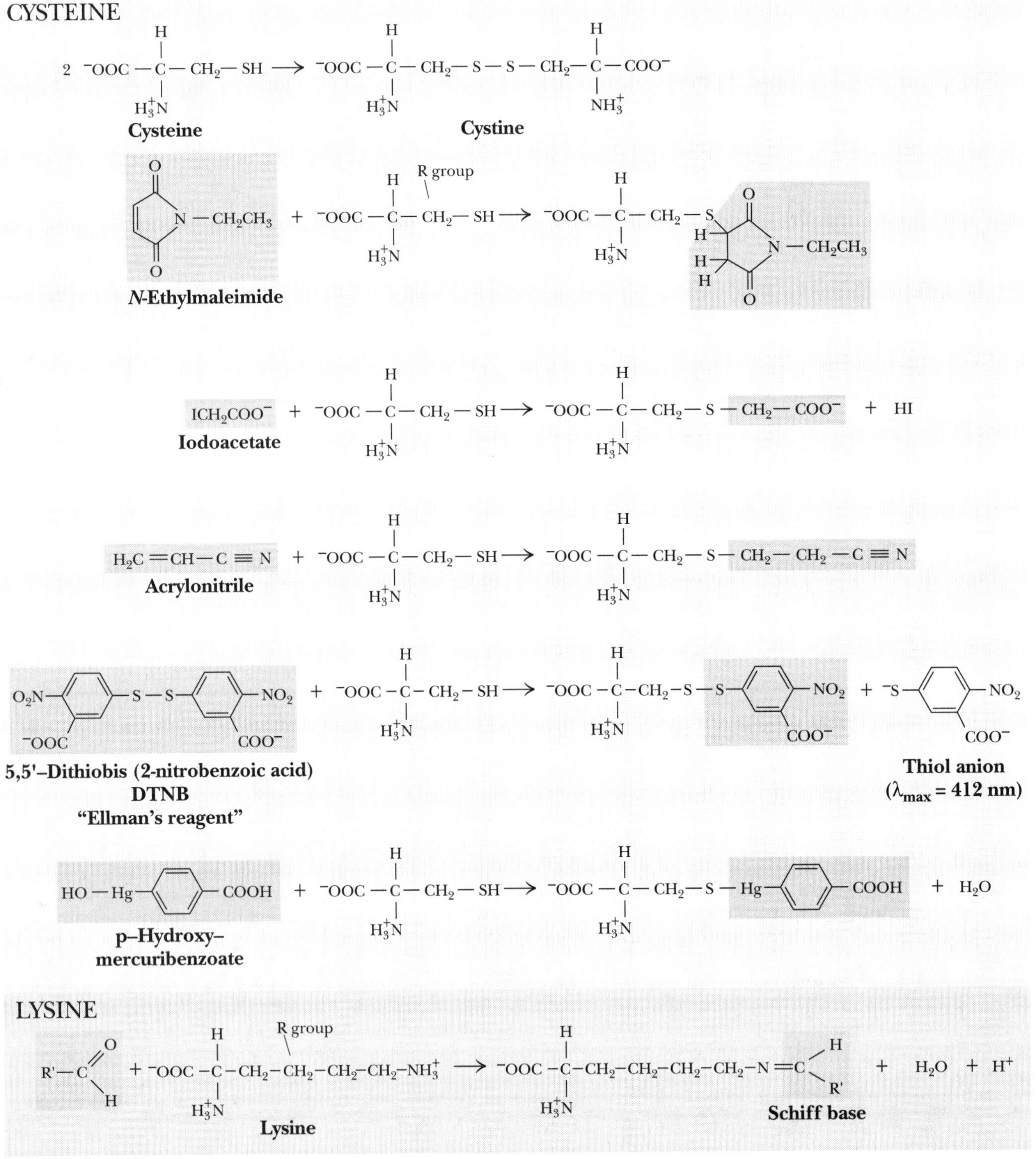

Figure 5.1   Fibrous and globular proteins.

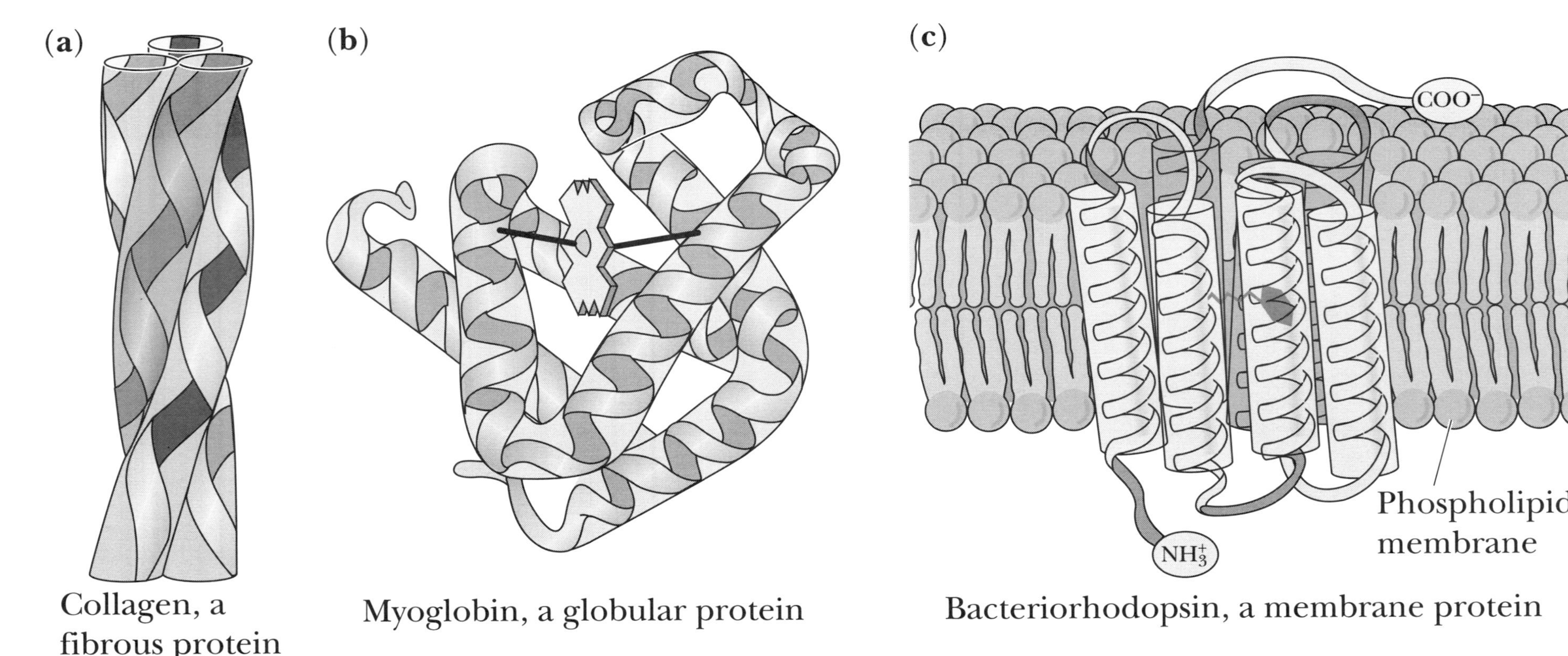

(a) Collagen, a fibrous protein

(b) Myoglobin, a globular protein

(c) Bacteriorhodopsin, a membrane protein

Figure 5.5   Configuration and confirmation.

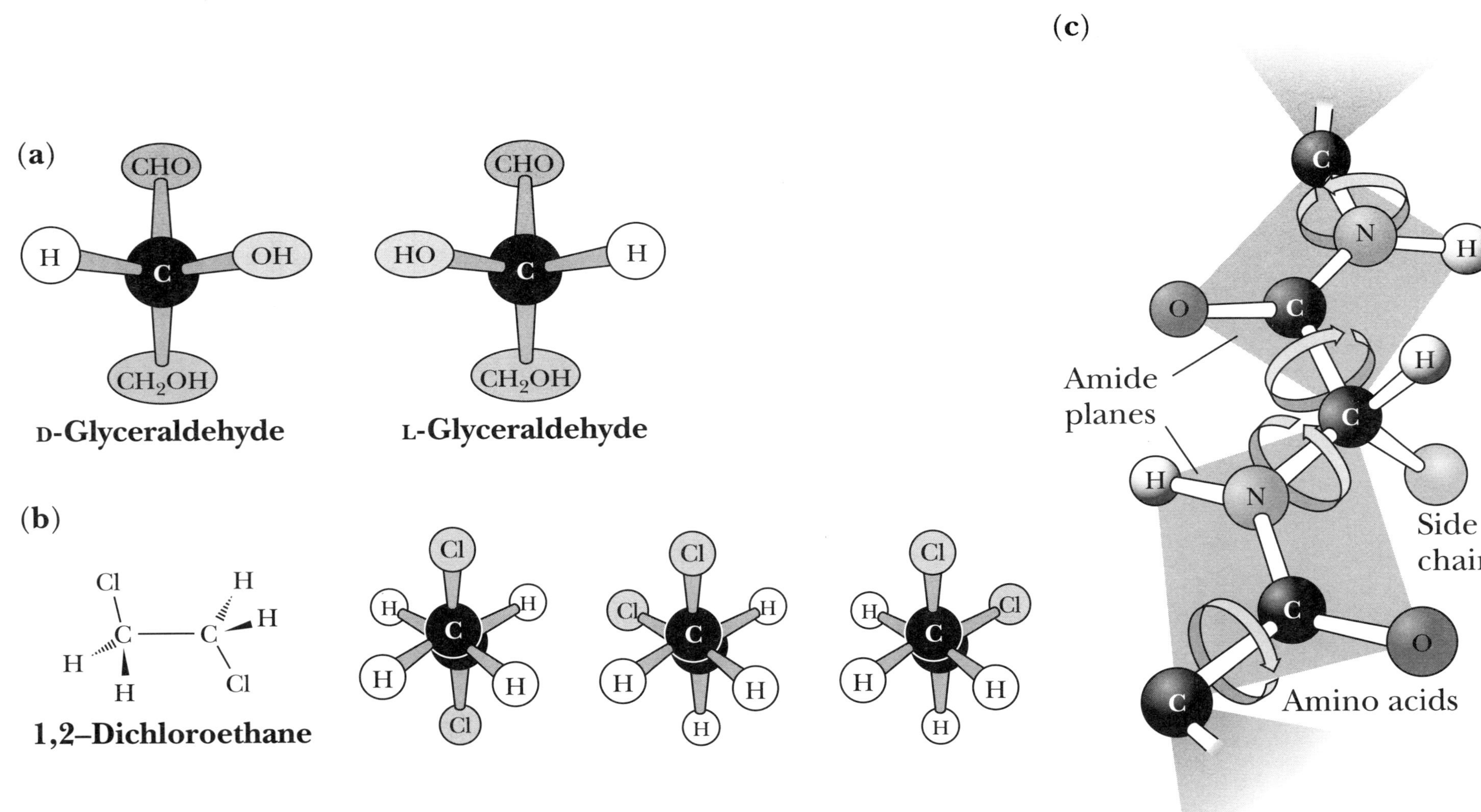

© Harcourt, Inc.

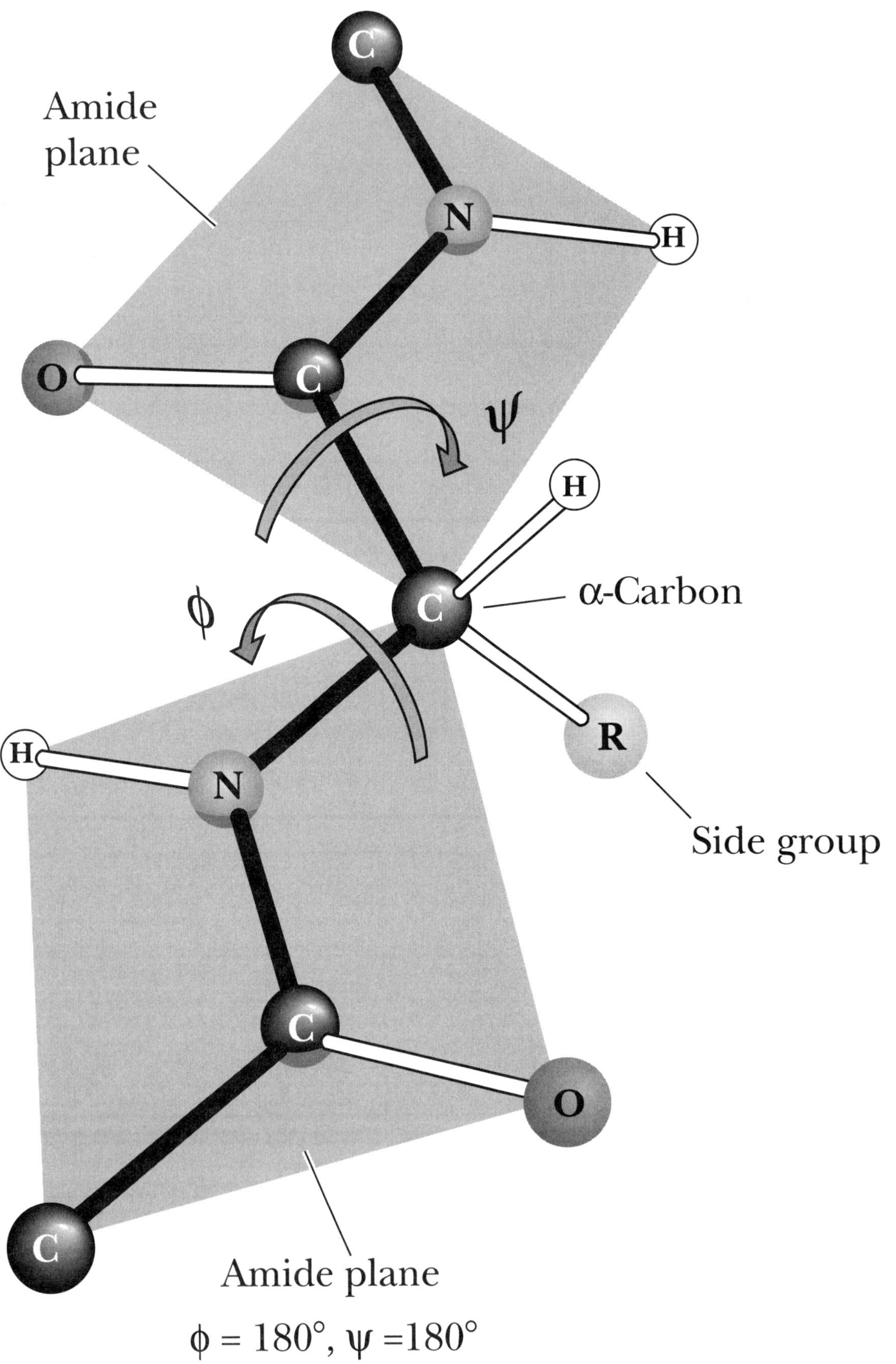
Amide plane
C
N
H
O
C
ψ
H
α-Carbon
φ
R
H
N
Side group
C
O
C
Amide plane
φ = 180°, ψ = 180°

Figure 5.7   Joined amide planes and forbidden conformations about an α-carbon between two peptide planes.

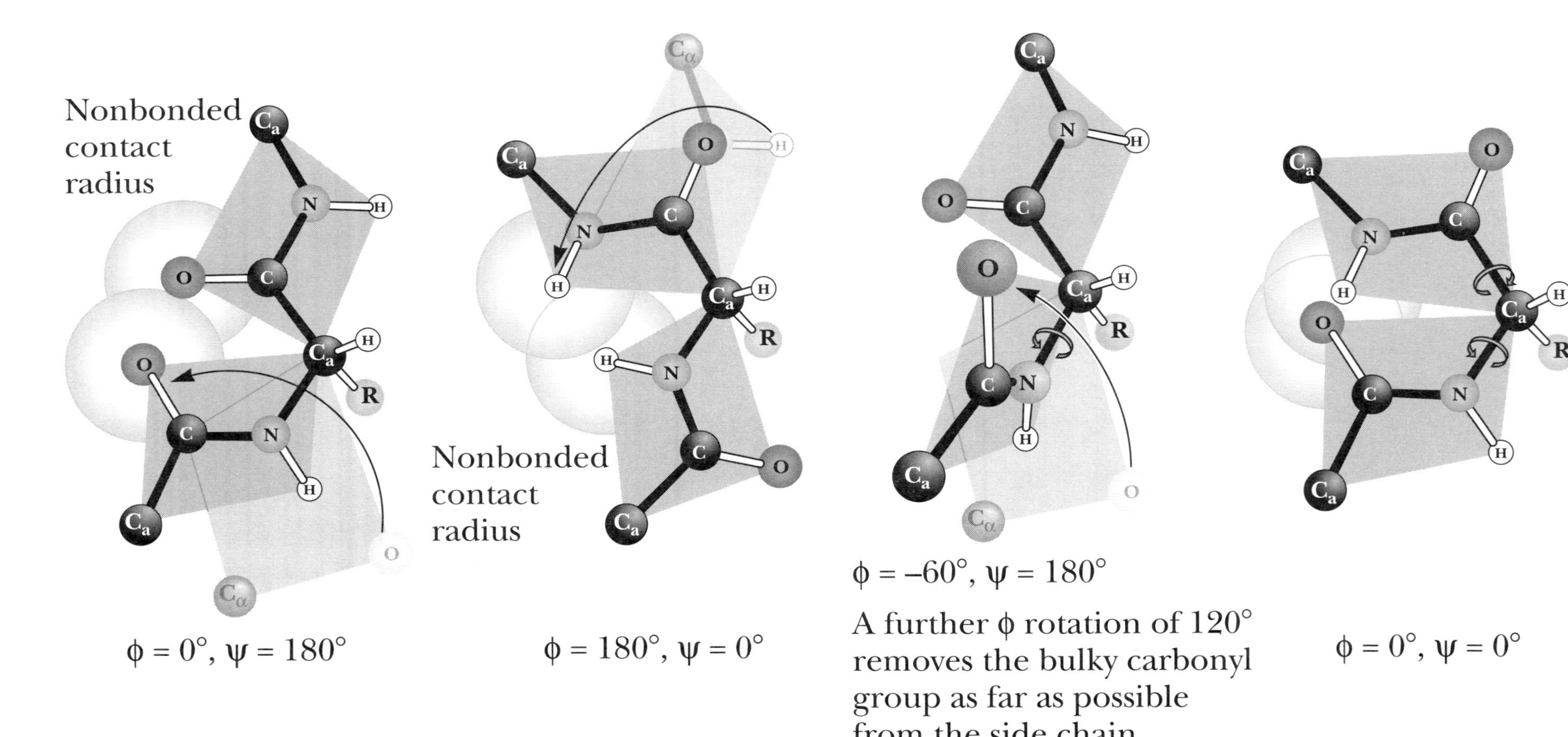

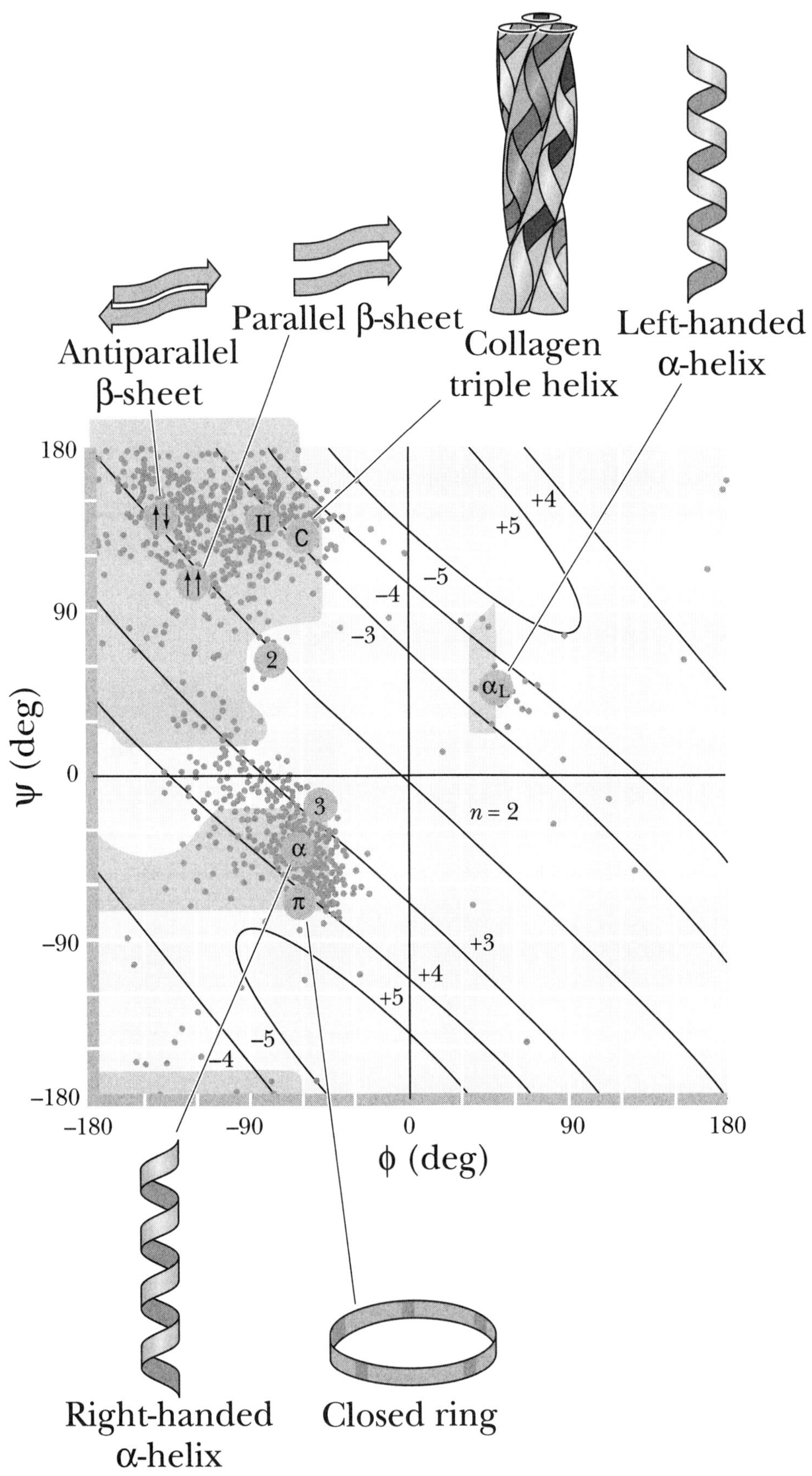
Antiparallel β-sheet
Parallel β-sheet
Collagen triple helix
Left-handed α-helix
180
+4
+5
II
C
−5
−4
90
−3
2
α_L
ψ (deg)
0
3
n = 2
α
π
−90
+3
+4
+5
−5
−4
−180
−180
−90
0
90
180
φ (deg)
Right-handed α-helix
Closed ring

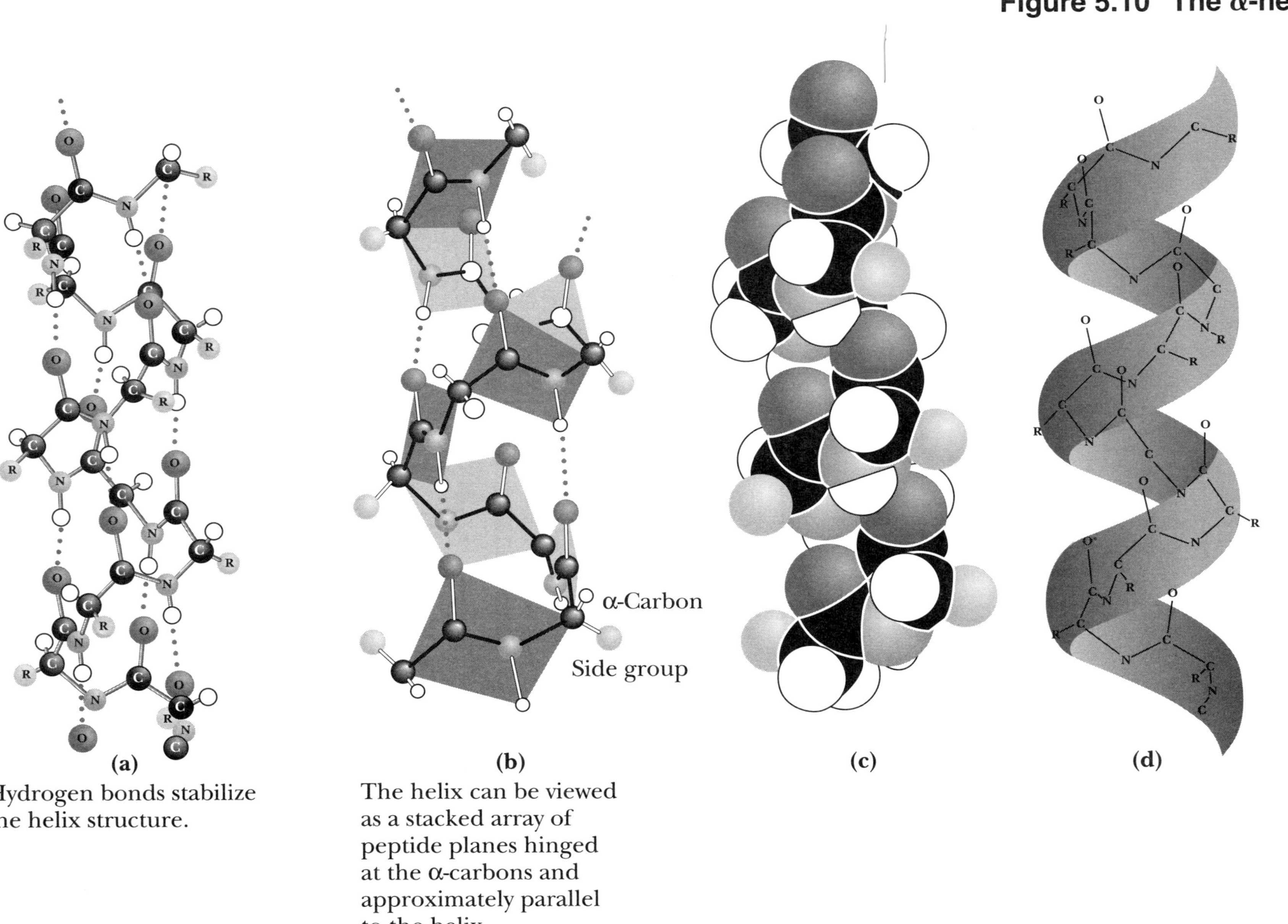

Figure 5.10   The α-helix.

Figure 5.12   An antiparallel-pleated sheet.

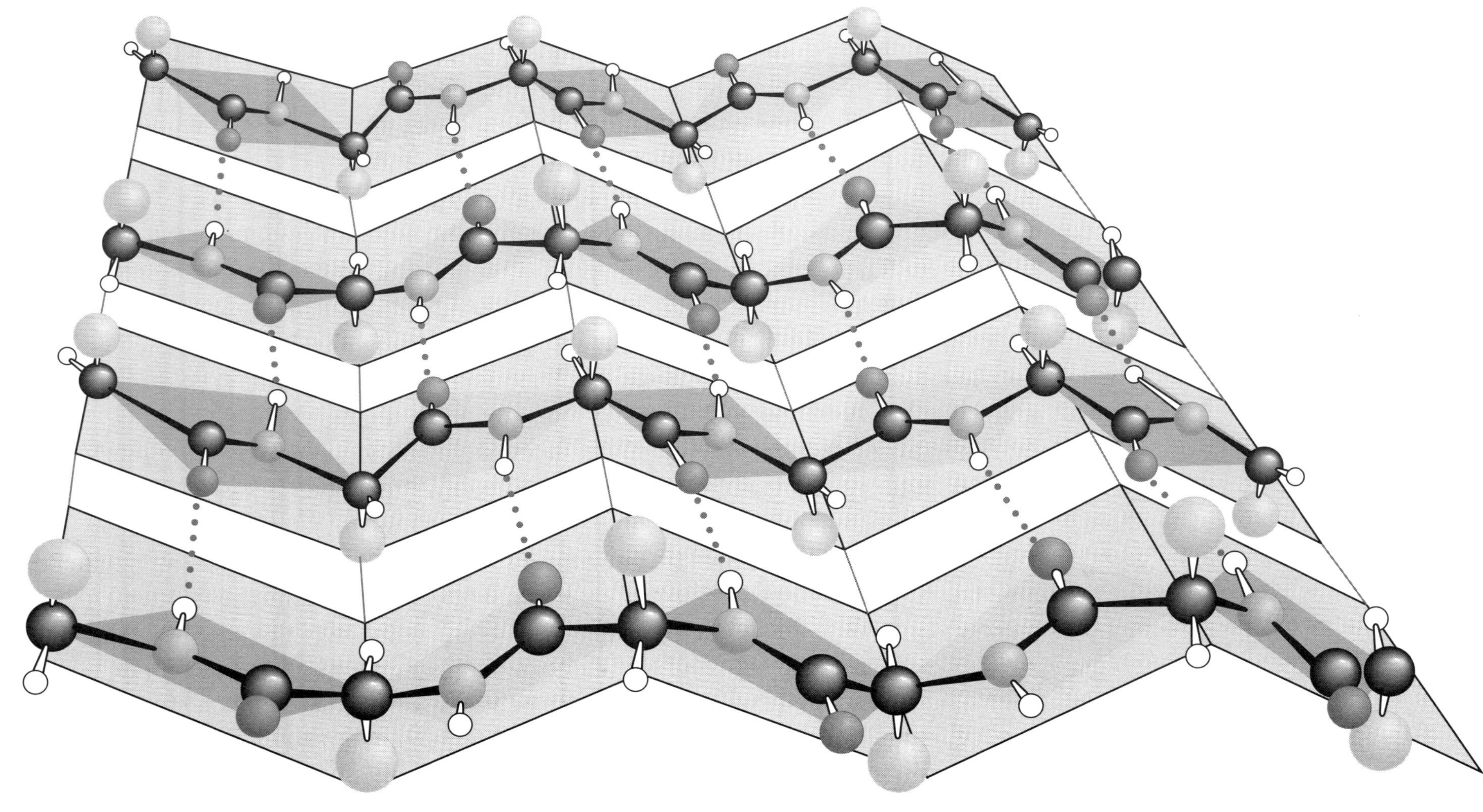

Figure 5.24 Examples of protein domains.

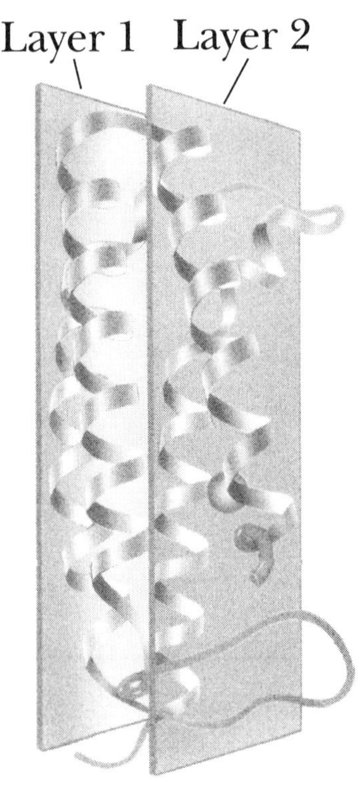

(a) Cytochrome *c′*

(b) Phosphoglycerate kinase
(Domain 2)

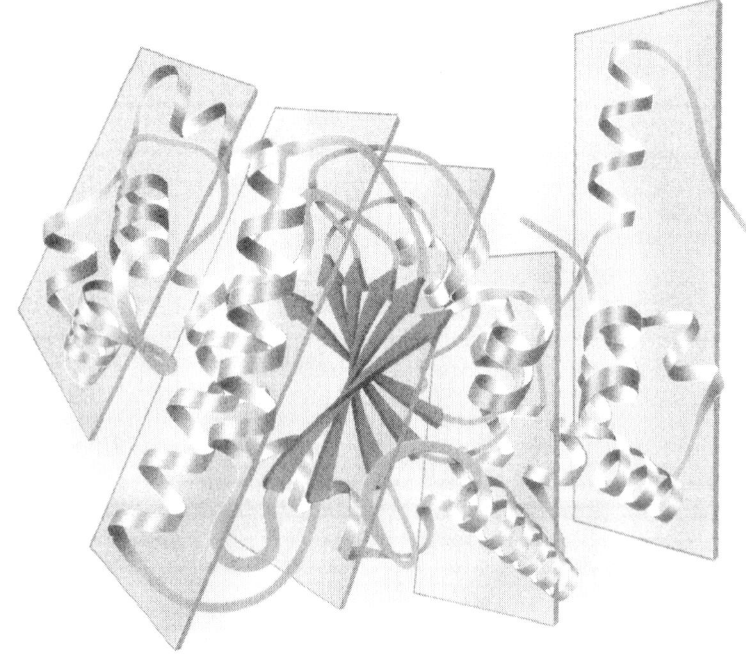

(c) Phosphorylase
(Domain 2)

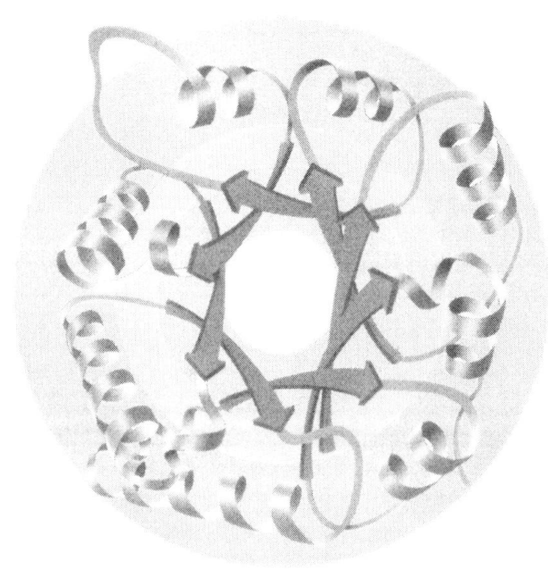

(d) Triose phosphate
isomerase

Figure 5.27 Protein mosaics.

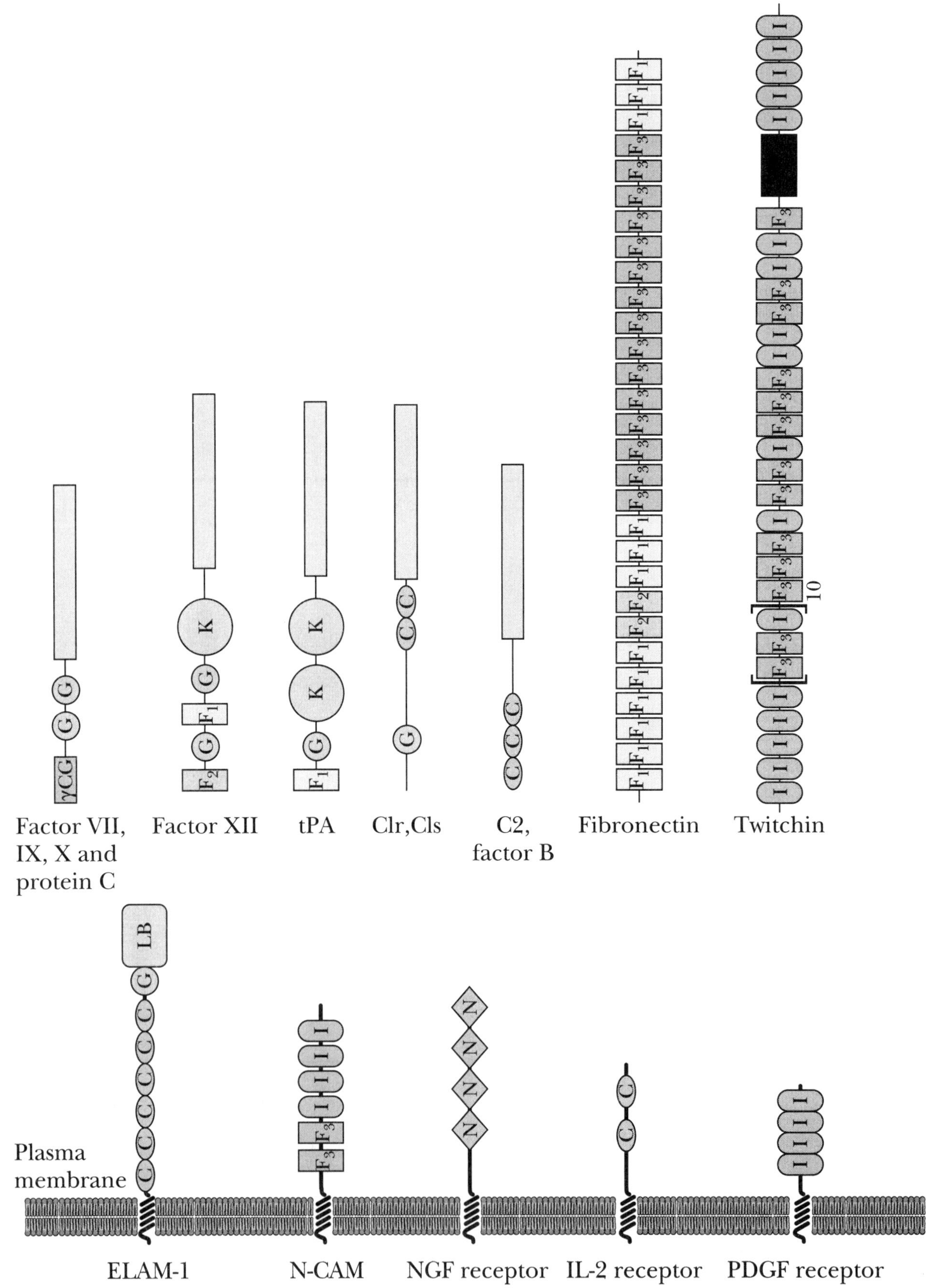

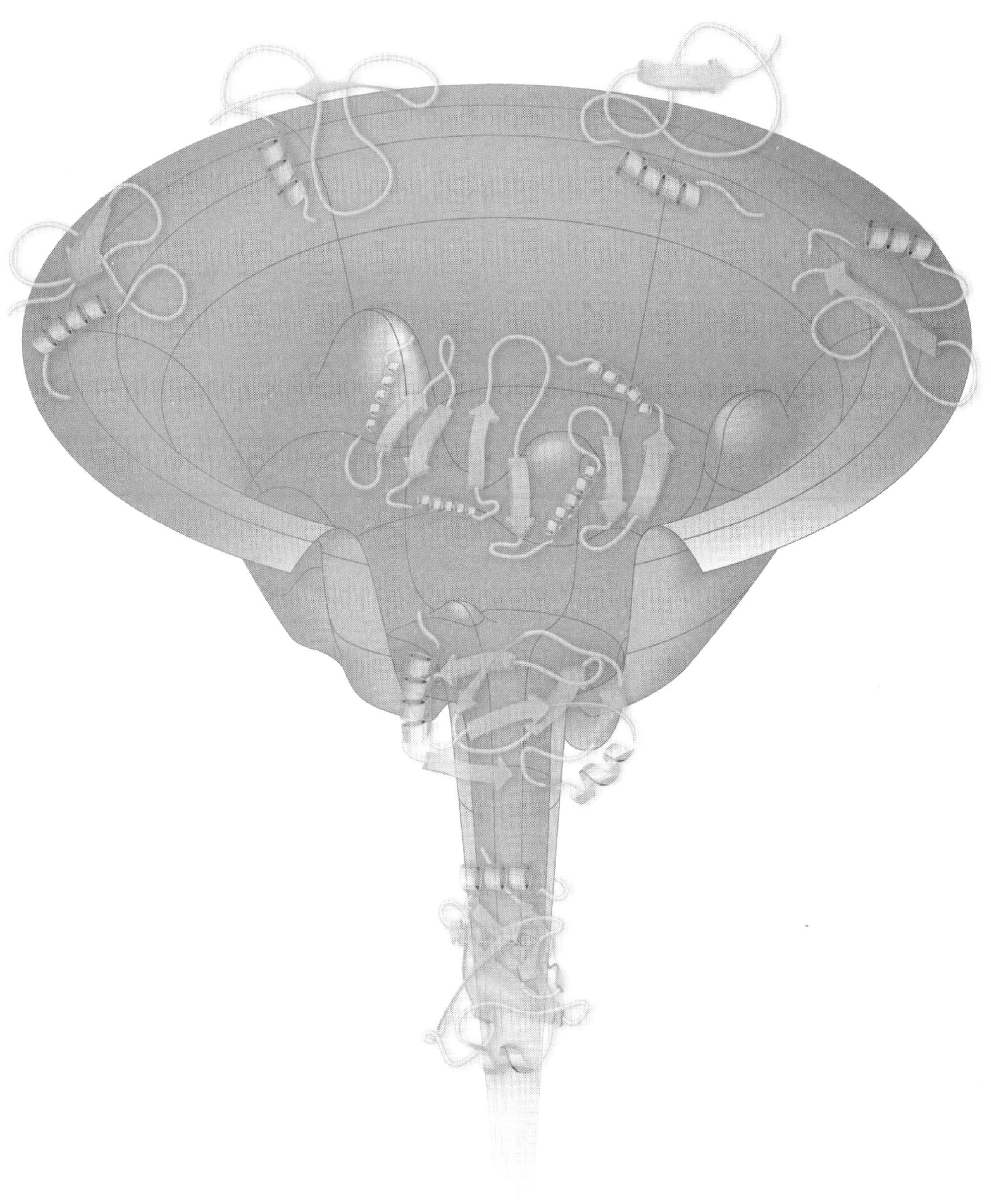

Figure 6.19   Several important sterols derived from cholesterol.

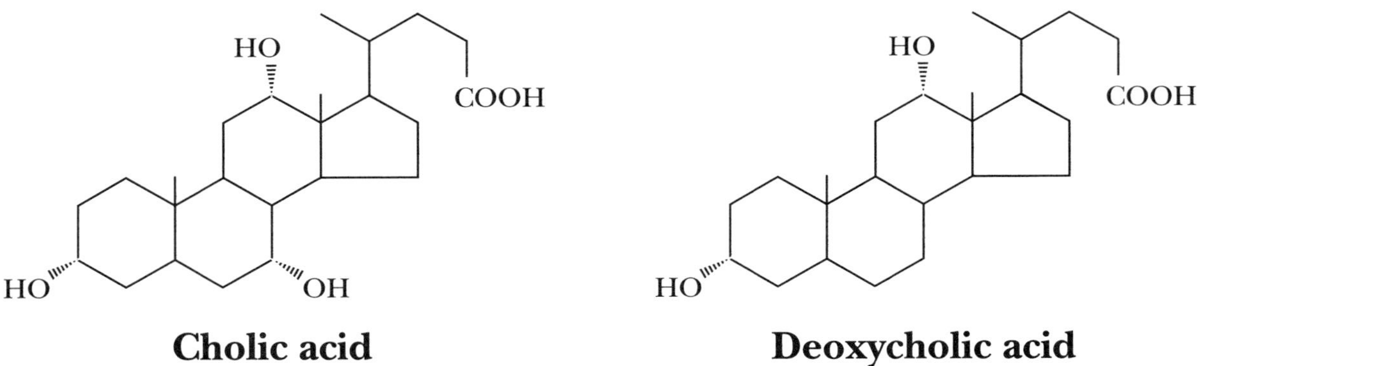

Figure 6.25   The fluid mosaic model.

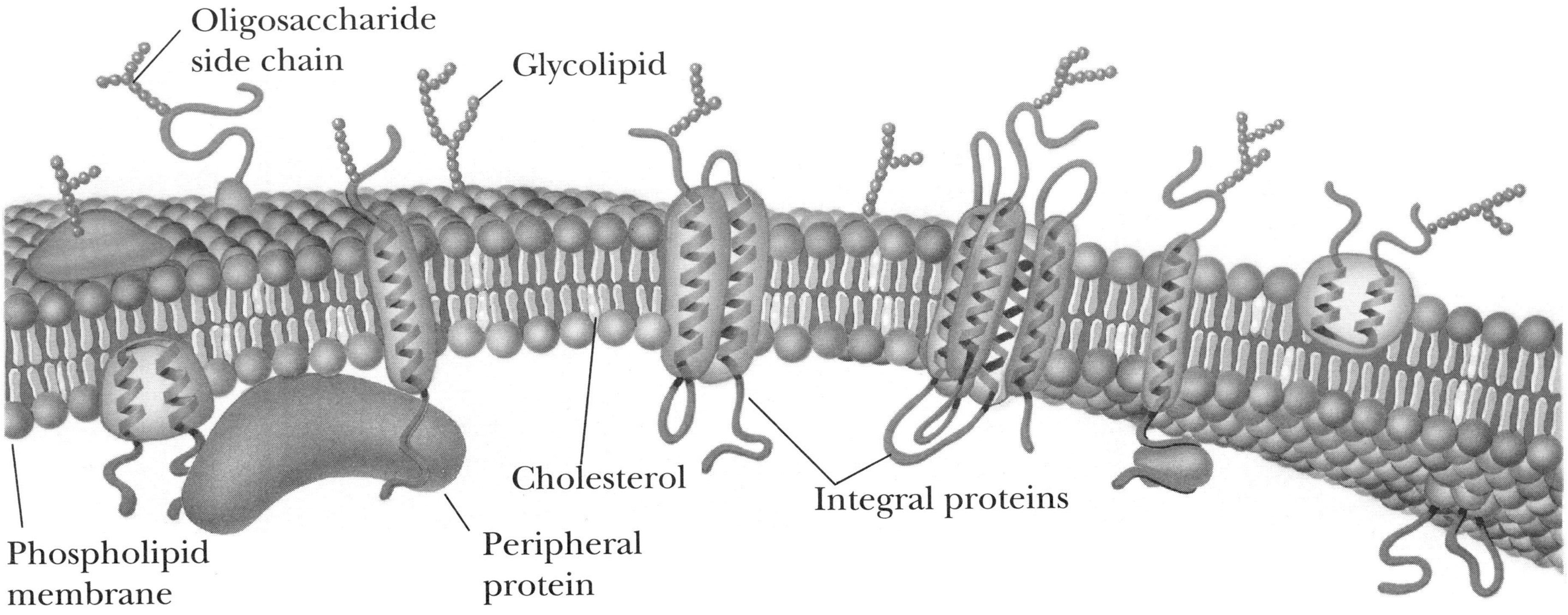

Figure 6.32 Bacteriorhodopsin: electron density profile.

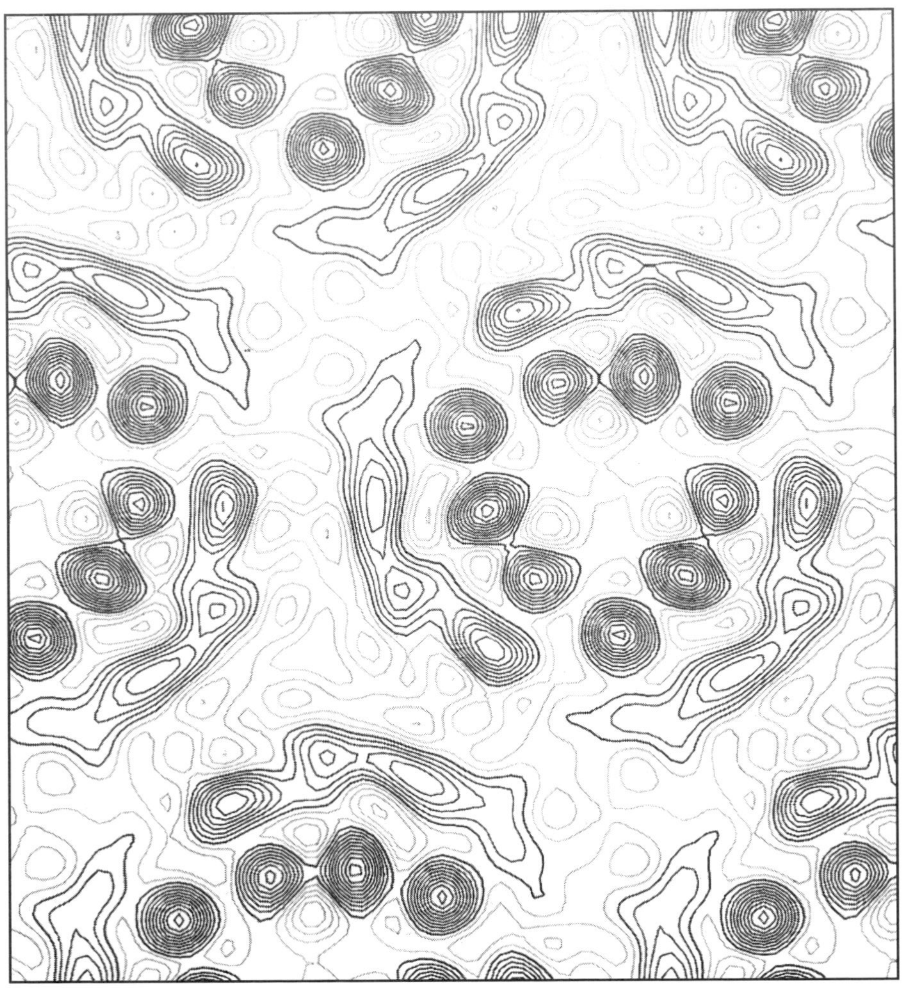

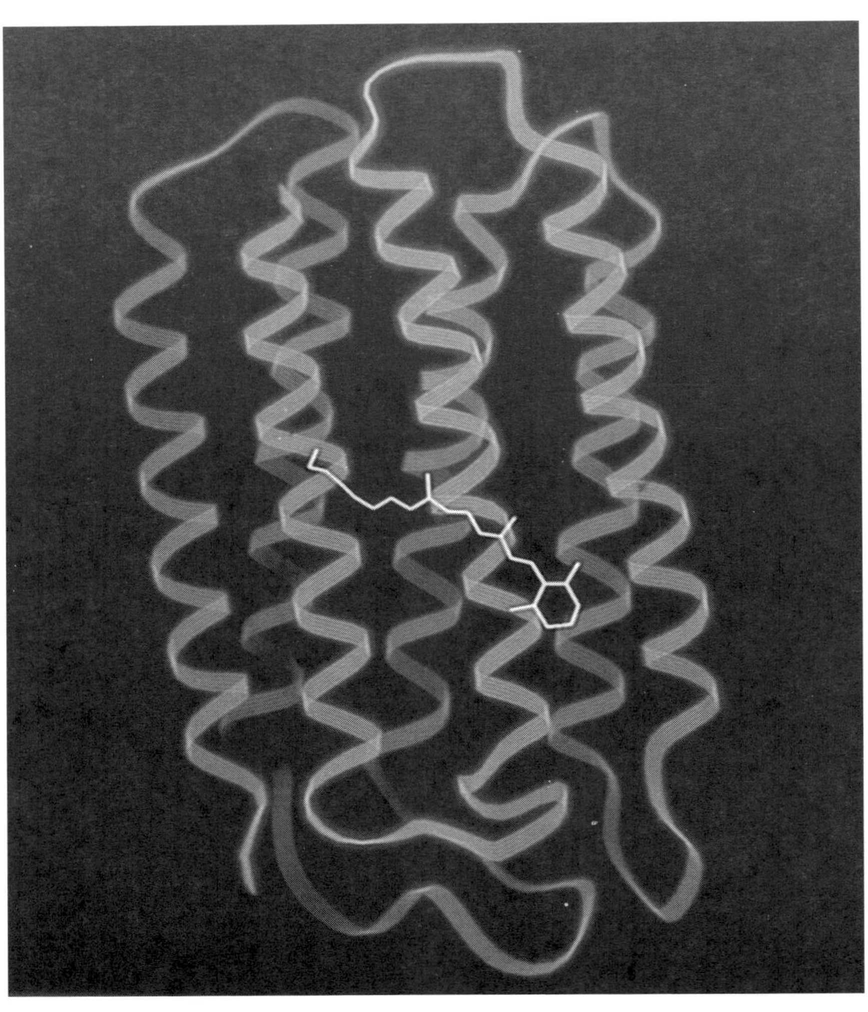

Figure 6.35   Proteins anchored to membranes by lipid anchors.

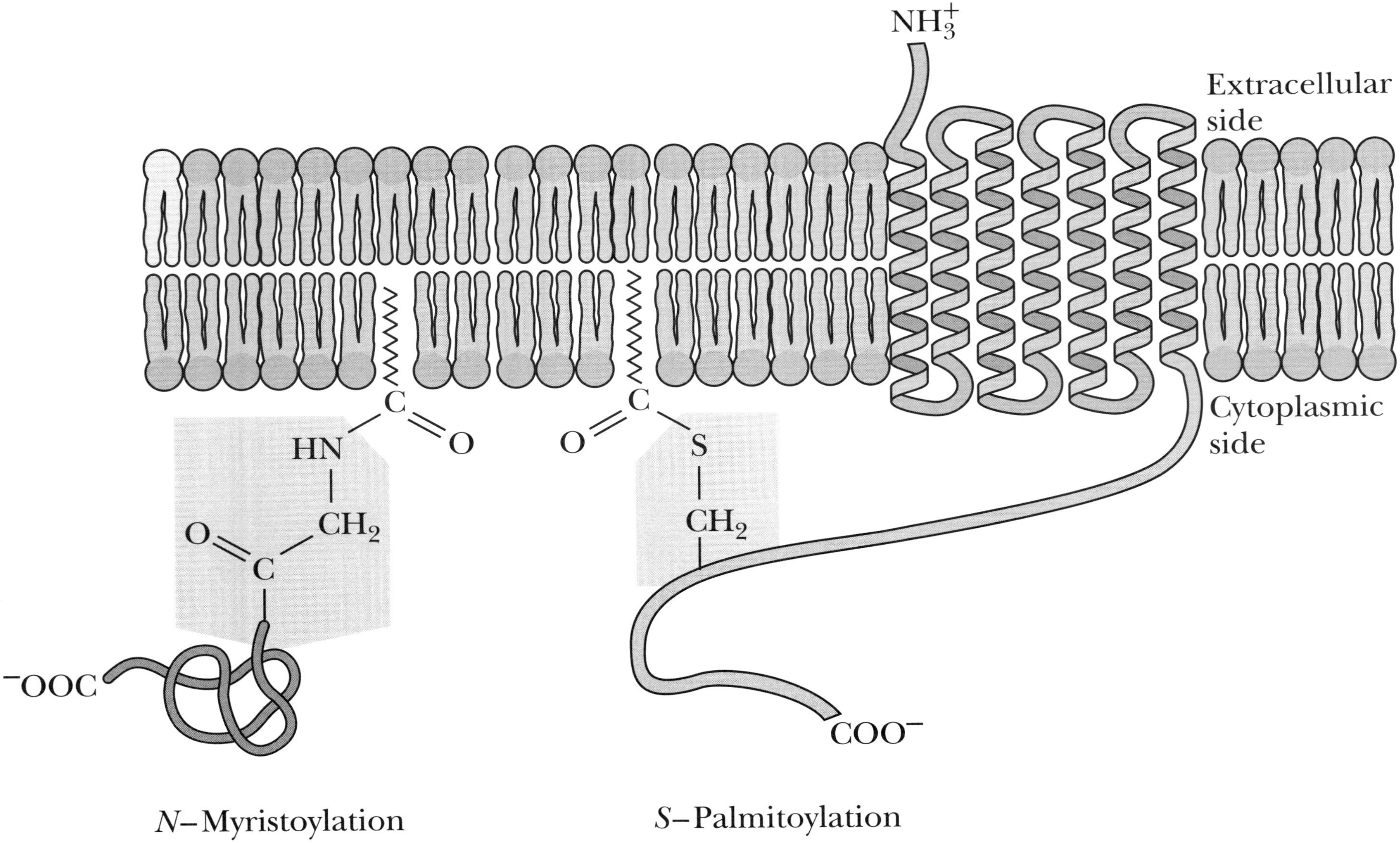

Figure 6.36   Prenylation of proteins destined for membrane anchoring.

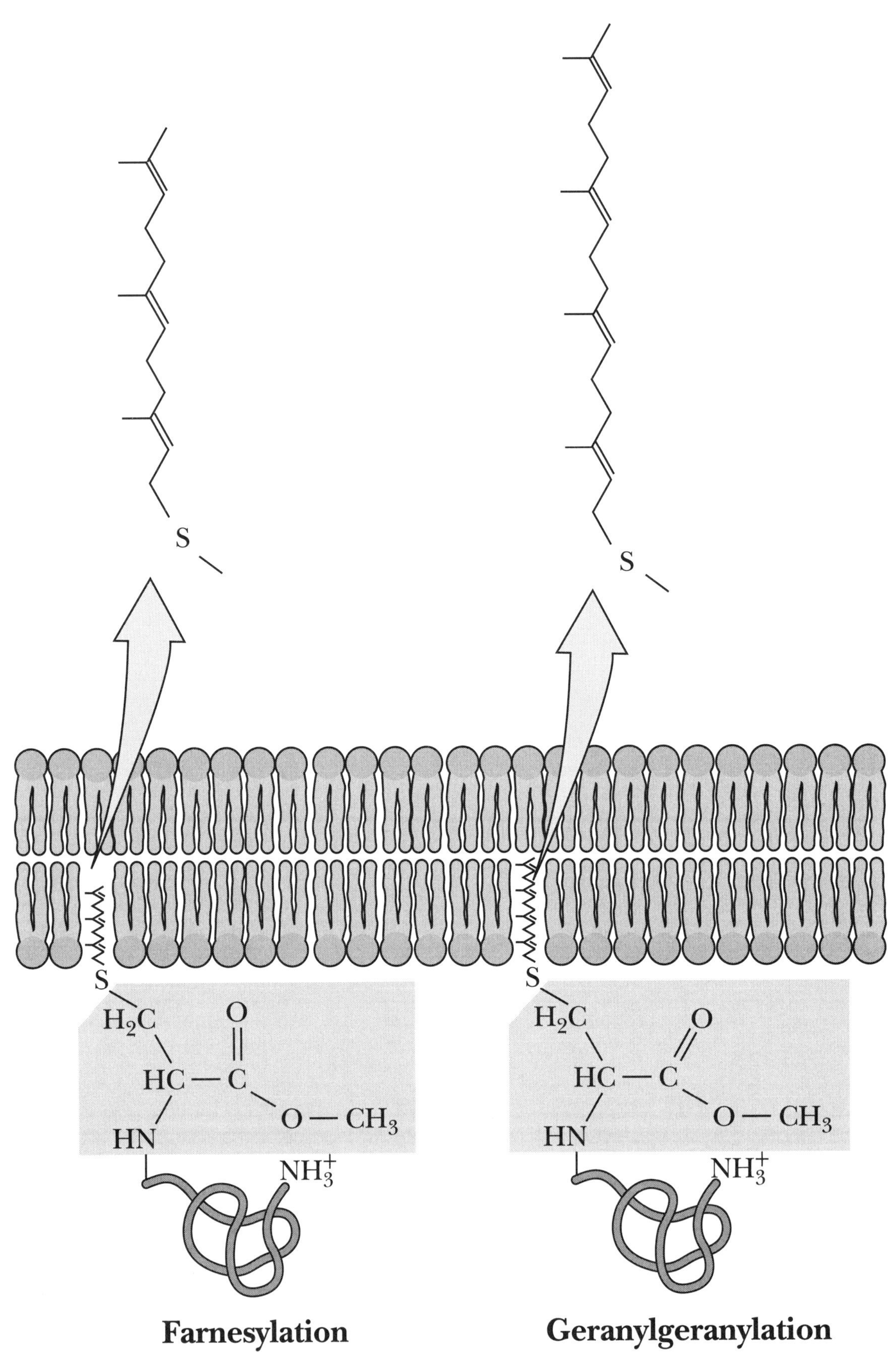

Figure 6.45   A model for the arrangement of Na^+, K^+-ATPase in the plasma membrane.

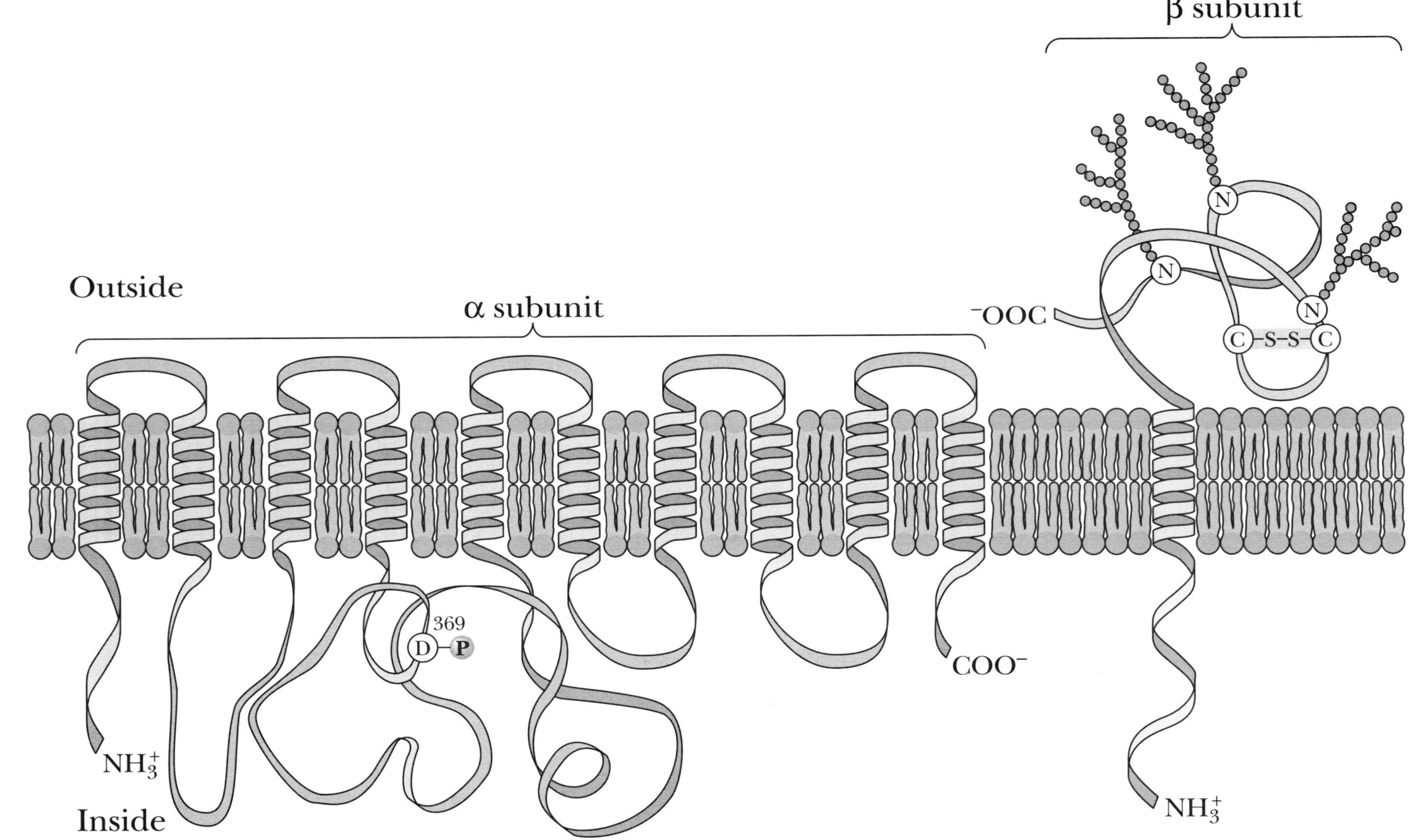

Figure 6.46   A mechanism for Na$^+$, K$^+$-ATPase.

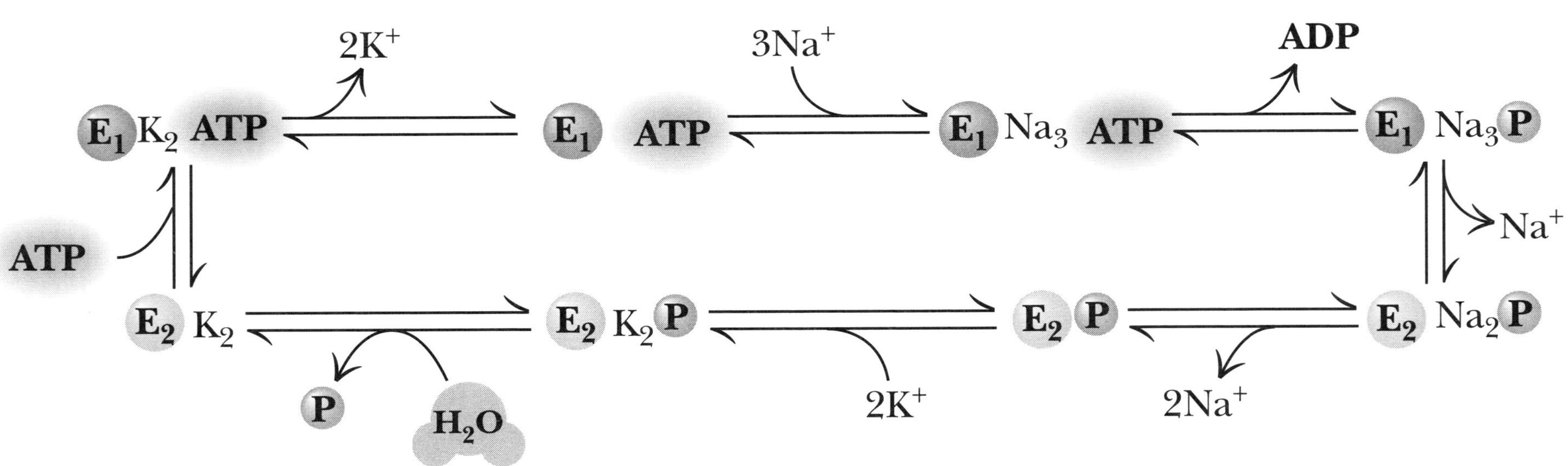

Figure 7.2   D-aldoses.

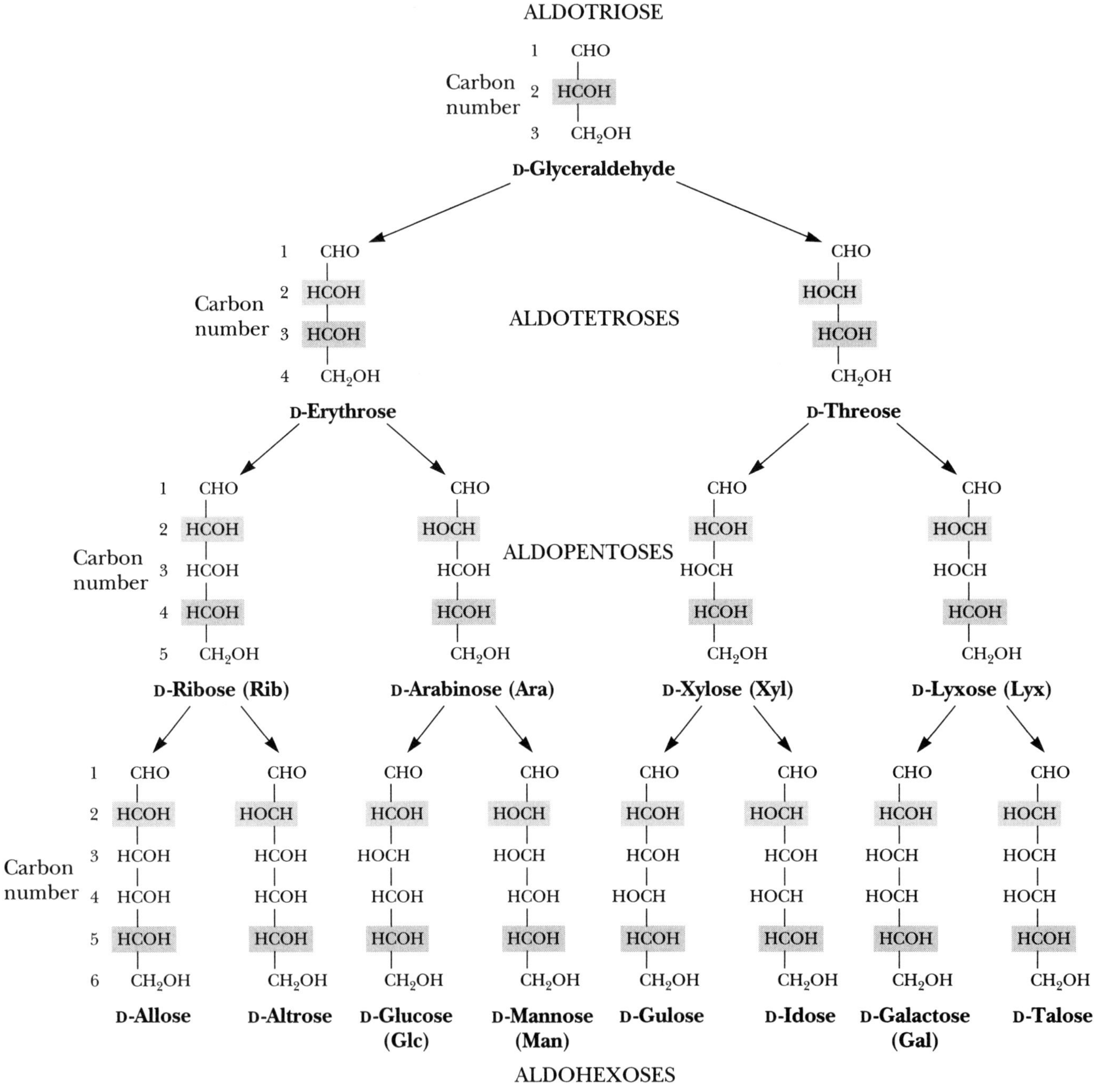

Figure 7.3   D-ketoses.

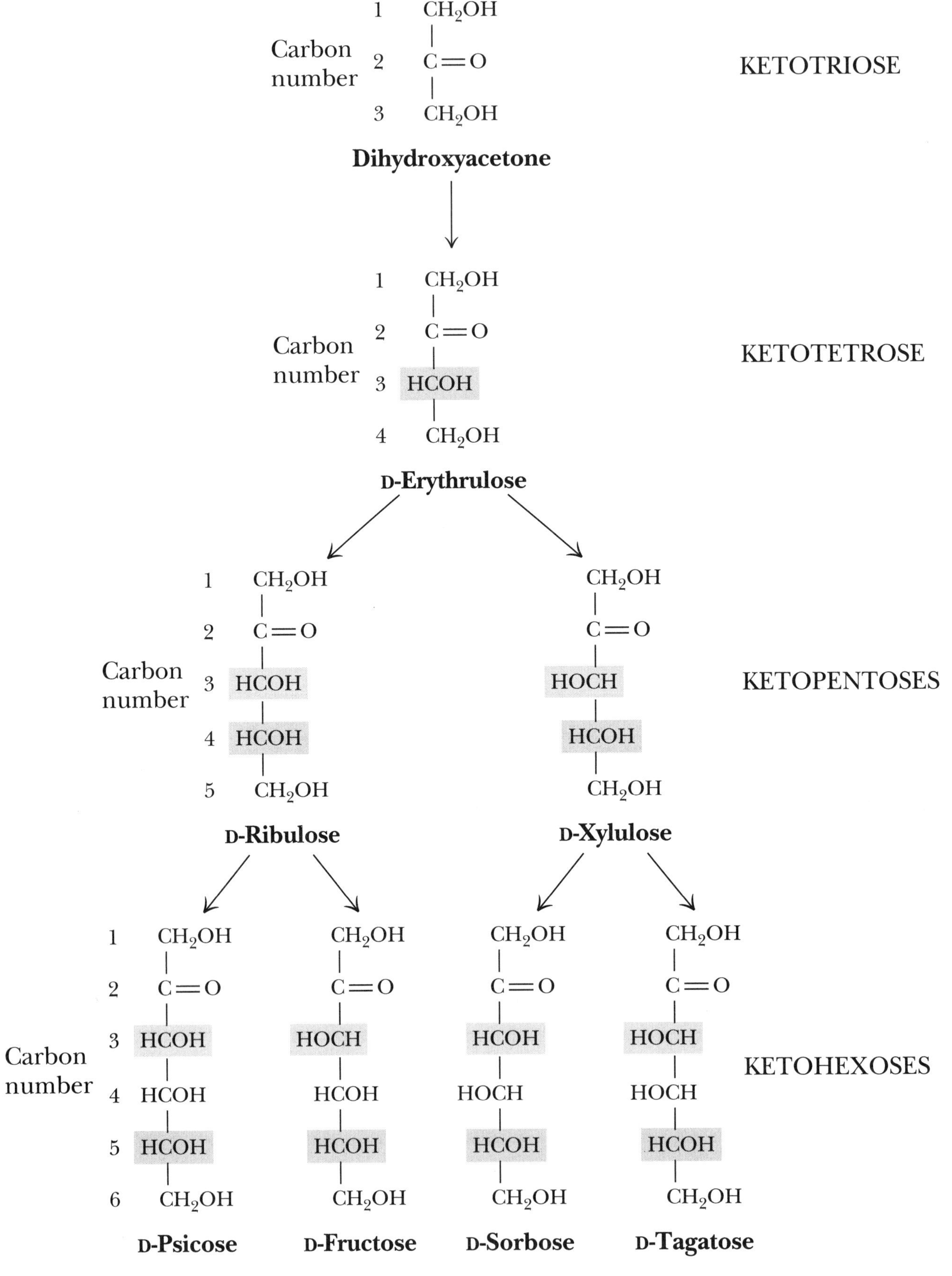

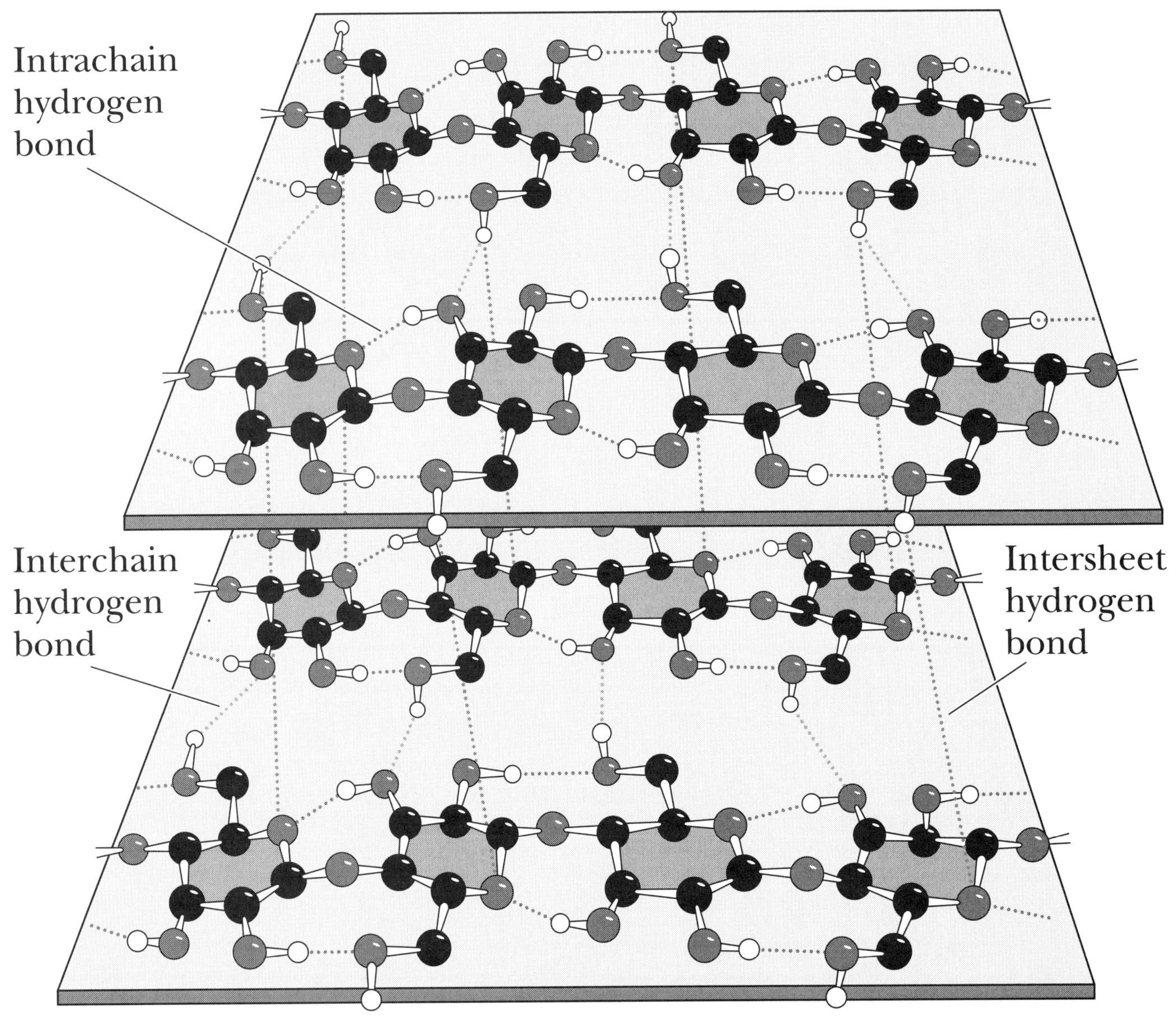

Intrachain
hydrogen
bond
Interchain
hydrogen
bond
Intersheet
hydrogen
bond

Figure 7.30 The structures of the cell wall and membrane(s) in Gram-positive and Gram-negative bacteria.

(**a**) Gram-positive bacteria

(**b**) Gram-negative bacteria

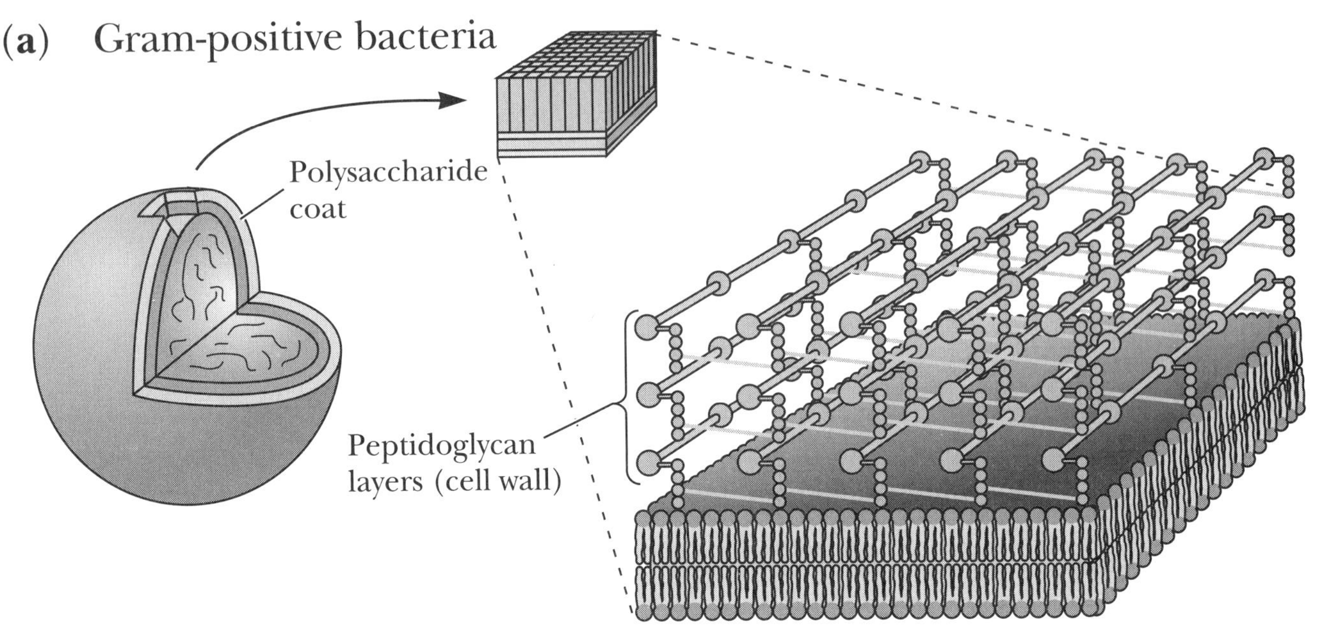

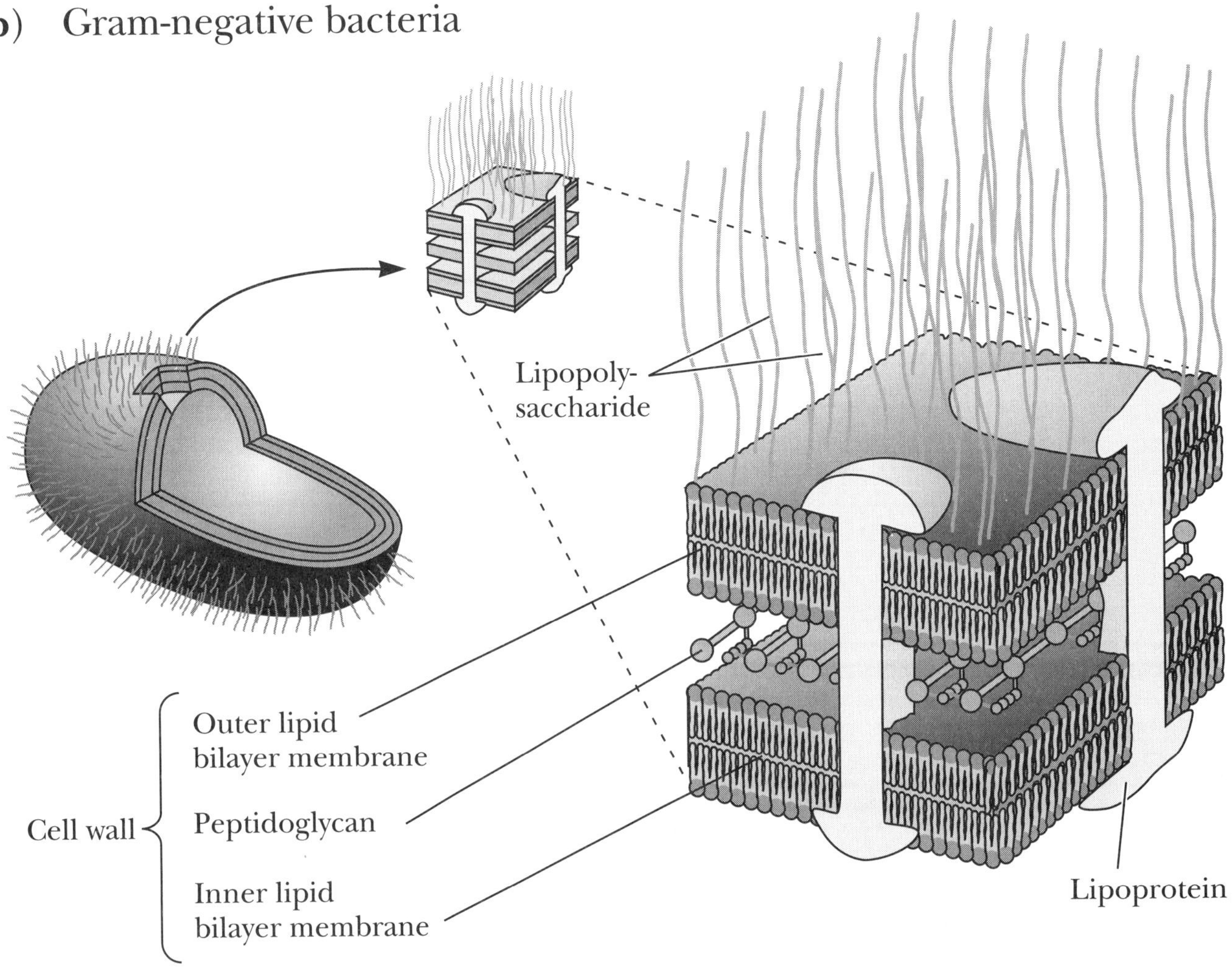

Figure 7.37 Structures of known proteoglycans.

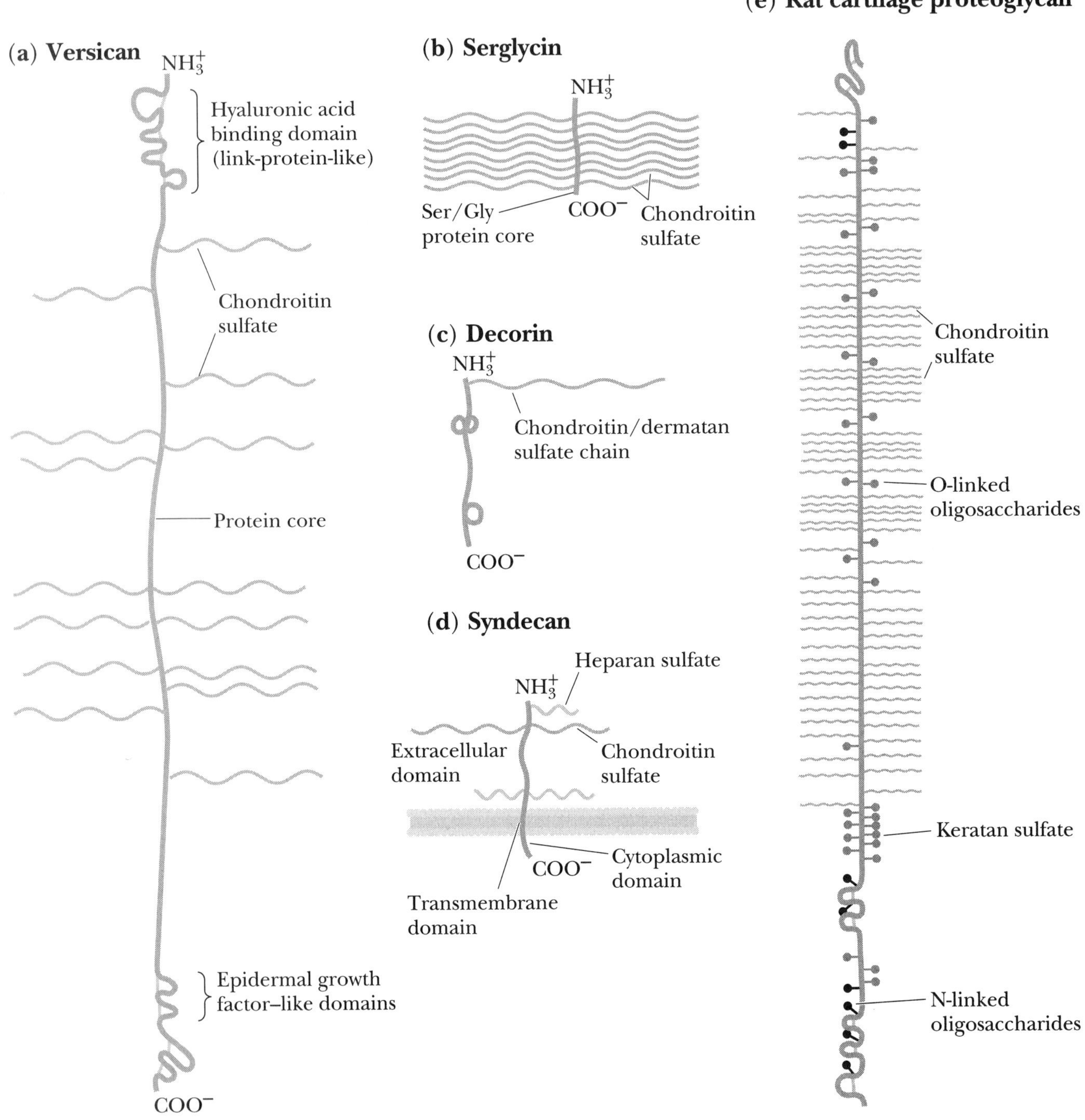

© **Harcourt, Inc.**

Figure 7.40 Hyaluronate forms the backbone of proteoglycan structures.

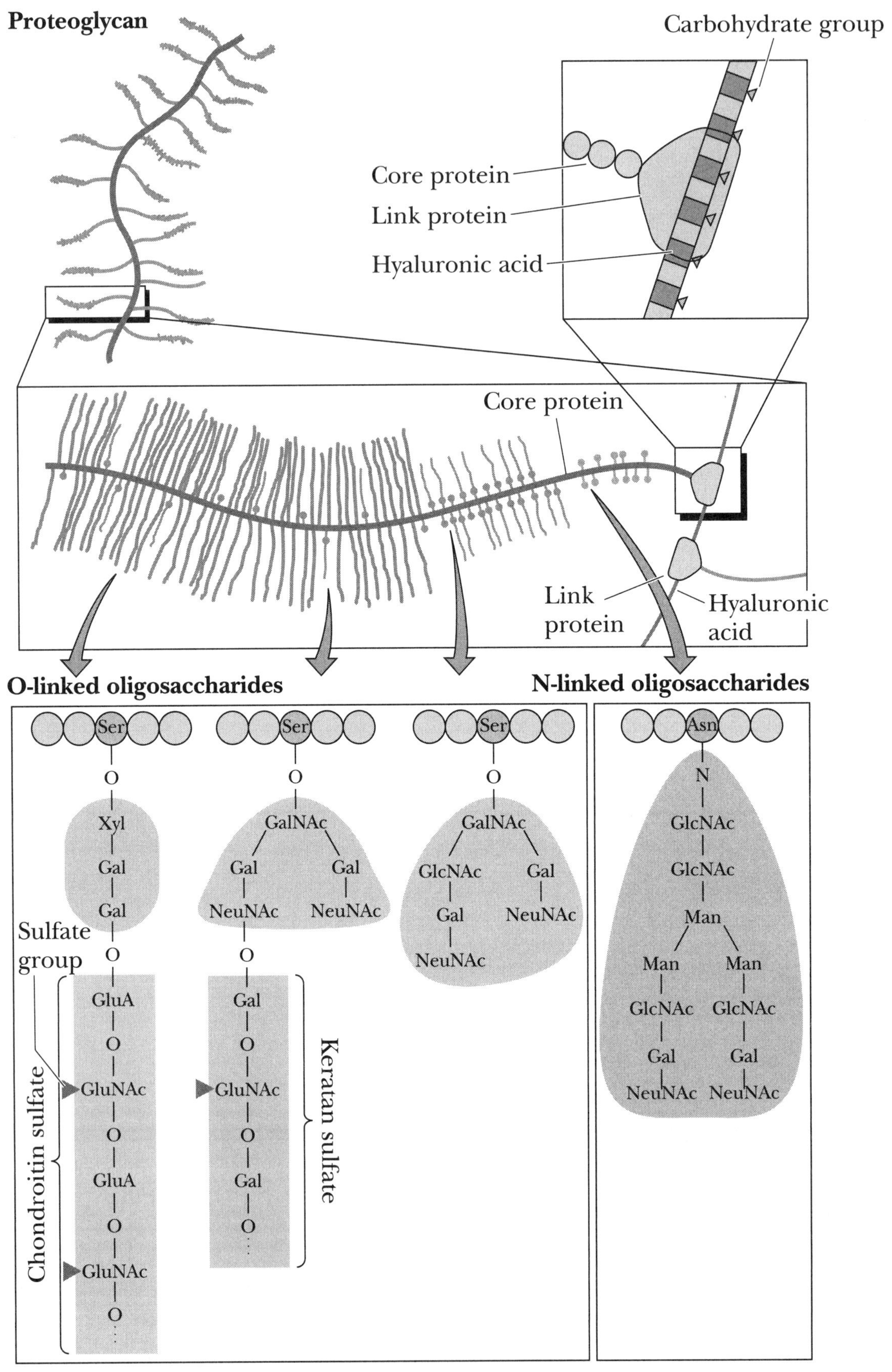

Figure 8.1   Information transfer in cells.

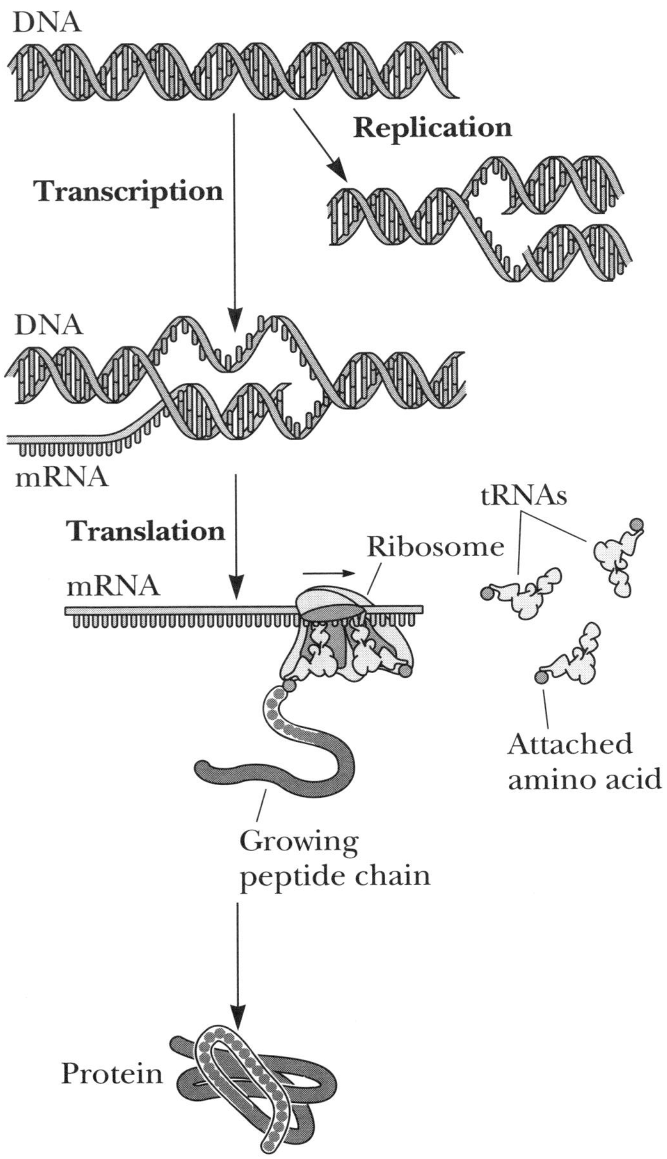

Replication

DNA replication yields two DNA molecules identical to the original one, ensuring transmission of genetic information to daughter cells with exceptional fidelity.

Transcription

The sequence of bases in DNA is recorded as a sequence of complementary bases in a single-stranded mRNA molecule.

Translation

Three-base codons on the mRNA corresponding to specific amino acids direct the sequence of building a protein. These codons are recognized by tRNAs (transfer RNAs) carrying the appropriate amino acids. Ribosomes are the "machinery" for protein synthesis.

Figure 8.11 The common ribonucleosides.

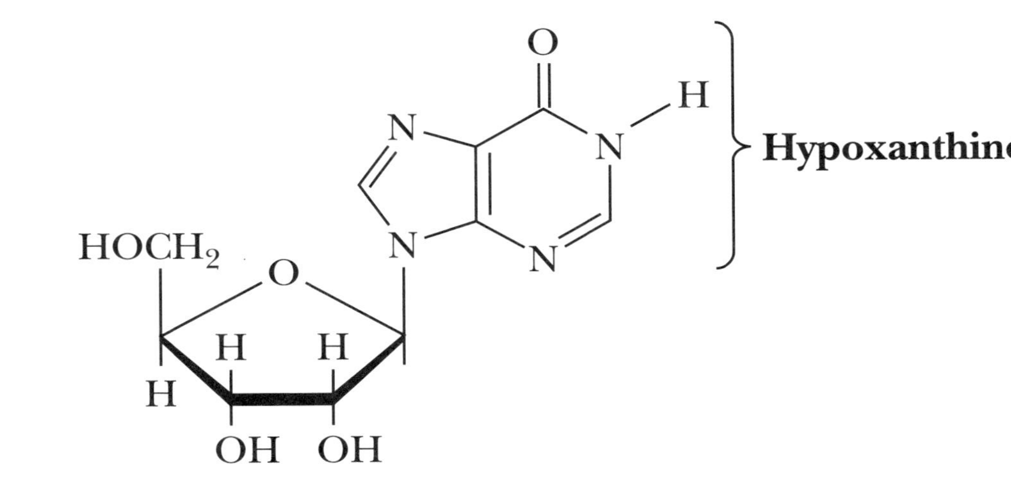
Cytidine
Uridine
Adenosine
Guanosine
Inosine, an uncommon nucleoside
Hypoxanthine

Figure 8.12   The four common ribonucleotides.

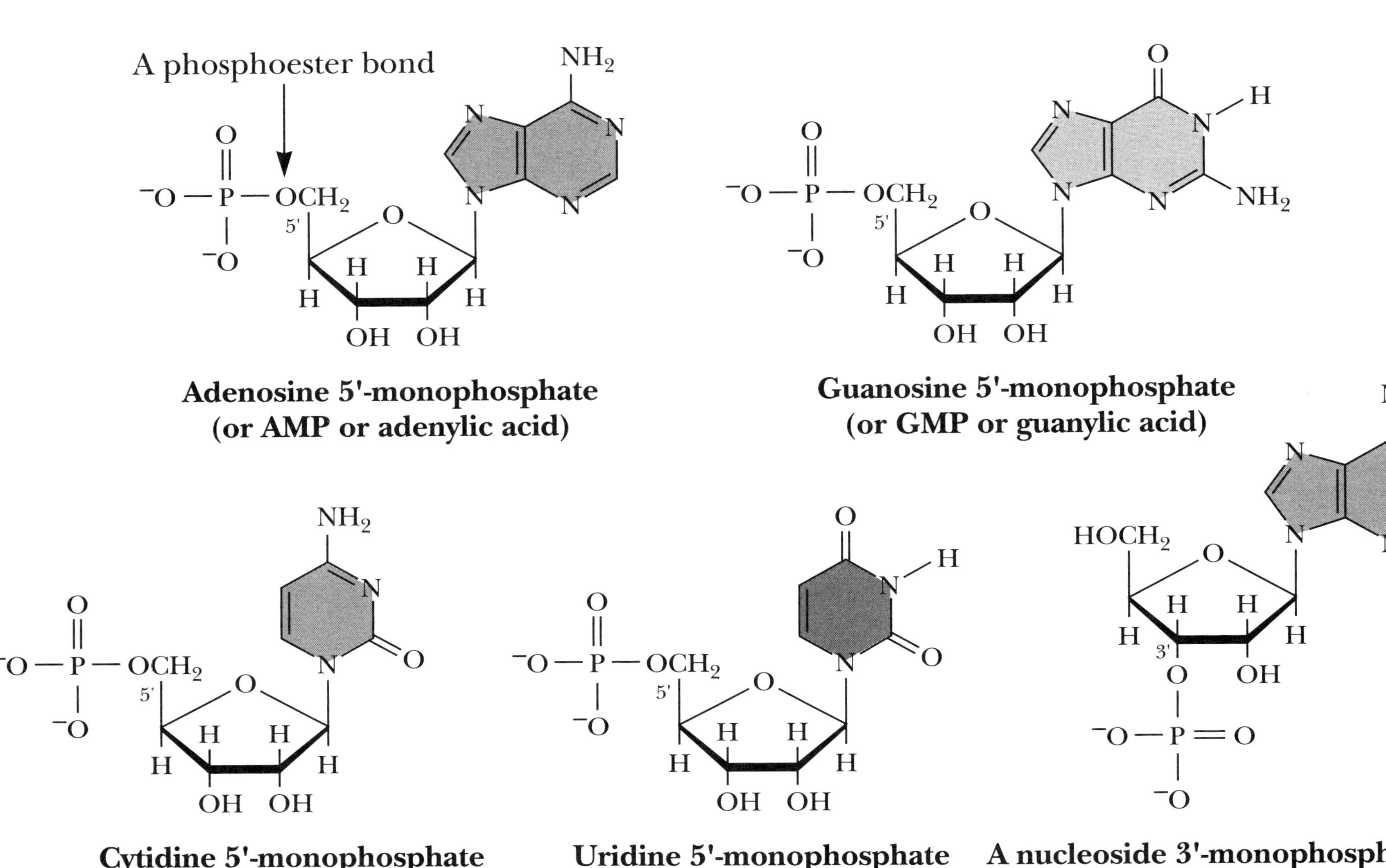

Figure 8.16 3′-5′-Phosphodiester bridges nucleotides to form polynucleotide chains.

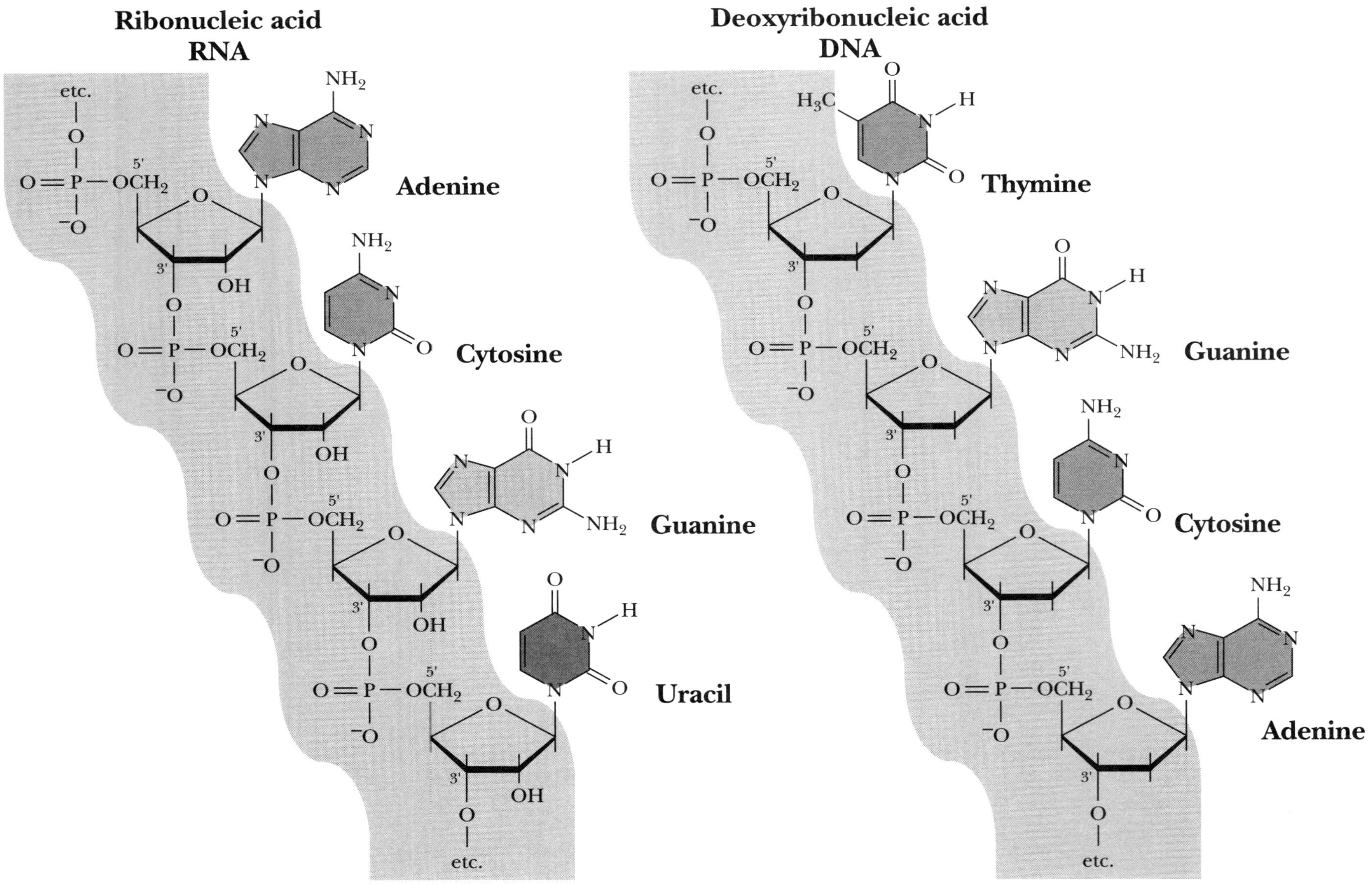

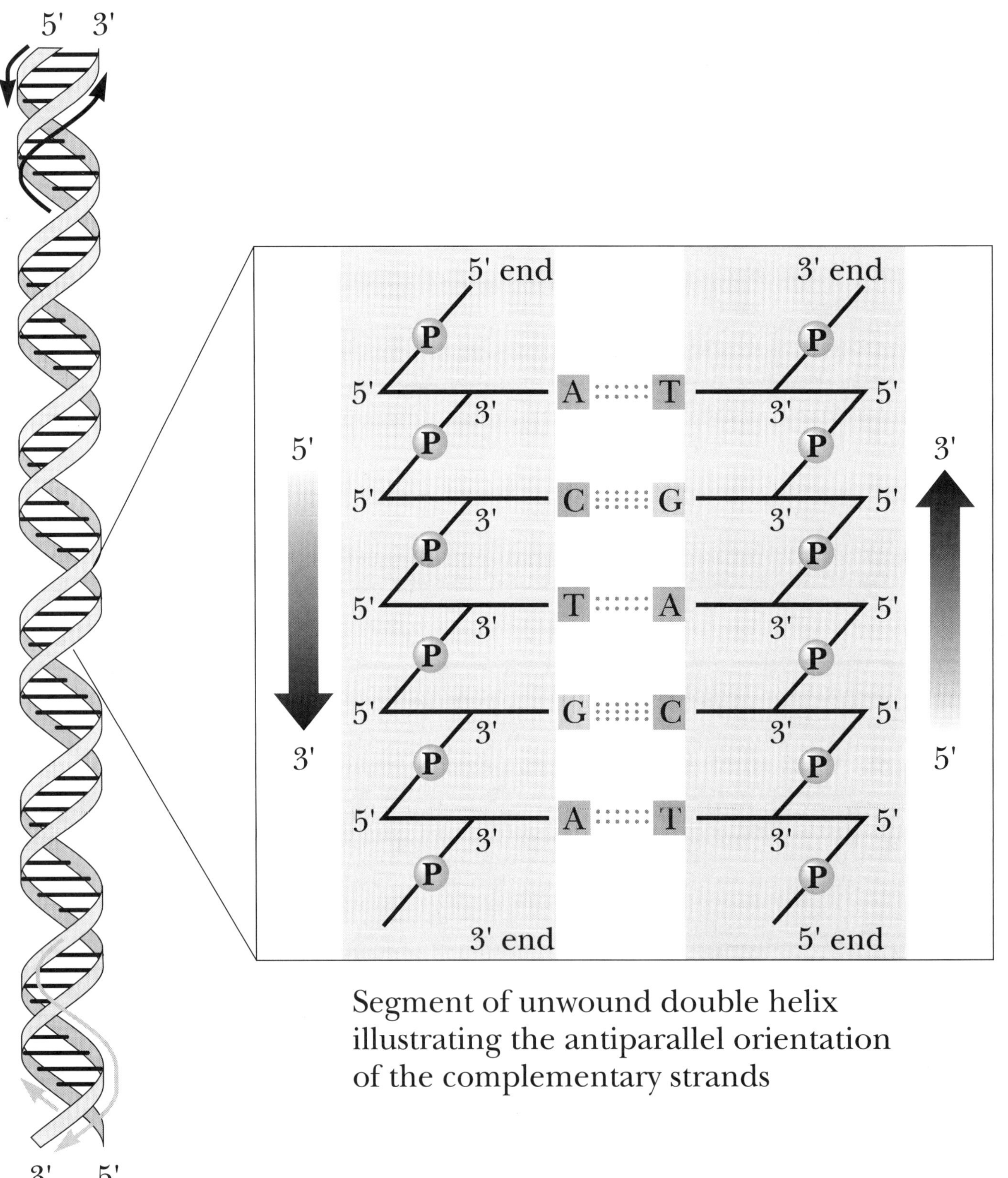

Segment of unwound double helix
illustrating the antiparallel orientation
of the complementary strands

Figure 8.19   The Watson-Crick base pairs A : T and G : C.

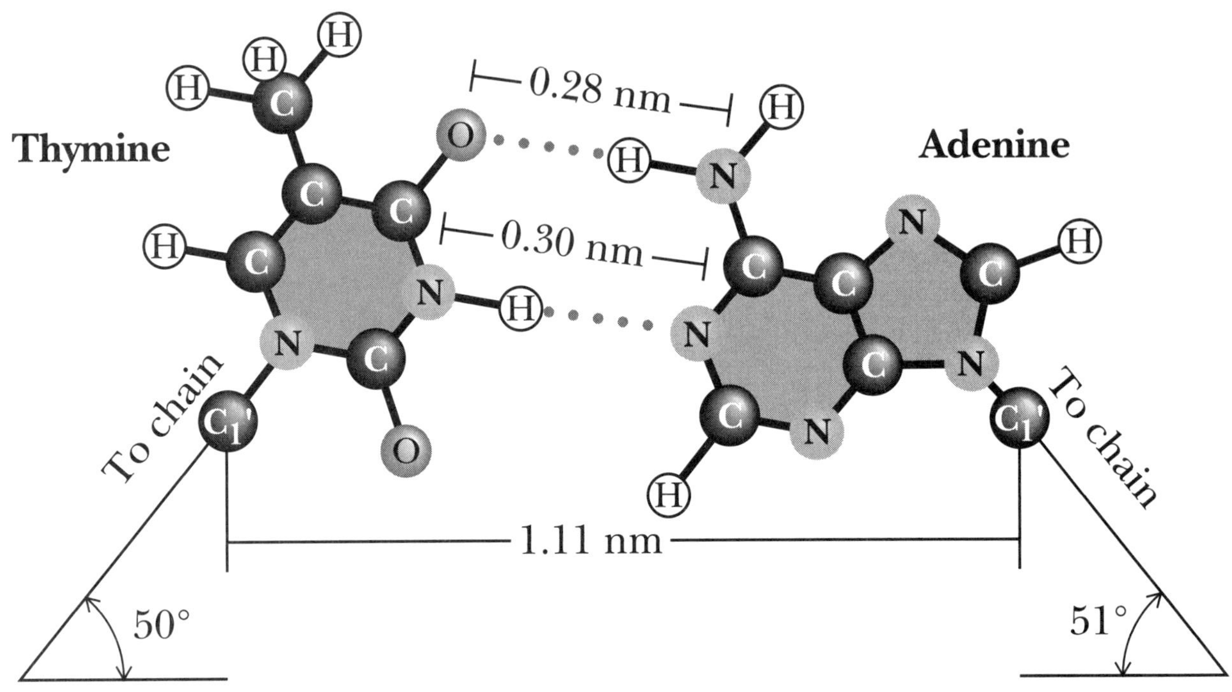

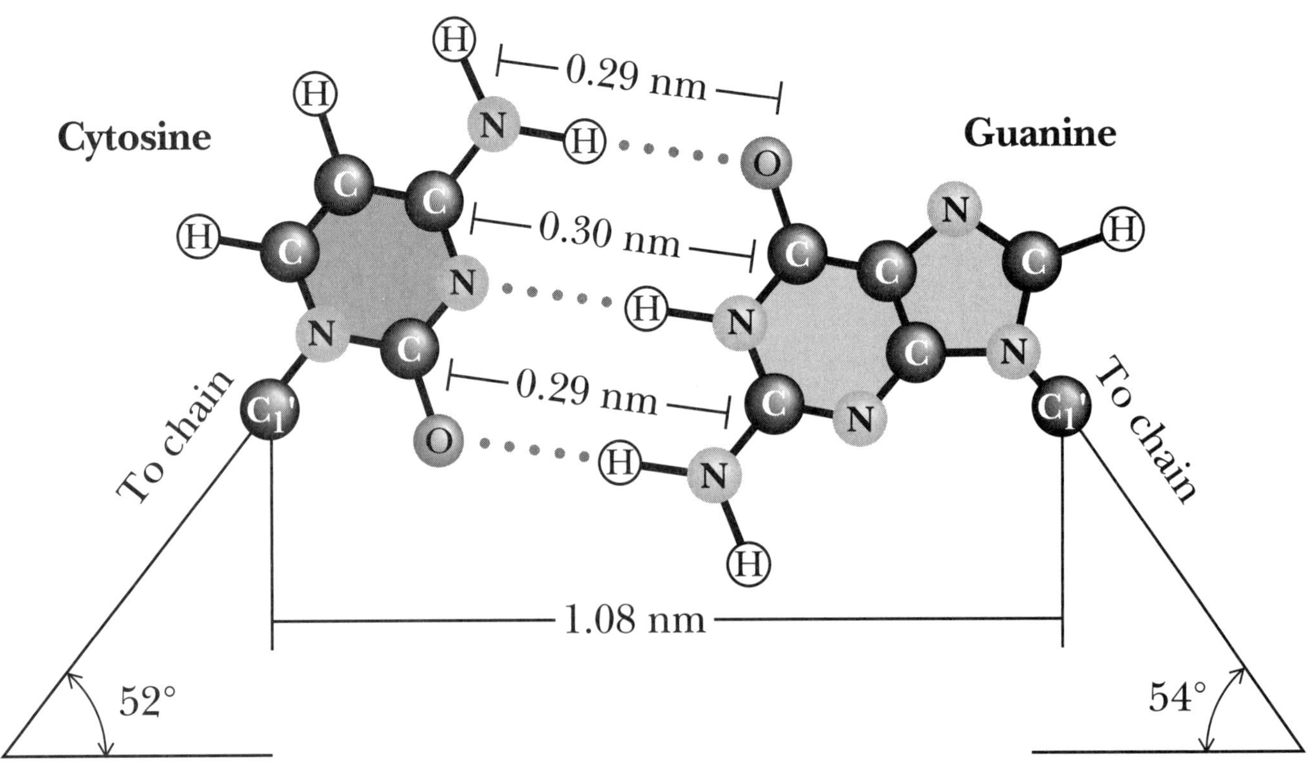

Figure 8.22 (part 1)   The properties of prokaryotic and eukaryotic mRNA.

Prokaryotes:

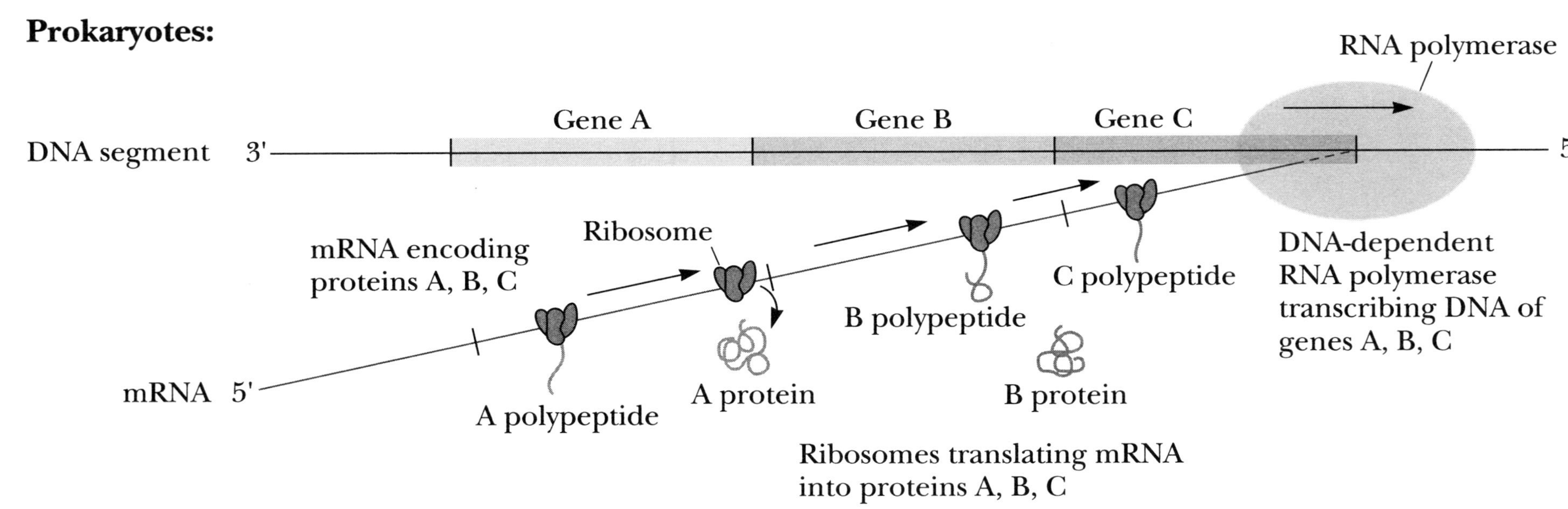

Figure 8.22 (part 2)   The properties of prokaryotic and eukaryotic mRNA.

Eukaryotes:

Exons are protein-coding regions that must be joined by removing introns, the noncoding intervening sequences. The process of intron removal and exon joining is called splicing.

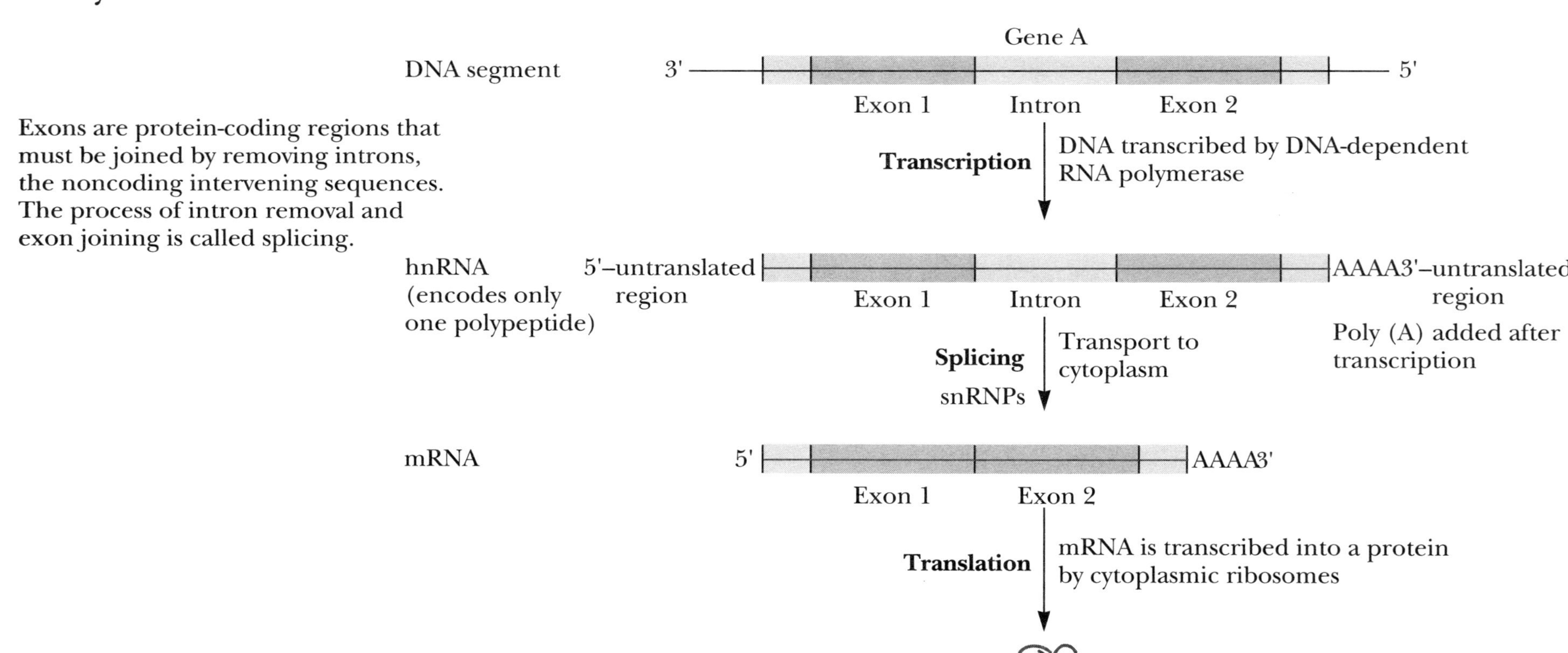

Figure 8.23   The organization and composition of prokaryotic and eukaryotic ribosomes.

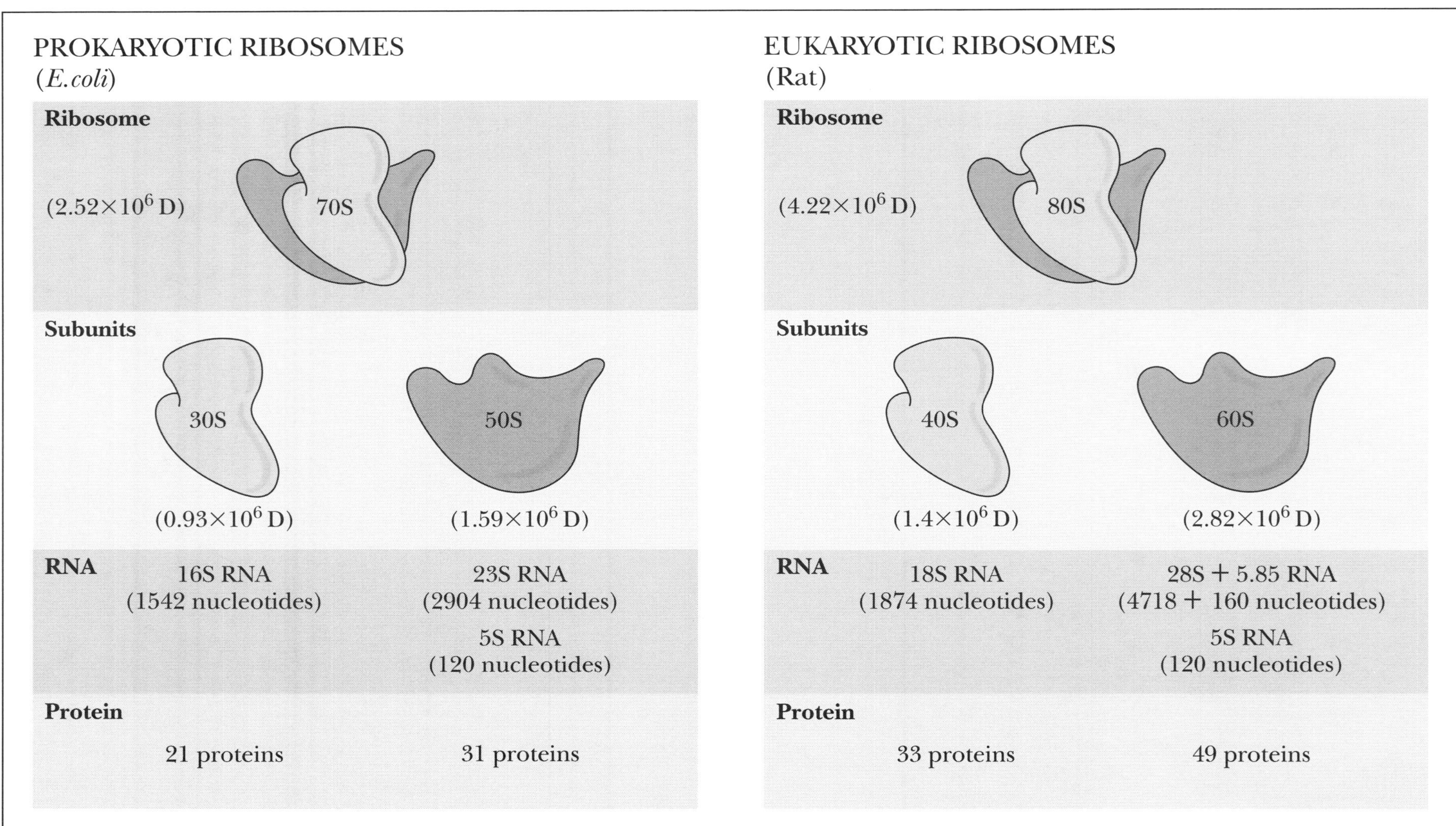

Table 8.3 Restriction endonucleases.

About 1000 restriction enzymes have been characterized. They are named by italicized three-letter codes, the first a capital latter denoting the genus of the organism of origin, while the next two letters are an abbreviation of the particular species. Because prokaryotes often contain more than one restriction enzyme, the various representatives are assigned letter and number codes as they are identified. Thus, *Eco*RI is the initial restriction endonuclease isolated from *Escherichia coli,* strain R. With one exception (*Nci*I), all known type II restriction endonucleases generate fragments with 5′-PO$_4$ and 3′-OH ends.

Enzyme	Common Isoschizomers	Recognition Sequence	Compatible Cohesive Ends
*Alu*I		AG↓CT	Blunt
*Ava*I		G↓PyCGPuG	*Sal*I, *Xho*I, *Xma*I
*Avr*II		C↓CTAGG	
*Bal*I		TGG↓CCA	Blunt
*Bam*HI		G↓GATCC	*Bcl*I, *Bgl*II, *Mbo*I, *Sau*3A, *Xho*II
*Bcl*I		T↓GATCA	*Bam*HI, *Bgl*II, *Mbo*I, *Sau*3A, *Xho*II
*Bgl*II		A↓GATCT	*Bam*HI, *Bcl*I, *Mbo*I, *Sau*3A, *Xho*II
*Bst*XI		CCANNNNN↓NTGG	
*Cla*I		AT↓CGAT	*Acc*I, *Acy*I, *Asy*II, *Hpa*II, *Taq*I
*Eco*RI		G↓AATTC	
*Eco*RII	*Atu*I, *Apy*I	↓CC (^A_T)GG	
*Hae*I		(^A_T)GG↓CC(^T_A)	Blunt
*Hae*II		PuGCGC↓Py	
*Hae*III		GG↓CC	Blunt
*Hinc*II		GTPy↓PuAC	Blunt
*Hind*III		A↓AGCTT	
*Hpa*I		GTT↓AAC	Blunt
*Hpa*II		C↓CGG	*Acc*I, *Acy*I, *Asu*II, *Cla*I, *Taq*I
*Kpn*I		GGTAC↓C	
*Mbo*I	*Sau*3A	↓GATC	*Bam*HI, *Bcl*I, *Bgl*II, *Xho*II
*Not*I		GC↓GGCCGC	
*Pst*I		CTGCA↓G	
*Sac*I	*Sst*I	GAGCT↓C	
*Sal*I		G↓TCGAC	*Ava*I, *Xho*I
*Sfi*I		GGCCNNNN↓NGGCC	
*Sma*I	*Xma*I	CCC↓GGG	Blunt
*Taq*I		T↓CGA	*Acc*I, *Acy*I, *Asu*II, *Cla*I, *Hpa*II
*Xba*I		T↓CTAGA	
*Xho*I		C↓TCGAG	*Ava*I, *Sal*I
*Xho*II		(^A_G)↓GATC(^T_C)	*Bam*HI, *Bcl*I, *Bgl*II, *Mbo*I, *Sau*3A

© **Harcourt, Inc.**

Figure 8.30   Restriction mapping of a DNA molecule.

Treatment of a linear 10kb DNA molecule with endonucleases gave the following results:

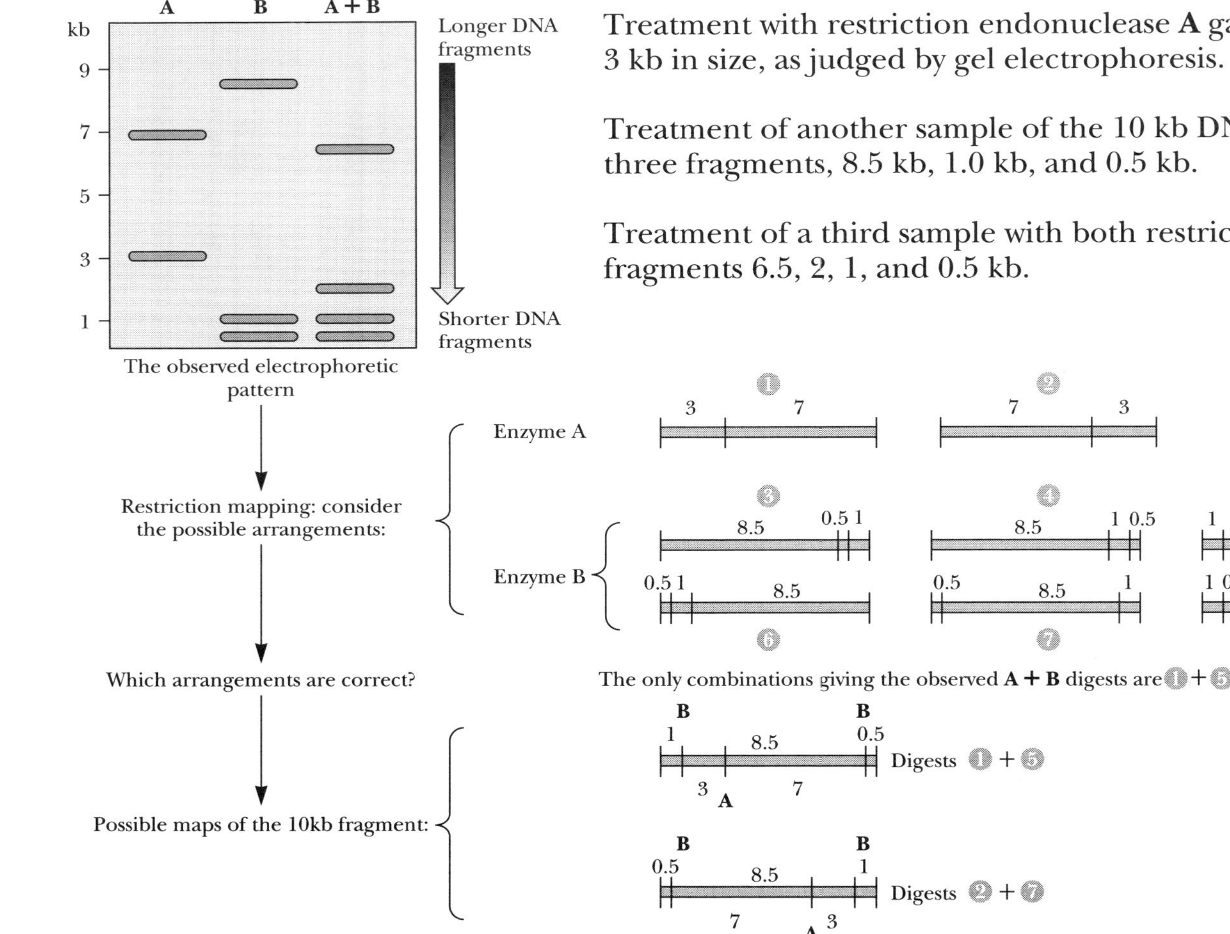

Treatment with restriction endonuclease **A** gave 2 fragments, one 7 kb in size and one 3 kb in size, as judged by gel electrophoresis.

Treatment of another sample of the 10 kb DNA with restriction endonuclease **B** gave three fragments, 8.5 kb, 1.0 kb, and 0.5 kb.

Treatment of a third sample with both restriction endonucleases **A** and **B** yielded fragments 6.5, 2, 1, and 0.5 kb.

To decide between these alternatives, a fixed point of reference, such as one of the ends of the fragment, must be identified or labeled. The task increases in complexity as DNA size, number of restriction sites, and/or number of restriction enzymes used increases.

Figure 8.33   Chain termination of dideoxy method of DNA sequencing.

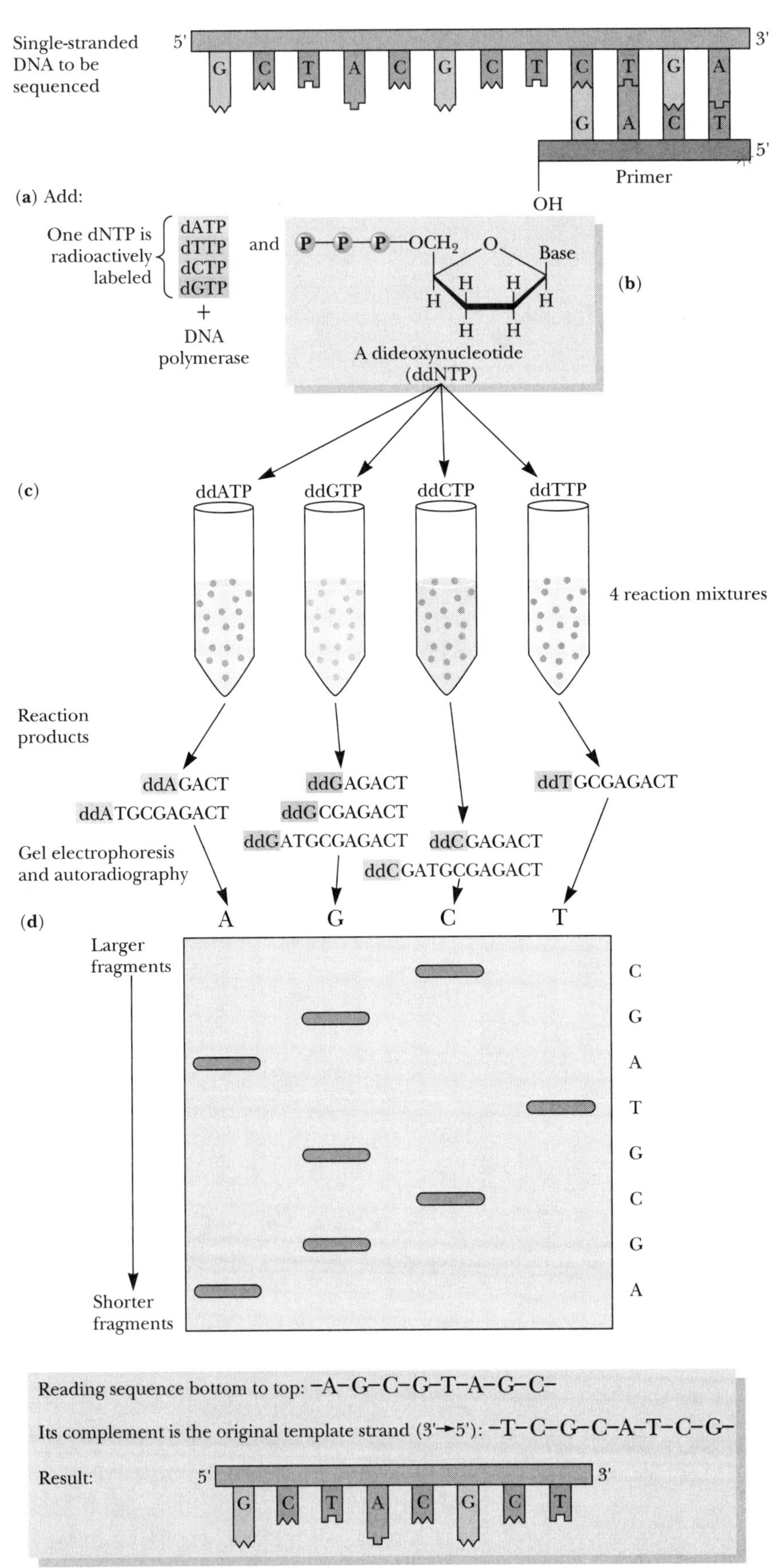

© Harcourt, Inc.

Figure 8.36   Bases in a base pair are not directly across the helix axis from one another.

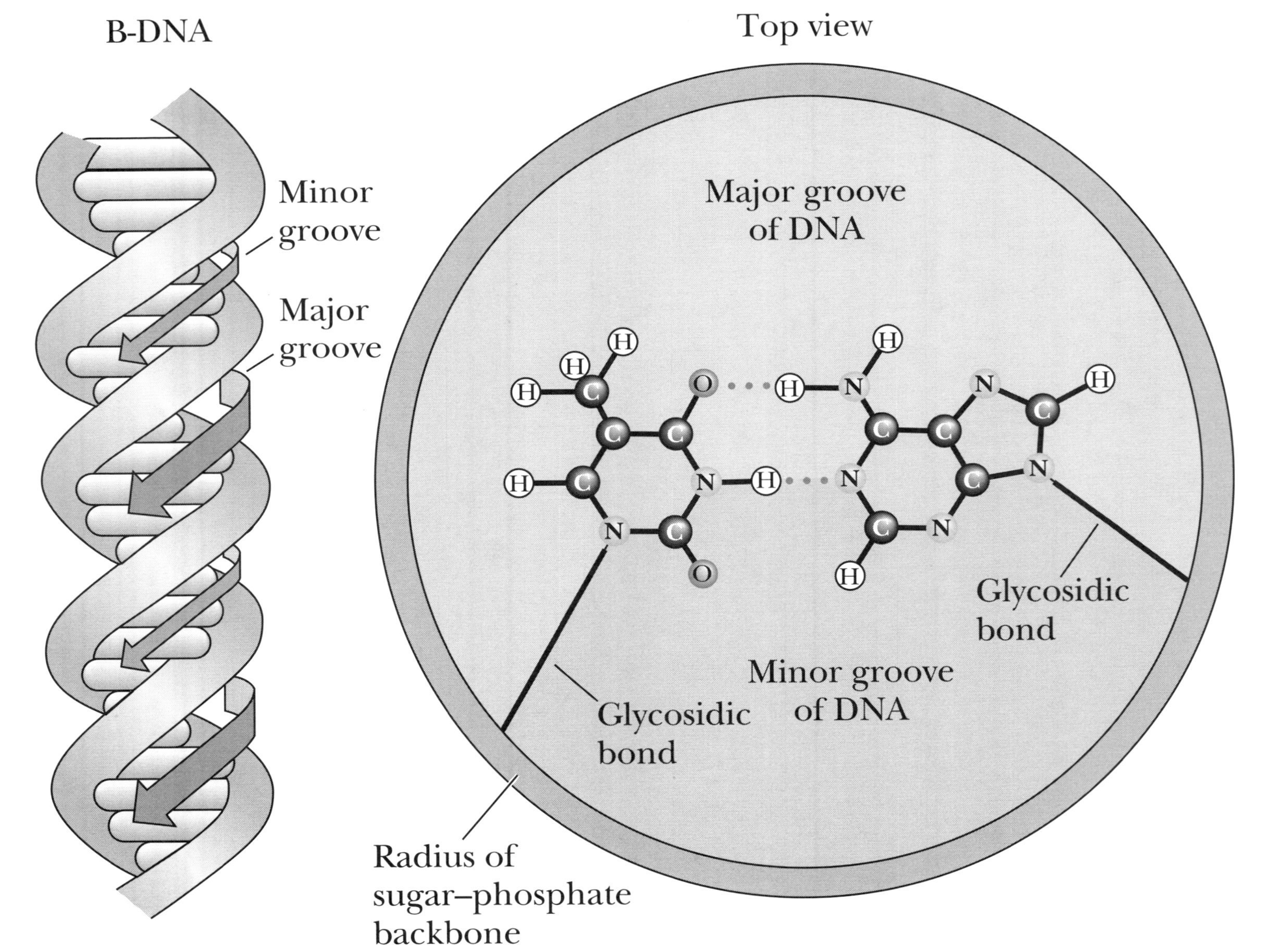

© Harcourt, Inc.

Figure 8.37 (part 1)   Comparison of the A-, B-, and Z-forms of the DNA double helix.

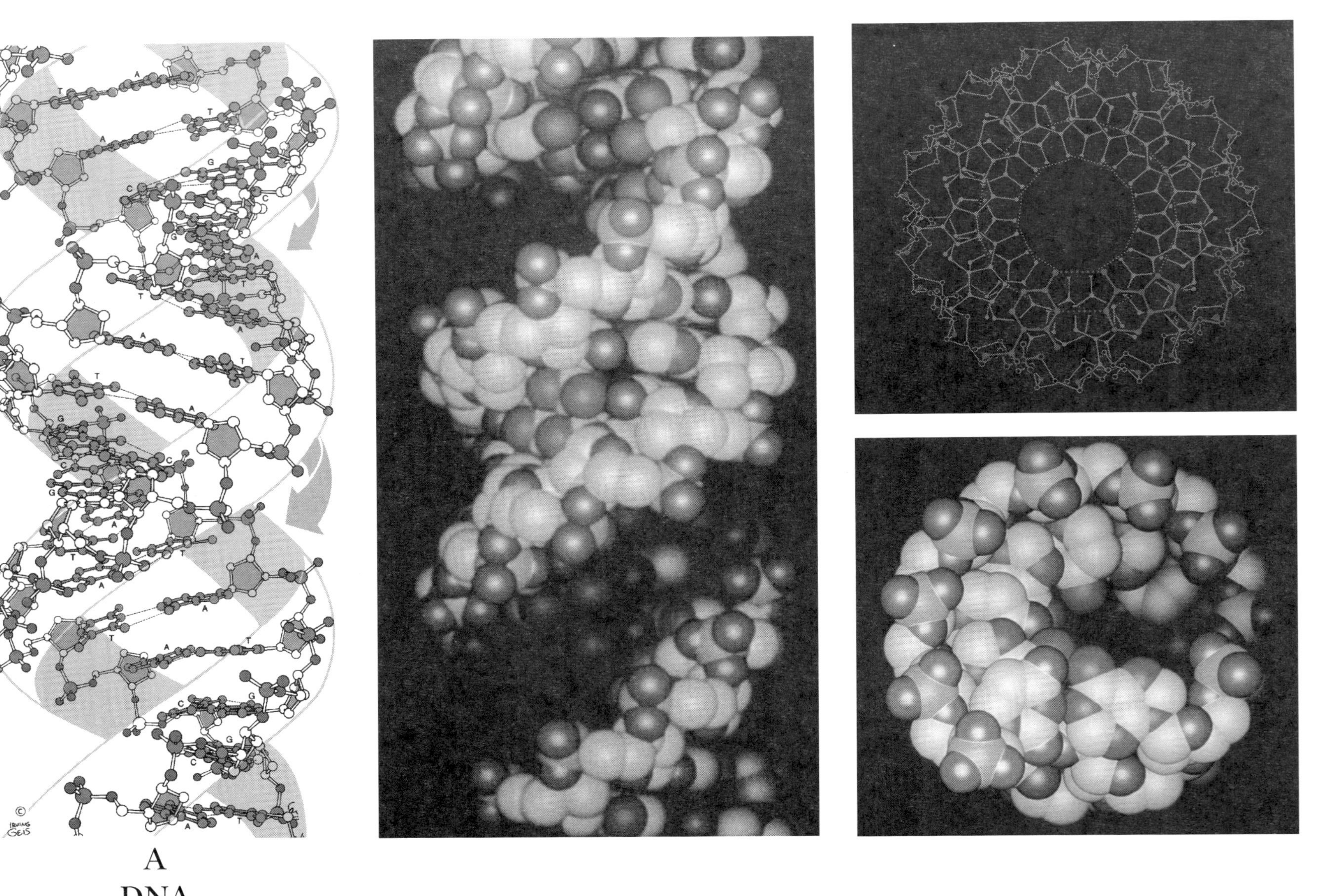

Figure 8.37 (part 2) Comparison of the A-, B-, and Z-forms of the DNA double helix.

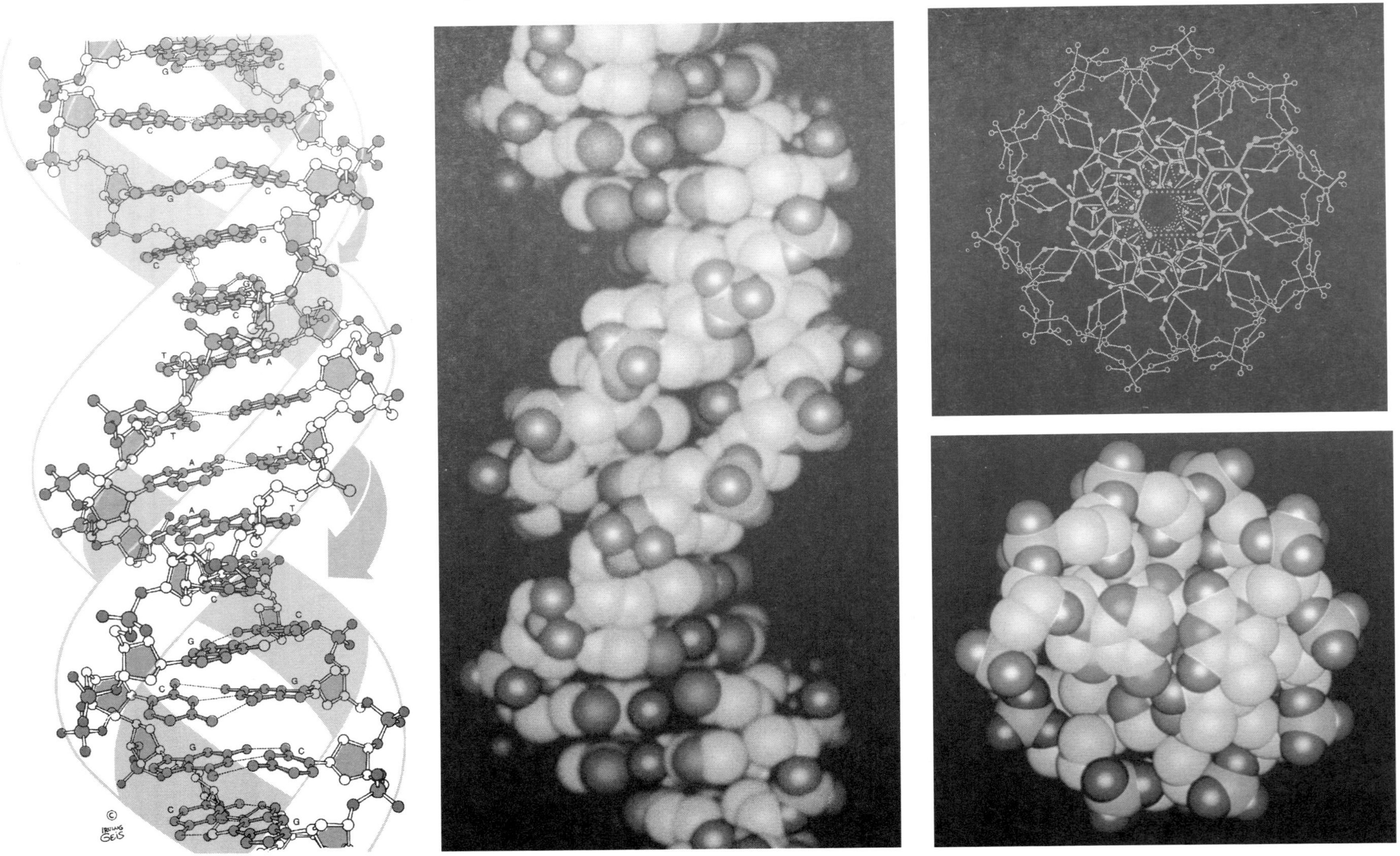

B
DNA

Figure 8.37 (part 3)   Comparison of the A-, B-, and Z-forms of the DNA double helix.

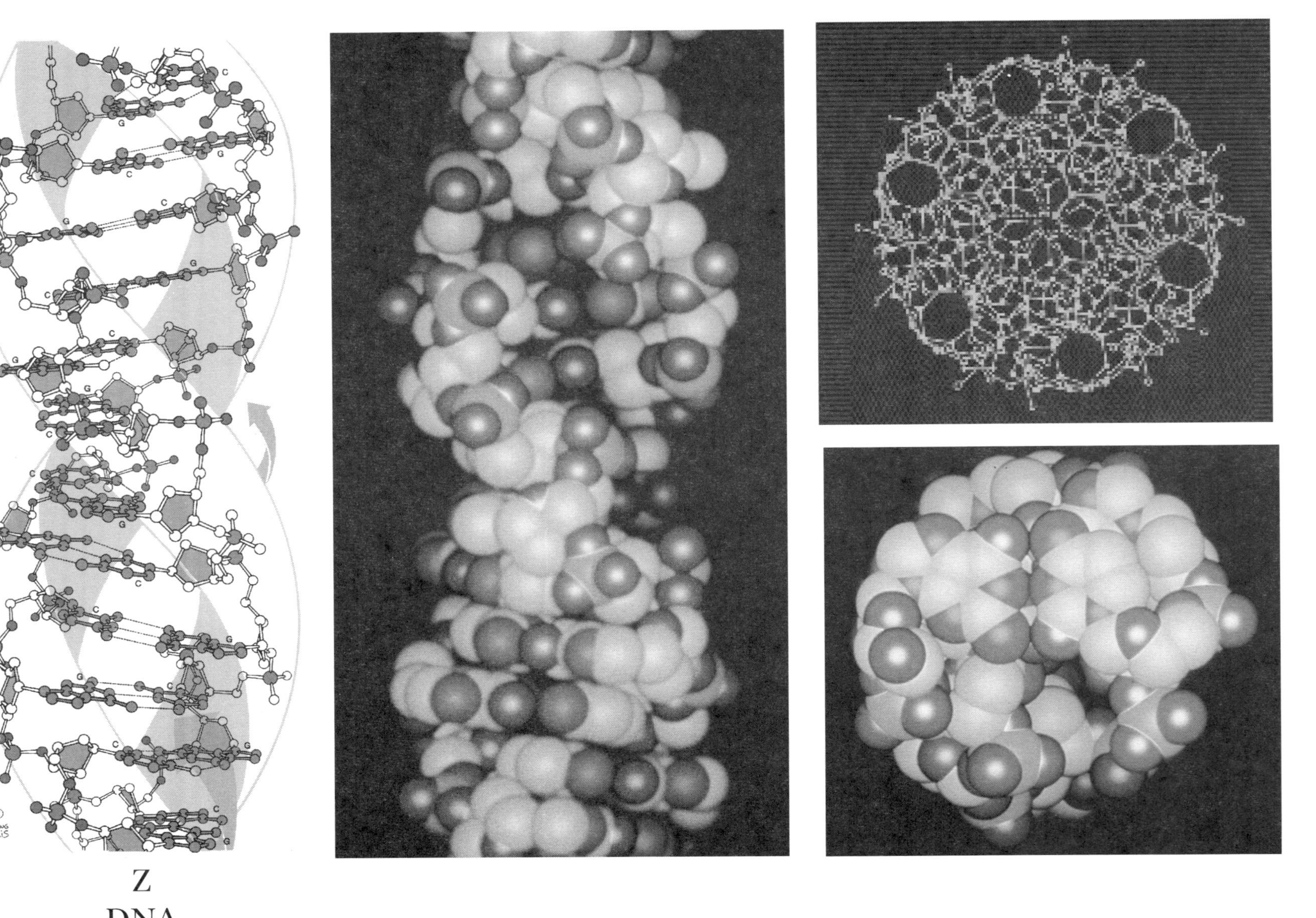

Table 8.4   Comparison of the structural properties of A-, B-, and Z-DNA.

Table 8.4 Comparison of the Structural Properties of A-, B-, and Z-DNA			
	Double Helix Type		
	A	**B**	**Z**
Overall proportions	Short and broad	Longer and thinner	Elongated and slim
Rise per base pair	2.3 Å	3.32 Å $\pm$ 0.19 Å	3.8 Å
Helix packing diameter	25.5 Å	23.7 Å	18.4 Å
Helix rotation sense	Right-handed	Right-handed	Left-handed
Base pairs per helix repeat	1	1	2
Base pairs per turn of helix	~11	~10	12
Mean rotation per base pair	33.6°	35.9° $\pm$ 4.2°	$-60°/2$
Pitch per turn of helix	24.6 Å	33.2 Å	45.6 Å
Base-pair tilt from the perpendicular	+19°	$-1.2°$ $\pm$ 4.1°	$-9°$
Helix axis location	Major groove	Through base pairs	Minor groove
Glycosyl bond conformation	anti	anti	anti at C, syn at G

Adapted from Dickerson, R. L., et al., 1982. *Cold Spring Harbor Symposium on Quantitative Biology* **47:** *14.*

Figure 8.43 Toroidal and interwound varieties of DNA supercoiling.

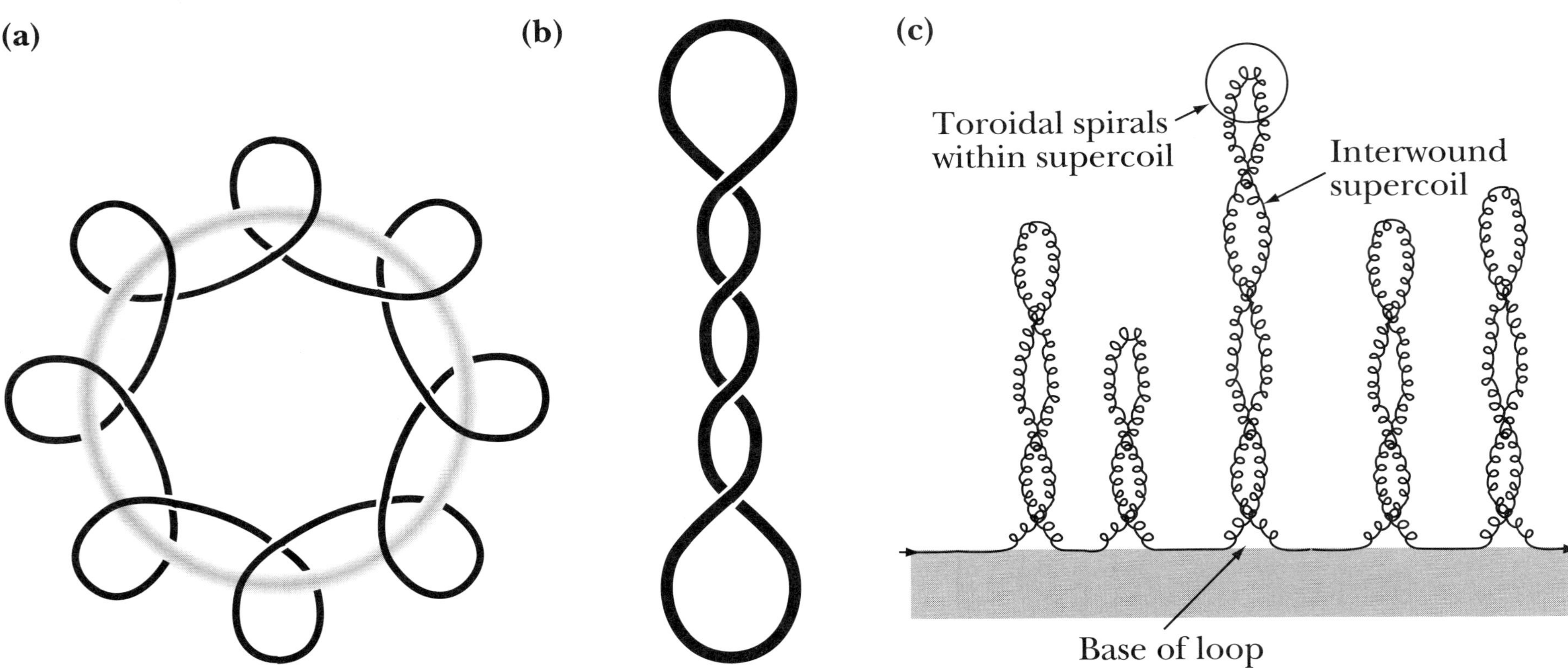

Figure 8.44 A model for the action of bacterial DNA gyrase.

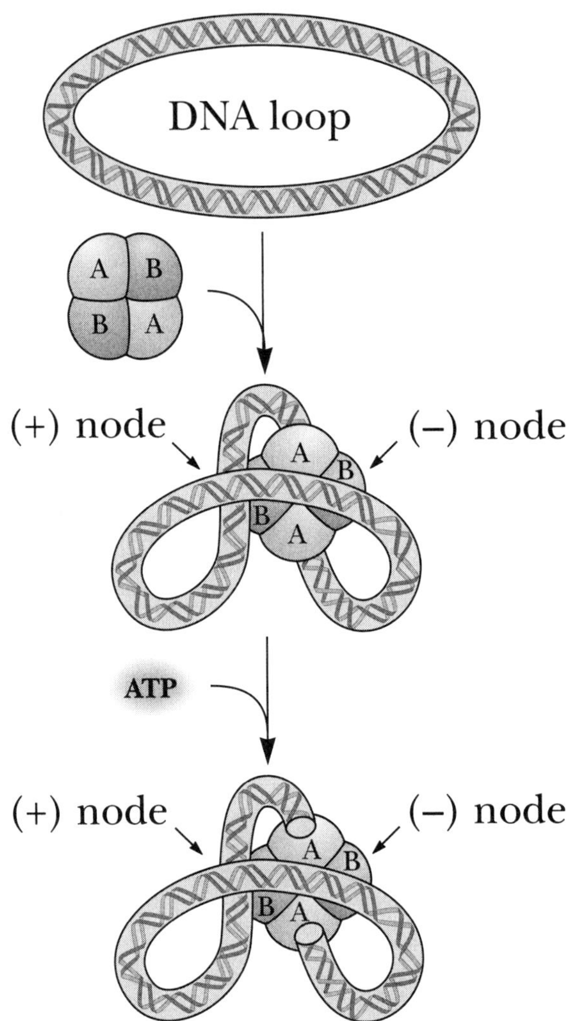

DNA is cut and a conformational change allows the DNA to pass through. Gyrase religates the DNA and then releases it.

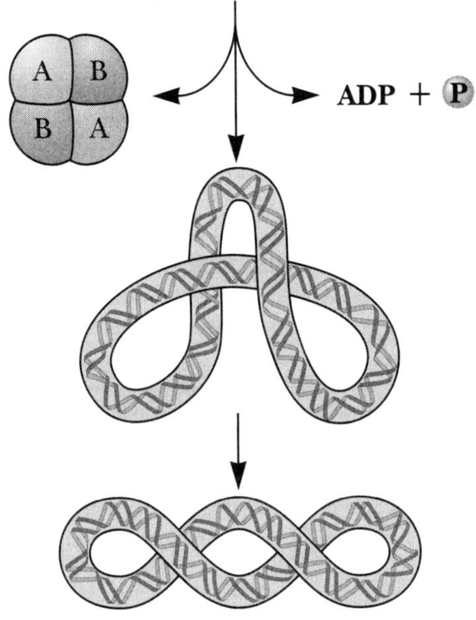

Figure 8.52 A general diagram for the structure of tRNA.

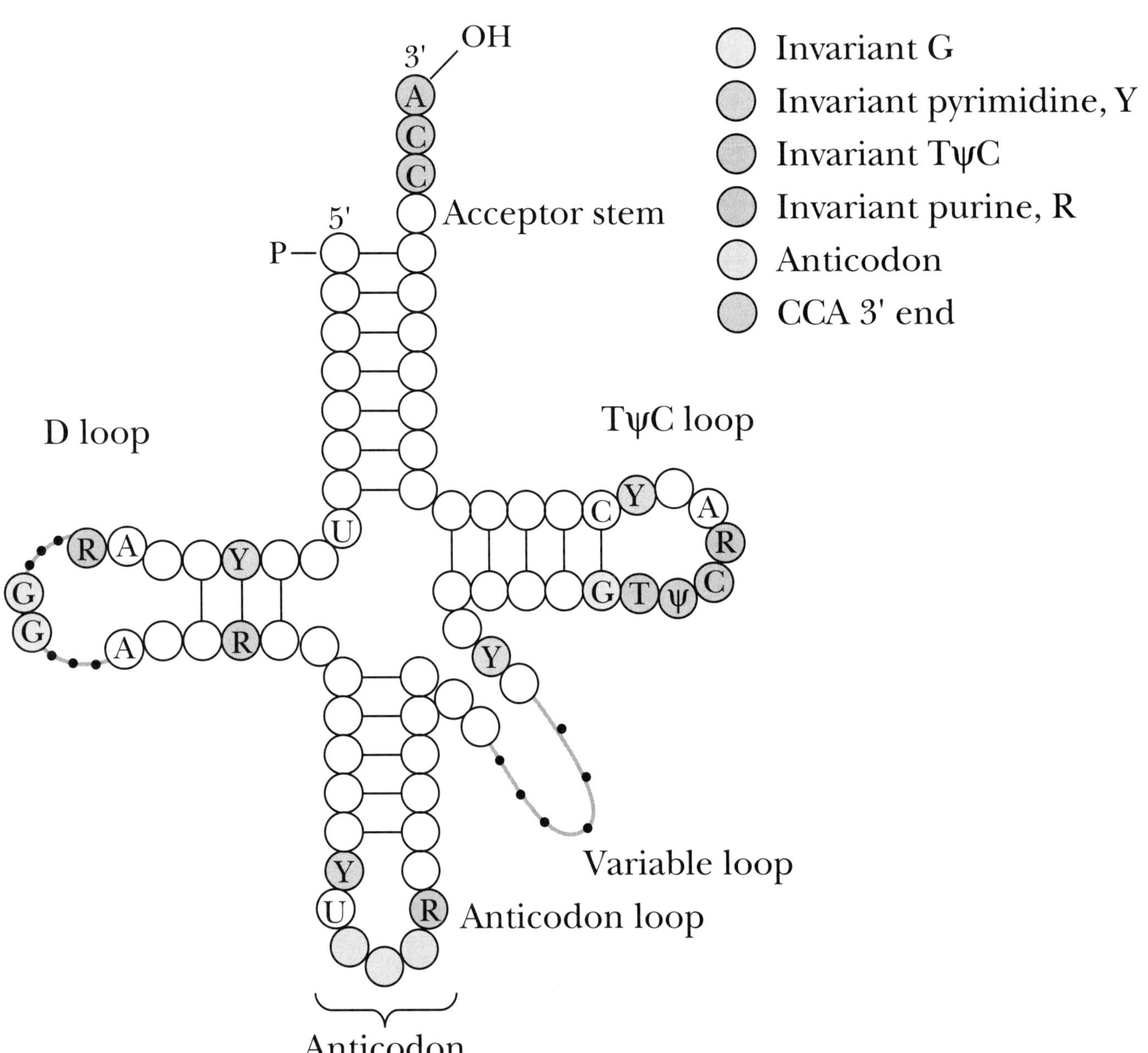

Figure 8.54 Tertiary interactions in yeast phenylalanine tRNA.

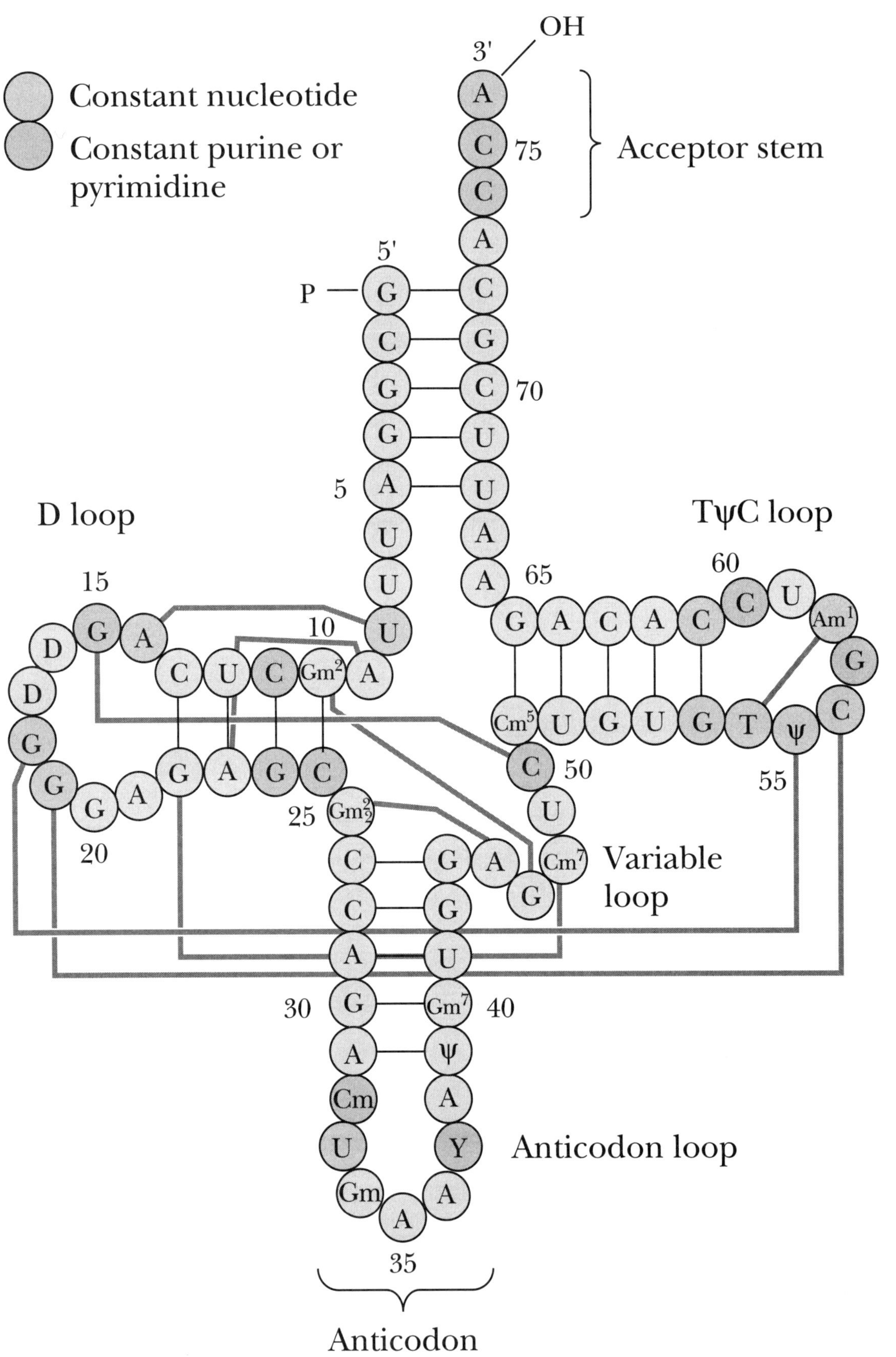

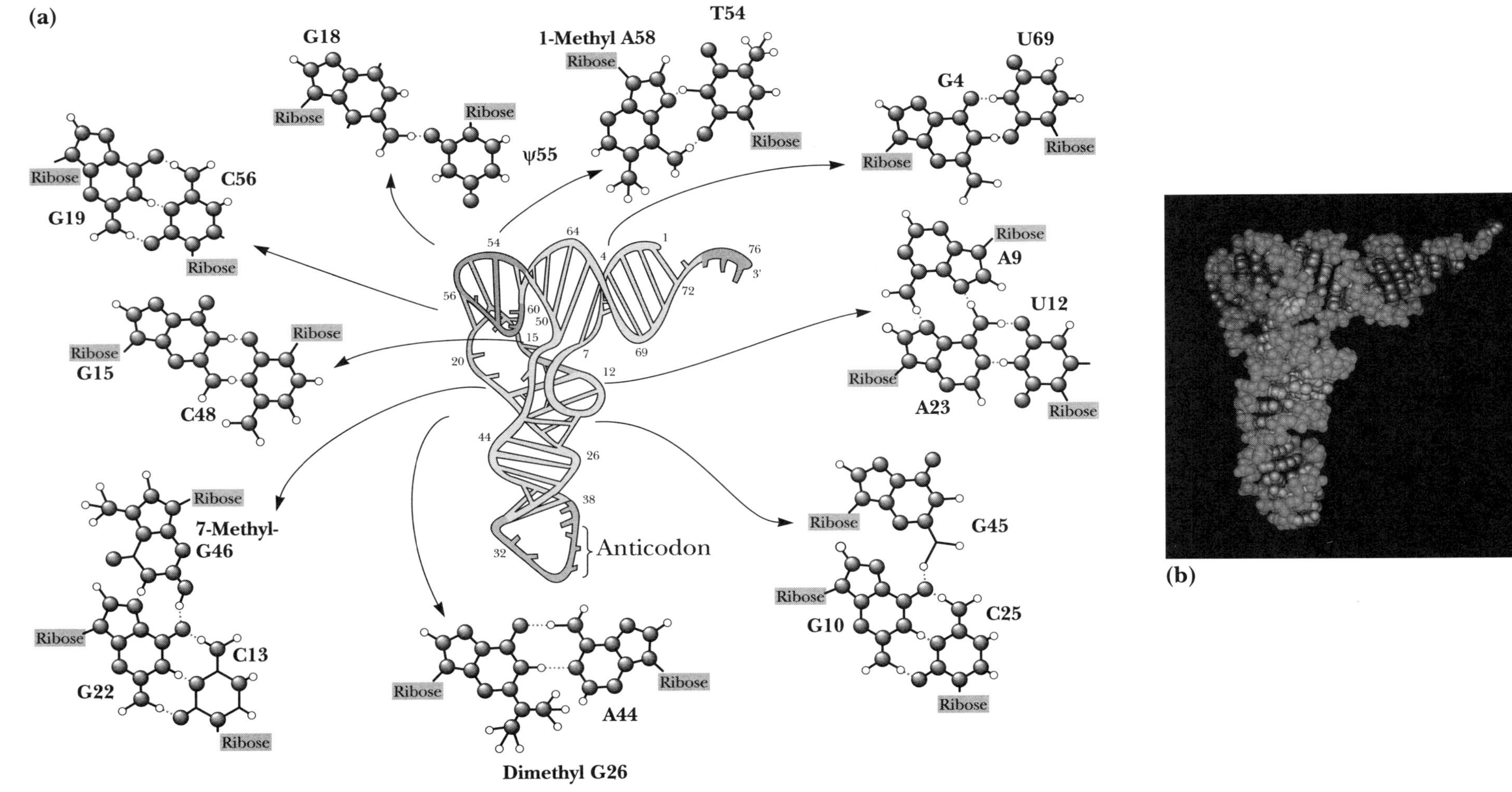

Figure 8.55   The 3-D structure of yeast phenylalanine tRNA.

Figure 9.3   Restriction endonuclease EcoRI leaves double-stranded DNA.

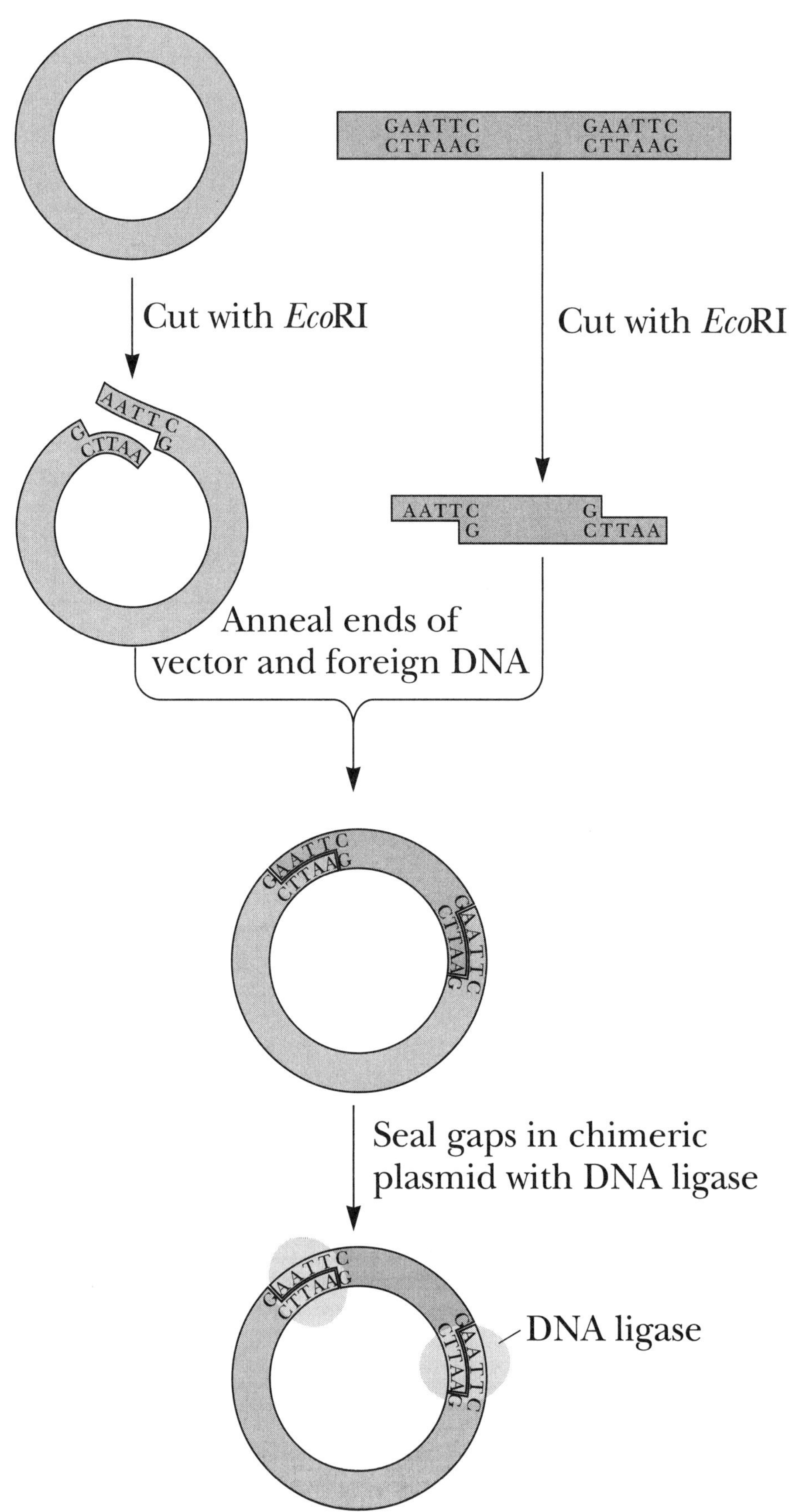

Figure 9.4   Blunt-end ligation using phage T4 DNA ligase.

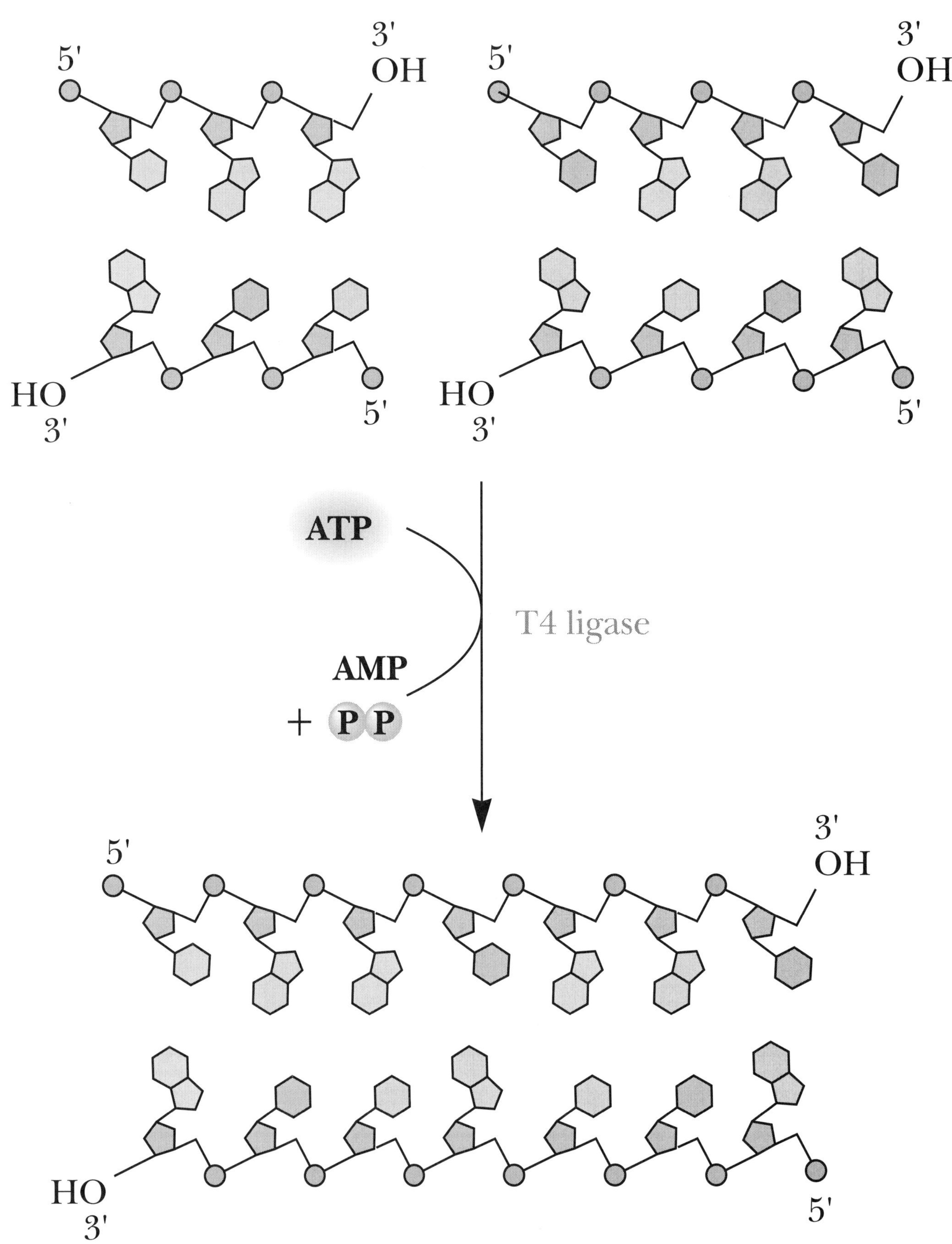

Figure 9.5 Linkers to create tailor-made ends on cloning fragments.

(a)

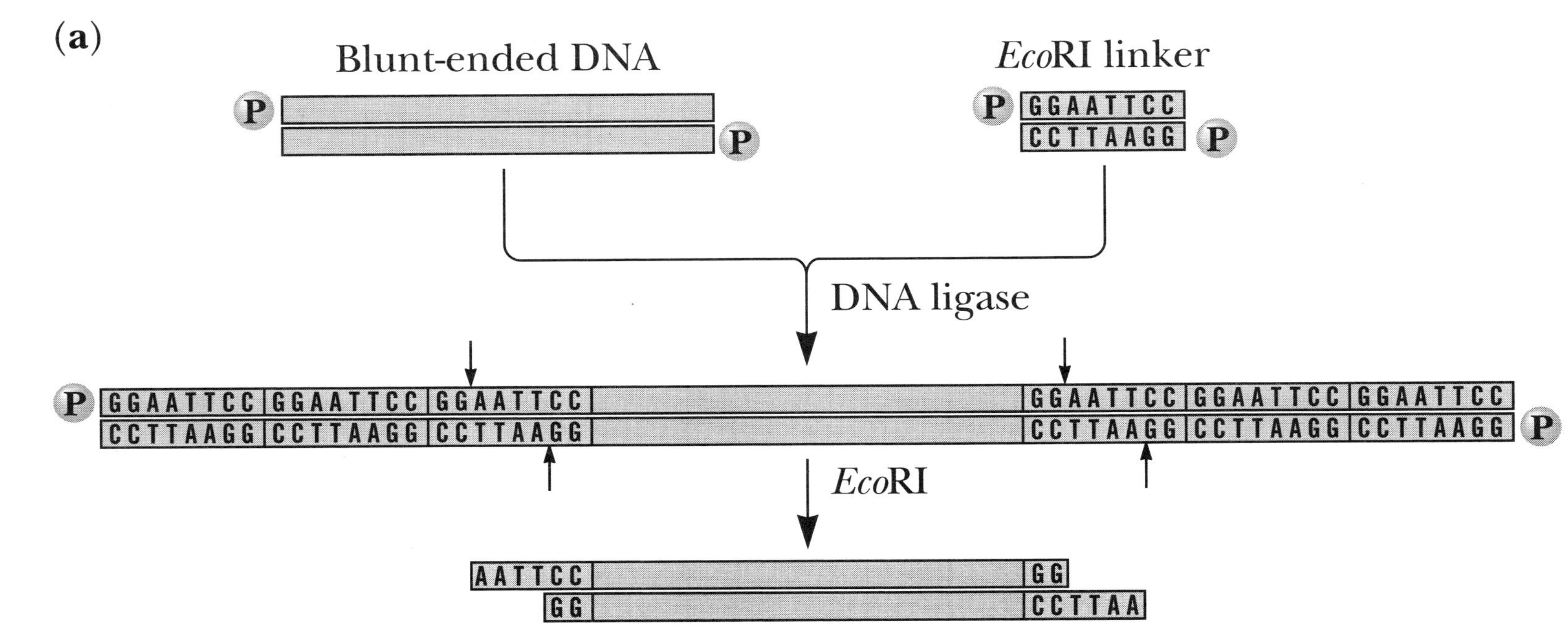

(b) A vector cloning site containing multiple restriction sites, a so-called *polylinker*.

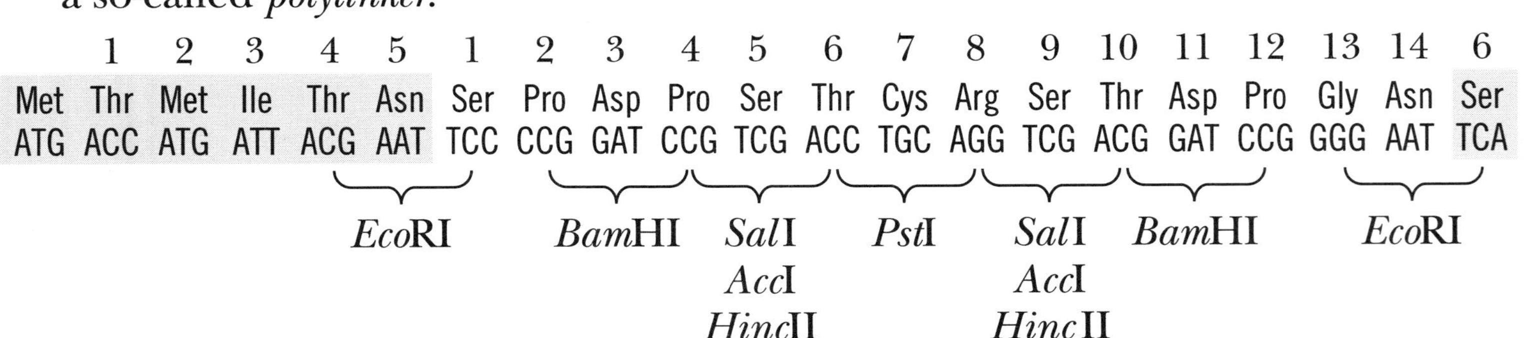

Figure 9.6 Directional cloning.

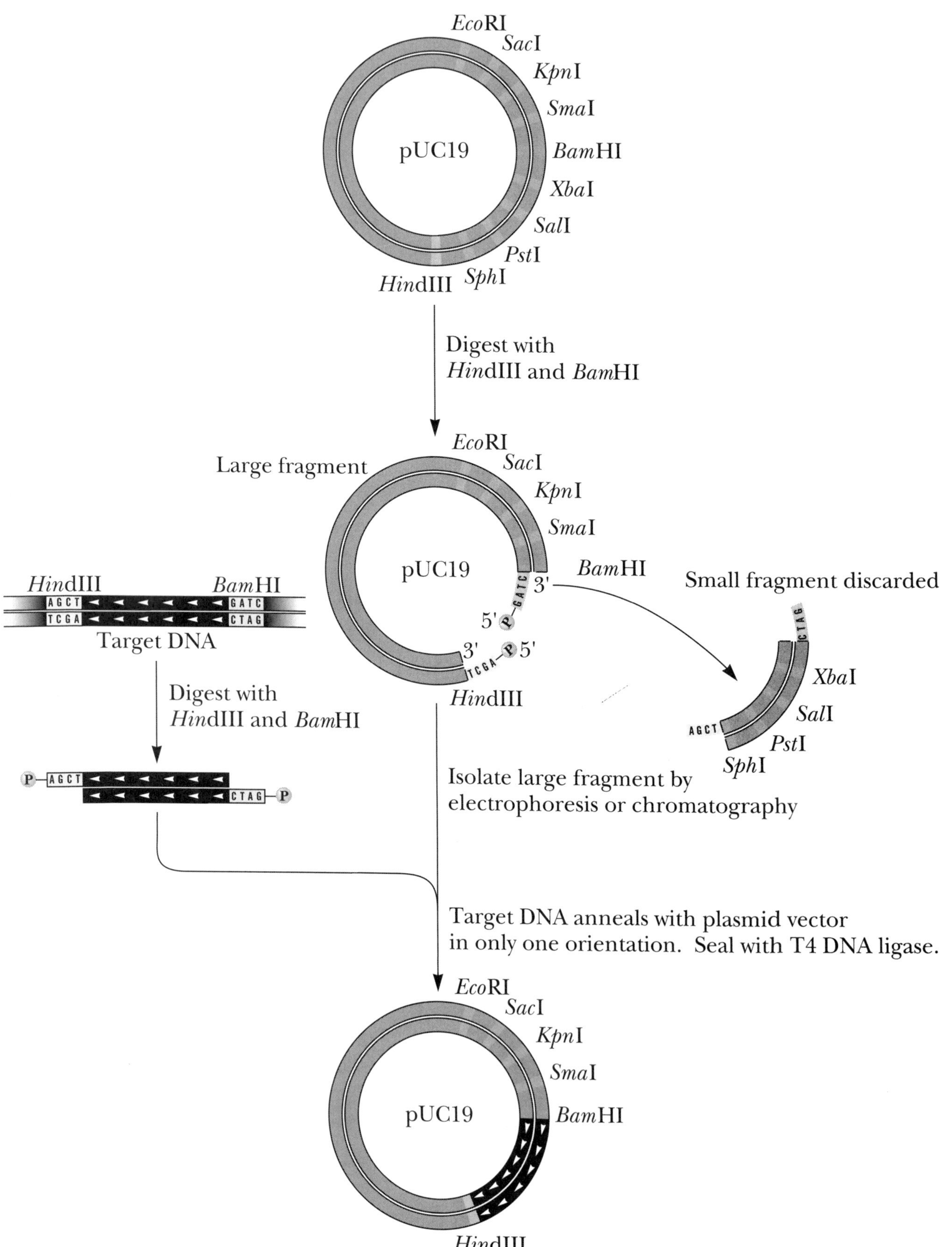

Figure 9.7　A typical bacterial transformation experiment.

pBR322
(4363 bases)
amp^r
tet^r
ori
EcoRI
HindIII
EcoRV
BamHI
SalI
AvaI
SalI
PvuII
PvuI
PstI

amp^r gene remains intact.

Chimeric plasmid

3 tet^r gene is split by the insertion of DNA fragment. amp^r gene remains intact.

1 A plasmid with genes for ampicillin resistance (amp^r) and tetracycline resistance (tet^r). A BamHI restriction site is located within the tet^r gene.

2 BamHI restriction fragment of DNA to be cloned is inserted into the BamHI site of tet^r.

4 Suspend 20 ng plasmid DNA + 10^7 E.coli cells in CaCl$_2$ solution.

42°C, 2 min

5 Plate bacteria on ampicillin media.

37°C, overnight

Ampicillin-containing medium

Tetracycline-containing medium

37°C, overnight

6 Only ampicillin-resistant (amp^r) bacterial colonies grow.

7 Using velvet-covered disk, bacterial colonies are lifted from surface of agar amp$^+$ plate and pressed briefly to surface of plate containing tetracycline media.

8 Only tet^r colonies appear; tet^s colonies can be recovered from amp^r plate by comparing two plates.

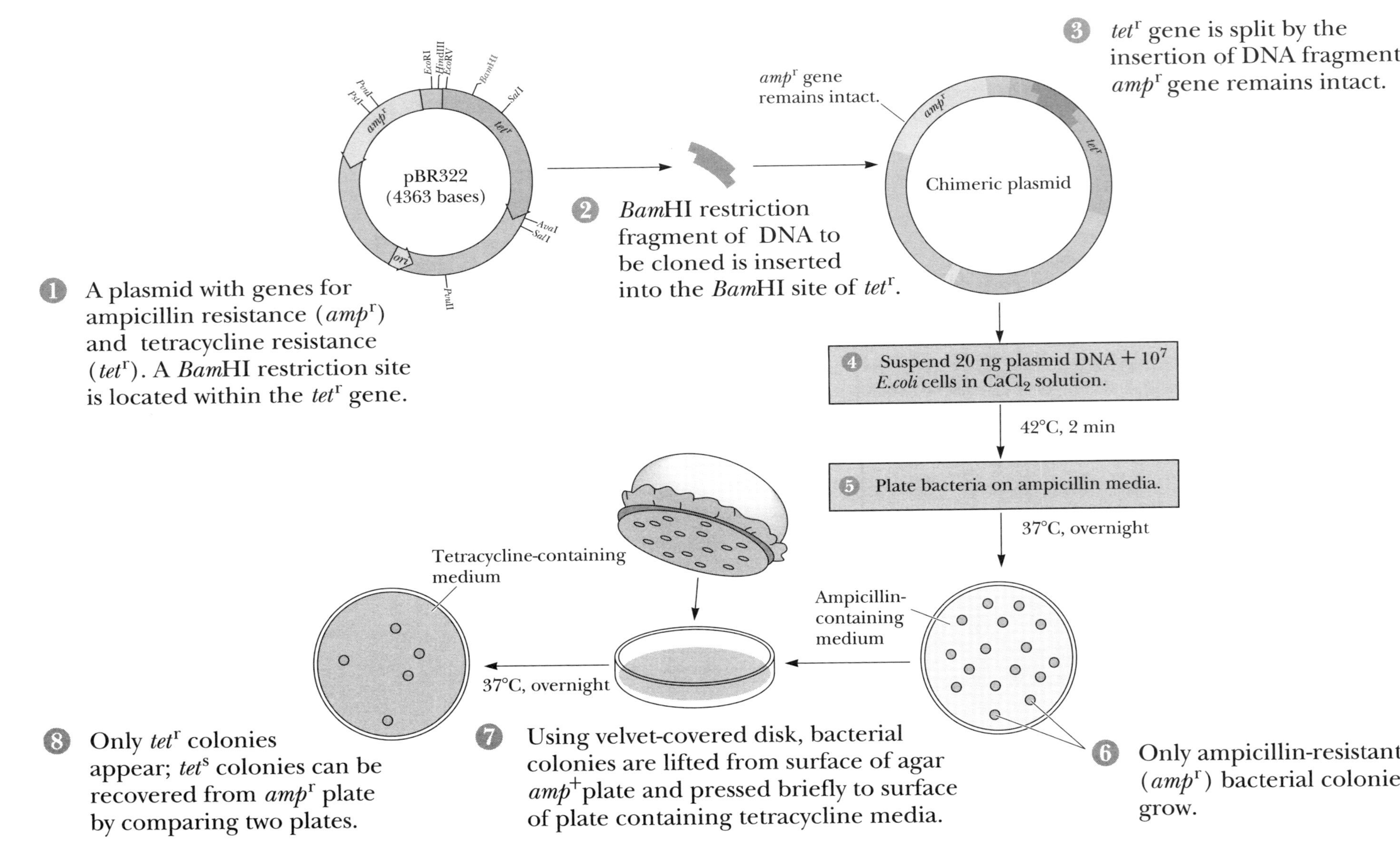

Figure 9.10 A typical shuttle vector.

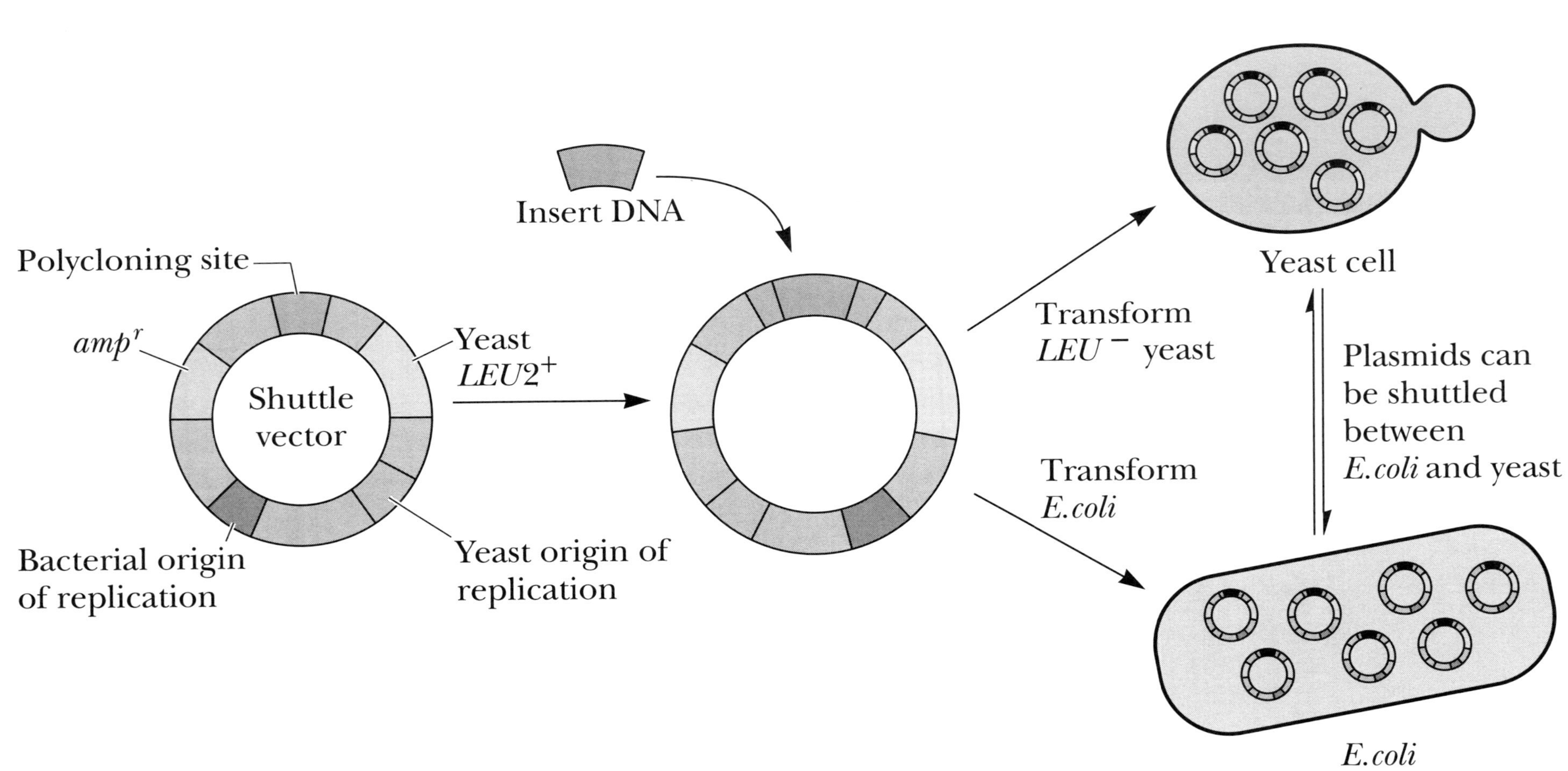

© Harcourt, Inc.

Figure 9.11   Screening a genomic library by colony hybridization.

Master plate of bacteria colonies (or phage plaques)
1 Replicate onto nitrocellulose disc
2 Treat with NaOH; neutralize, dry
Denatured DNA bound to nitrocellulose
3 Place nitrocellulose filter in sealable plastic bag with solution of labeled DNA probe
4 Wash filter, prepare autoradiograph, and compare with master plate
Radioactive probe will hybridize with its complementary DNA
5 Darkening identifies colonies (plaques) containing the DNA desired
Autoradiograph film

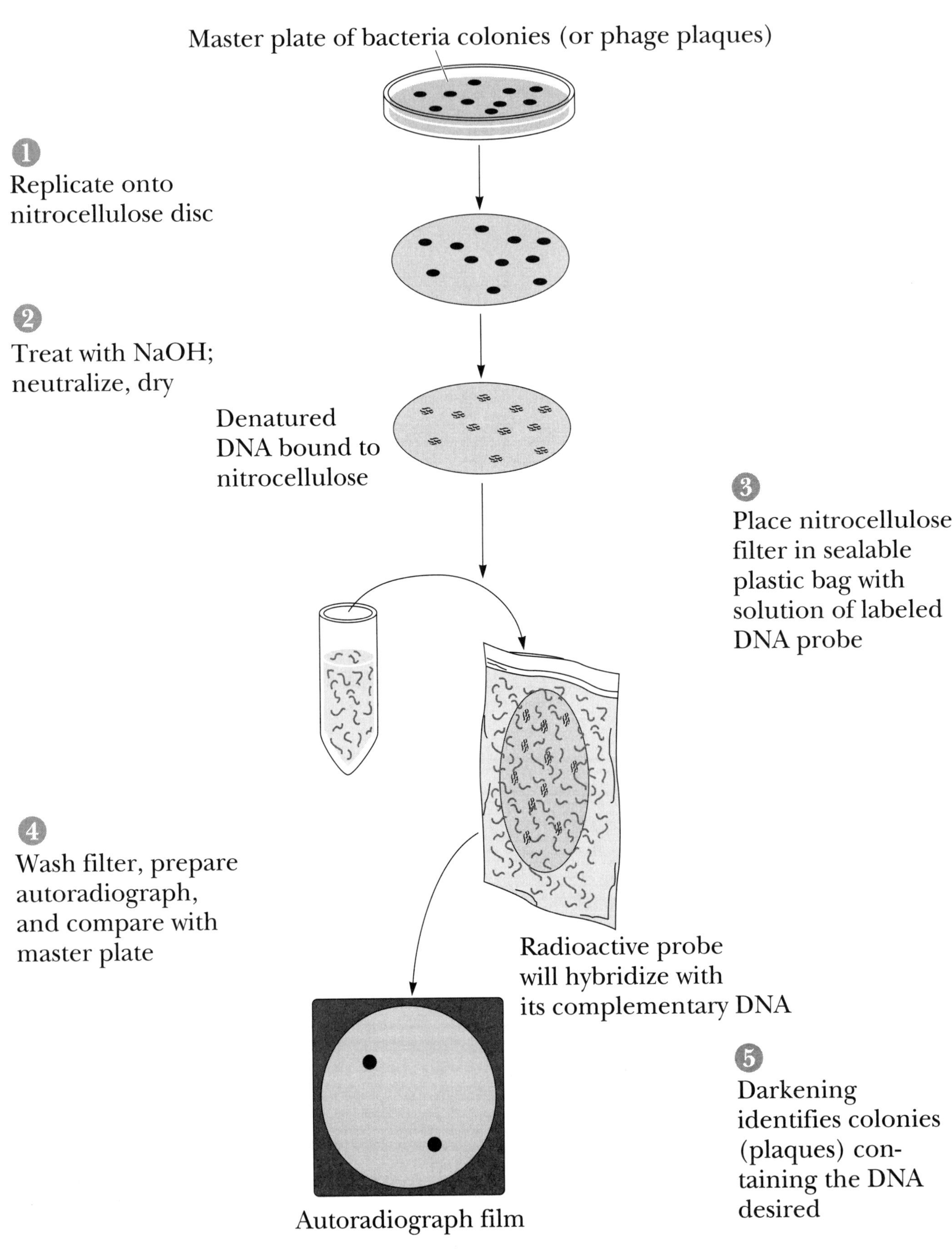

122

© Harcourt, Inc.

Figure 9.12 Cloning genes using oligonucleotide probes.

Known amino acid sequence:

Phe	Met	Glu	Trp	His	Lys	Asn

Possible mRNA sequence:

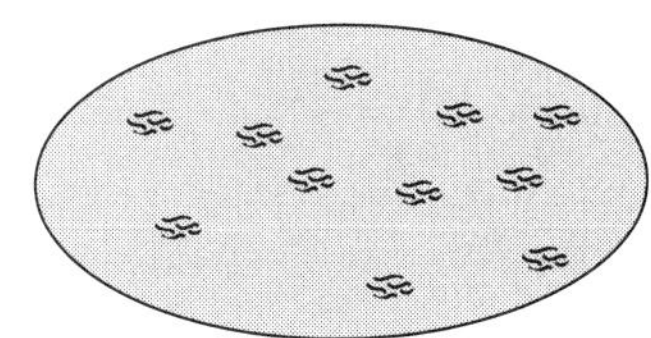

1 Nitrocellulose filter replica of bacterial colonies carrying different DNA fragments

2 Synthesize 32 possible DNA oligonucleotides and end label with radioactive ^{32}P

3 Incubate nitrocellulose filter with probe solution in plastic bag

4 Hybridization of the correct oligonucleotide to the DNA

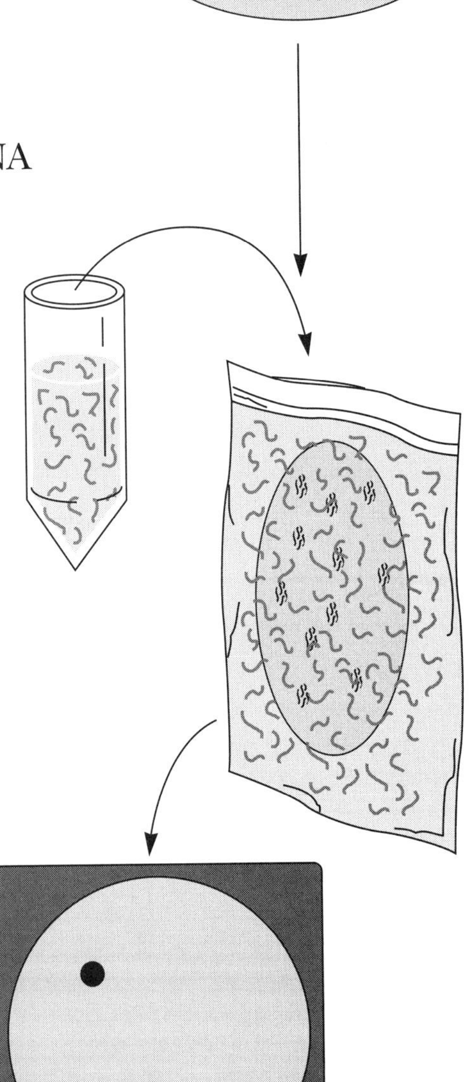

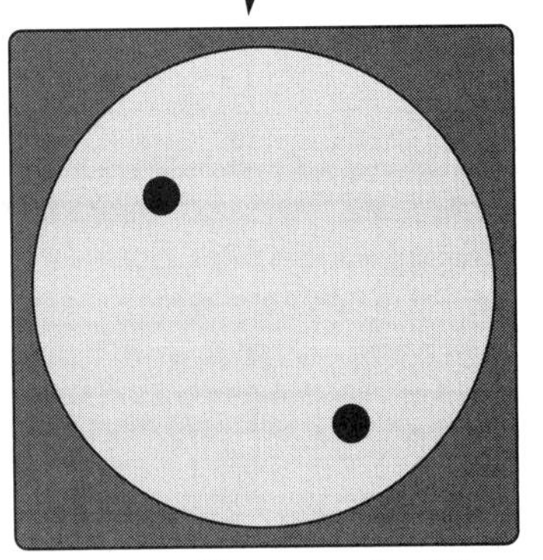

Autoradiograph film

5 Detection by autoradiography

Figure 9.14 Reverse transcriptase.

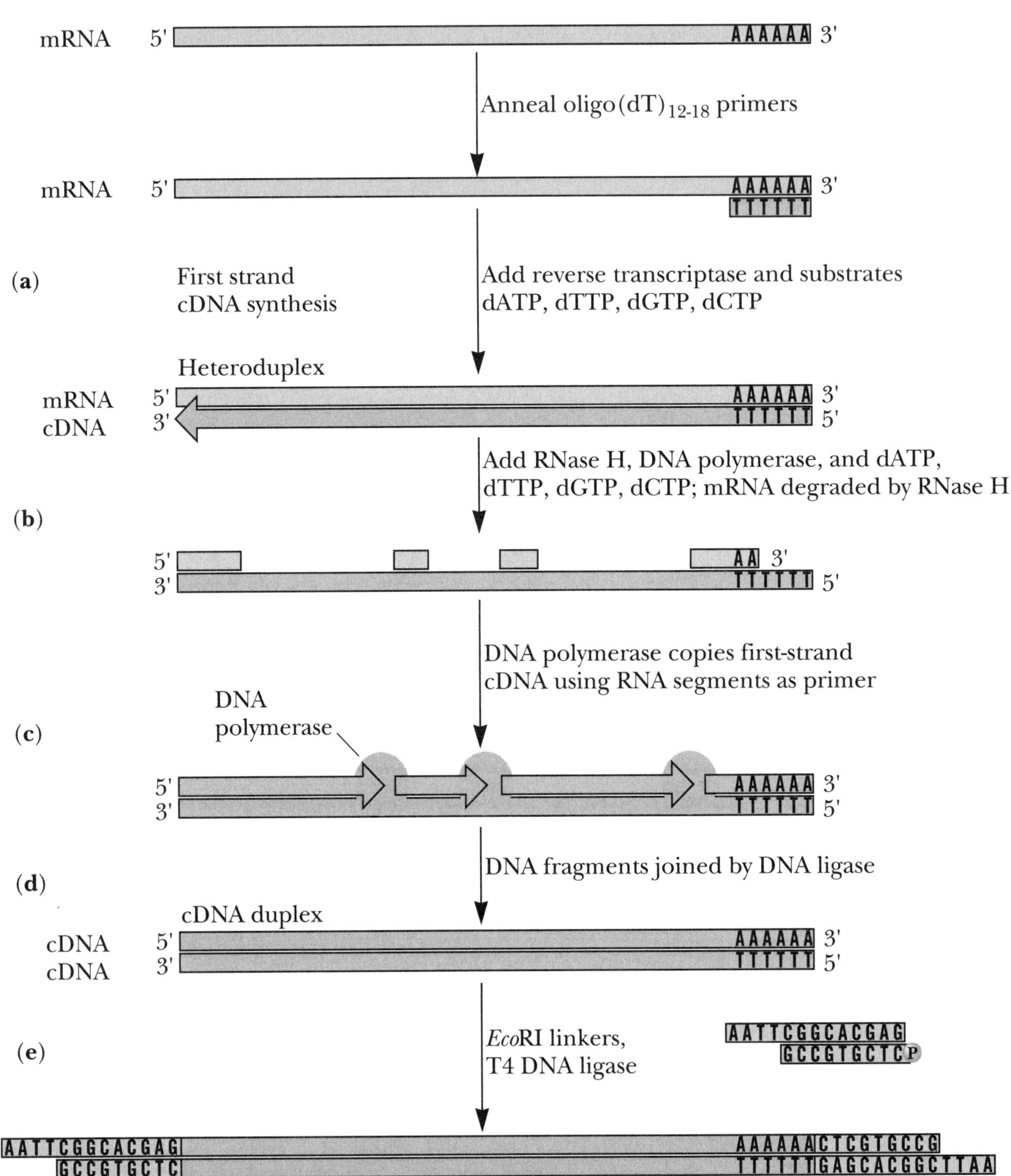

*Eco*RI ended–cDNA duplexes for cloning

© Harcourt, Inc.

Unnumbered Figure Page 308 The Southern blotting technique.

Digest DNA with restriction endonucleases
DNA
DNA restriction fragments
Perform agarose gel electrophoresis on the DNA fragments from different digests
Buffer solution
Agarose gel
DNA fragments fractionated by size (visible under UV light if gel is soaked in ethidium bromide)
Longer DNA fragments
Shorter DNA fragments
Transfer (blot) gel to nitrocellulose filter using Southern blot technique
Soak gel in NaOH, neutralize
Weight
Absorbent paper
Garrett & Grisham
Principles of Biochemistry
Nitrocellulose filter
Gel
Wick
Buffer
DNA fragments are bound to the filter in positions identical to those on the gel
Hybridize filter with radioactively labeled probe
Radioactive probe solution
Expose filter to X-ray film. Resulting autoradiograph shows hybridized DNA fragments

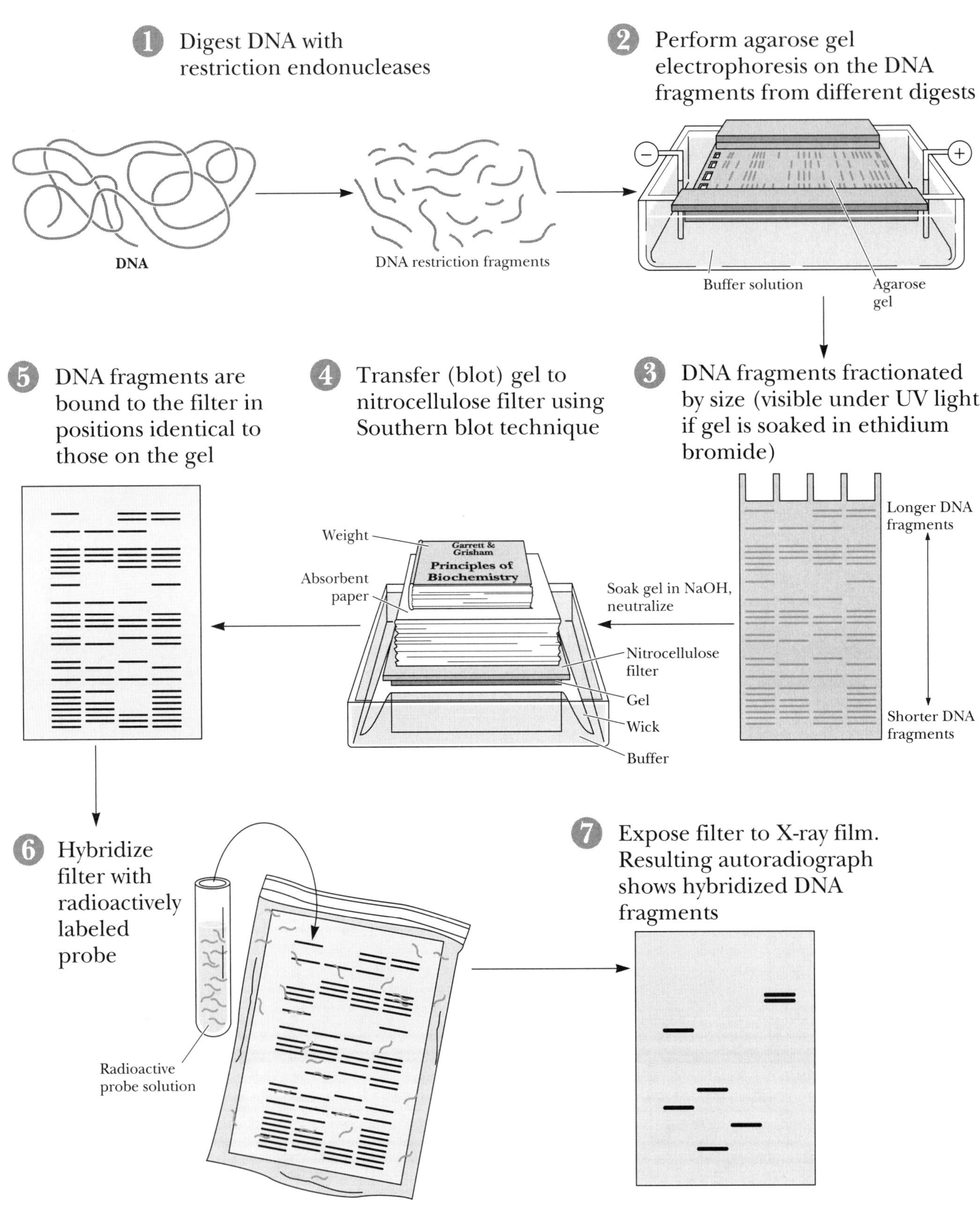

Figure 9.20 The polymerase chain reaction.

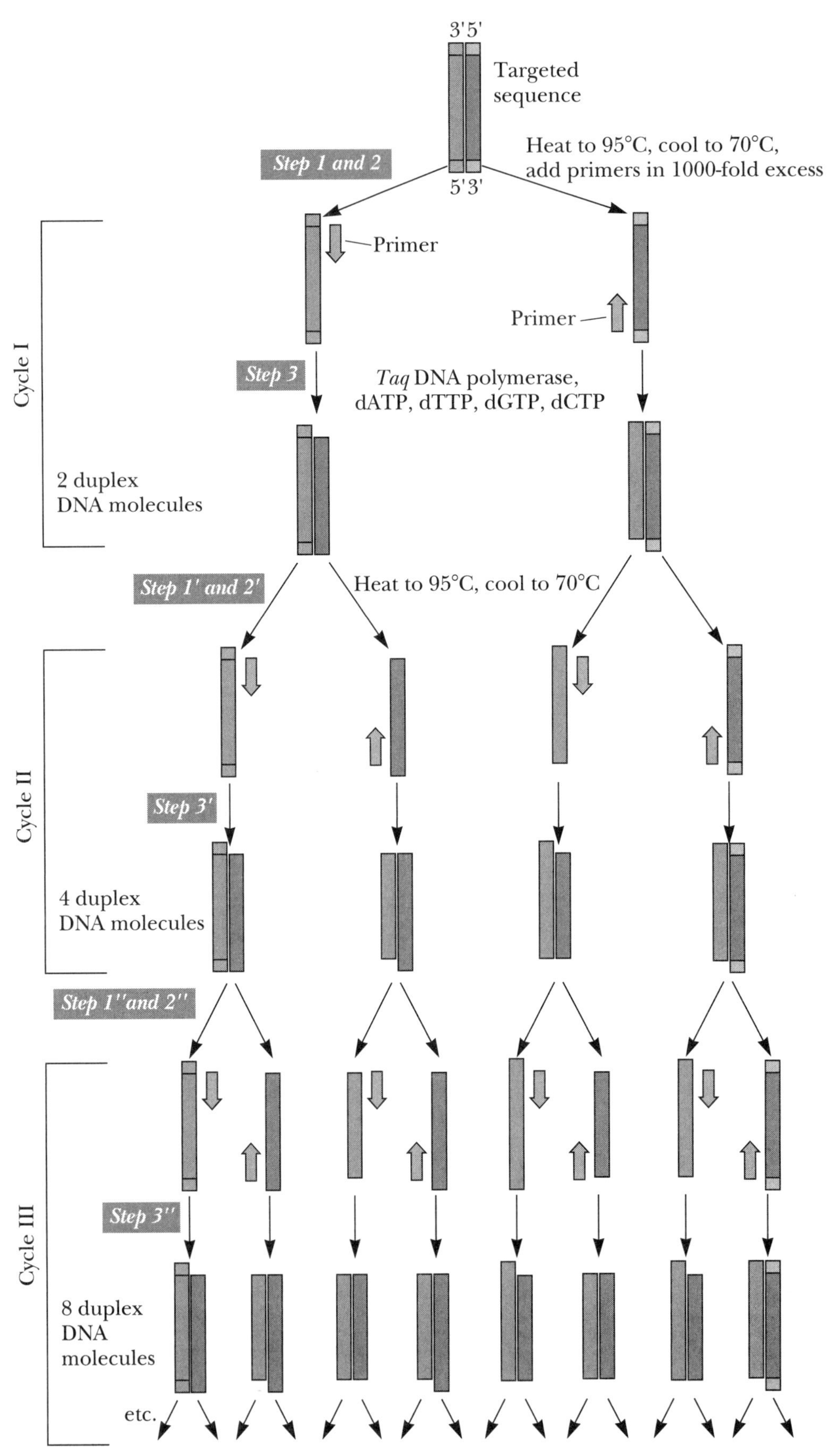

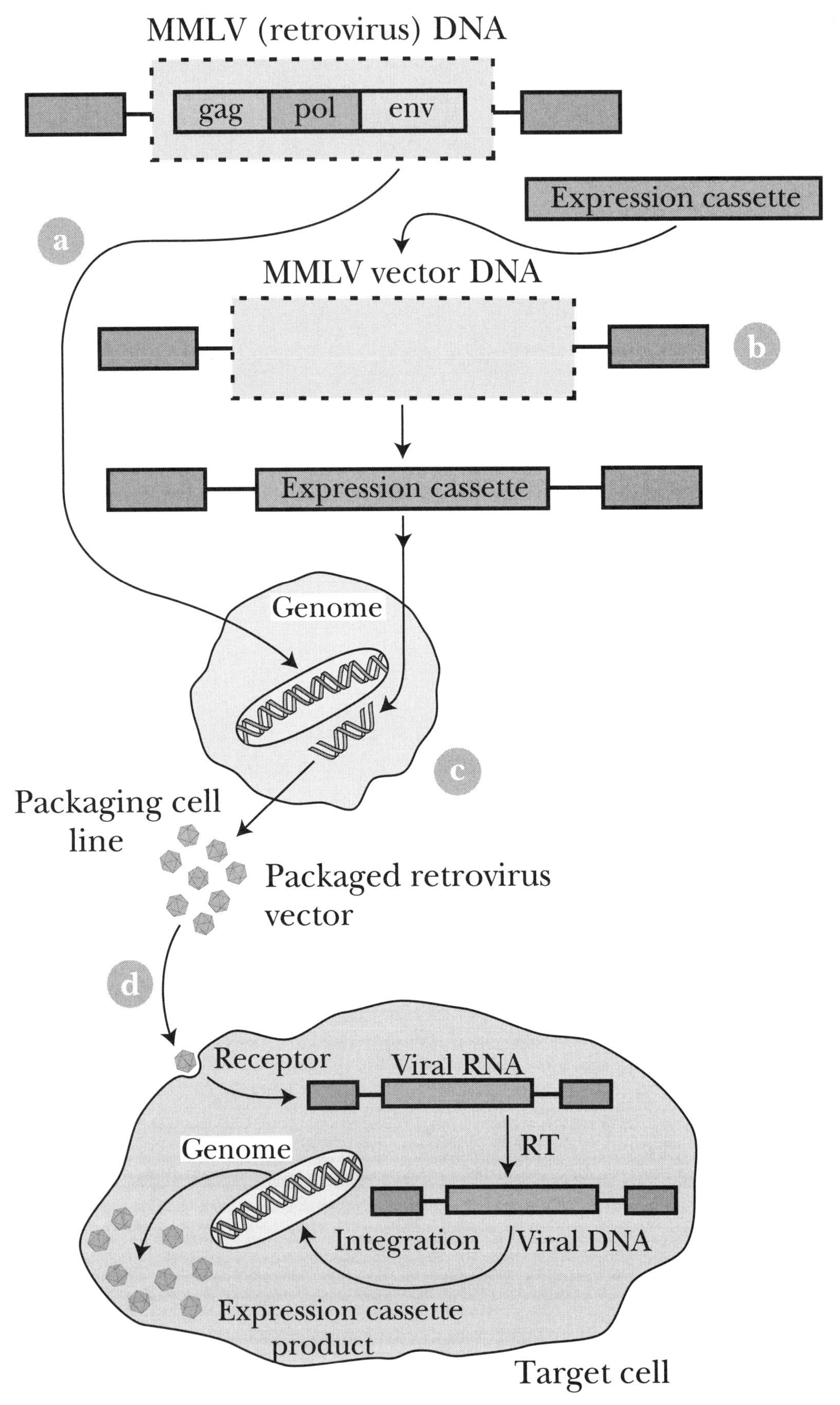
MMLV (retrovirus) DNA
gag
pol
env
Expression cassette
a
MMLV vector DNA
b
Expression cassette
Genome
c
Packaging cell line
Packaged retrovirus vector
d
Receptor
Viral RNA
RT
Genome
Integration
Viral DNA
Expression cassette product
Target cell

Figure 9.23   Adenovirus-mediated gene delivery *in vivo*.

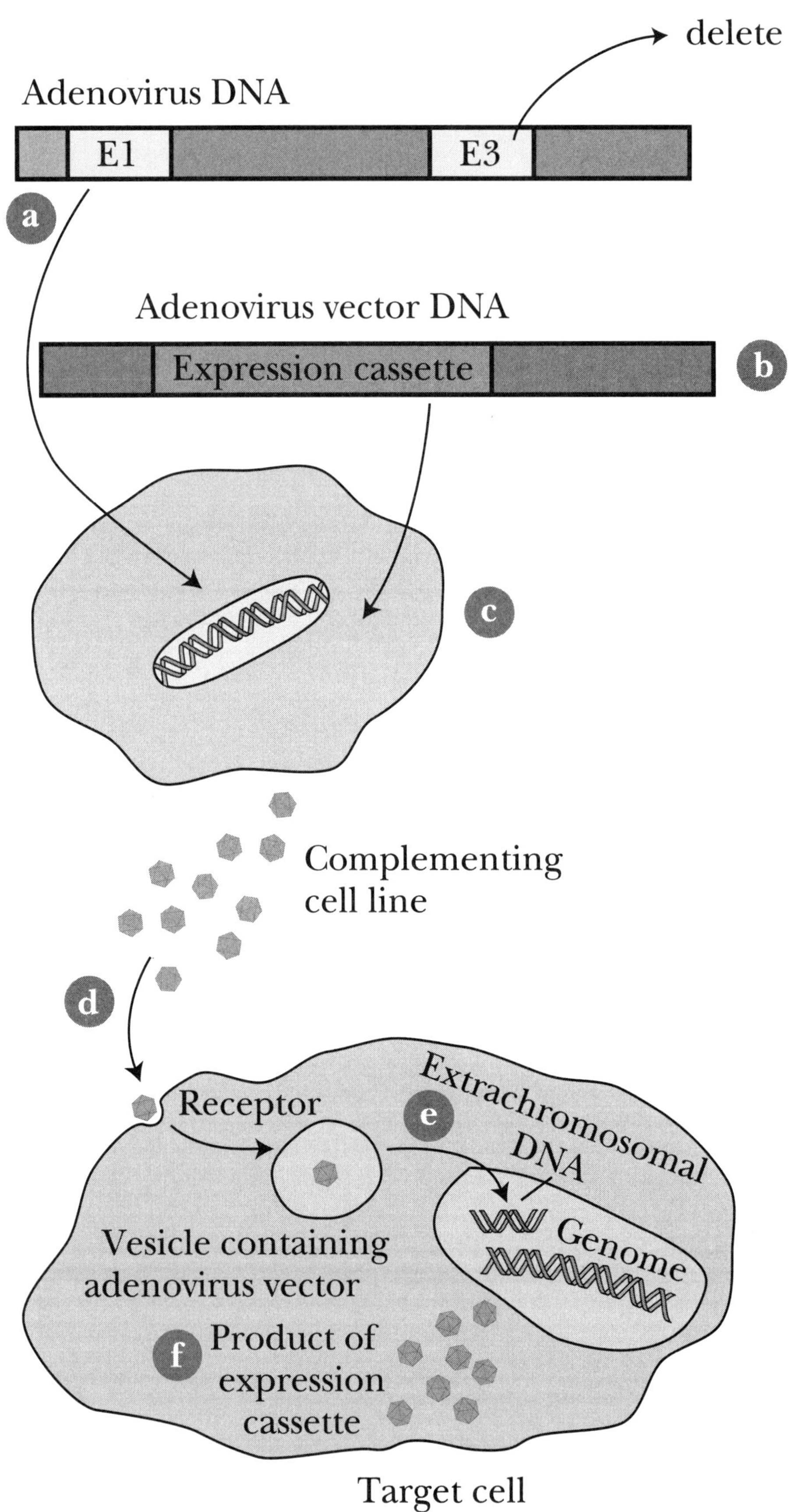

Table 10.1   Enzyme cofactors: metal ions and coenzymes.

Table 10.1   Enzyme Cofactors: Some Metal Ions and Coenzymes and the Enzymes with Which They Are Associated

Metal Ions and Some Enzymes That Require Them		Coenzymes Serving As Transient Carriers of Specific Atoms or Functional Groups		Representative Enzymes Using Coenzymes
Metal Ion	Enzyme	Coenzyme	Entity Transferred	
Fe^{2+} or Fe^{3+}	Cytochrome Oxidase	Thiamine pyrophosphate (TPP)	Aldehydes	Pyruvate dehydrogenase
	Catalase	Flavin adenine dinucleotide (FAD)	Hydrogen atoms	Succinate dehydrogenase
	Peroxidase	Nicotinamide adenine dinucleotide (NAD)	Hydride ion (H^-)	Alcohol dehydrogenase
Cu^{2+}	Cytochrome oxidase			
Zn^{2+}	DNA polymerase	Coenzyme A (CoA)	Acyl groups	Acetyl-CoA carboxylase
	Carbonic anhydrase	Pyridoxal phosphate (PLP)	Amino groups	Aspartate aminotransferase
	Alcohol dehydrogenase	5′-Deoxyadenosylcobalamin (vitamin B_{12})	H atoms and alkyl groups	Methylmalonyl-CoA mutase
Mg^{2+}	Hexokinase			
	Glucose-6-phosphatase	Biotin (biocytin)	CO_2	Propionyl-CoA carboxylase
Mn^{2+}	Arginase	Tetrahydrofolate (THF)	Other one-carbon groups	Thymidylate synthase
K^+	Pyruvate kinase (also requires Mg^{2+})			
Ni^{2+}	Urease			
Mo	Nitrate reductase			
Se	Glutathione peroxidase			

Figure 10.8 Time course for formation of the ES complex and establishment of the steady-state.

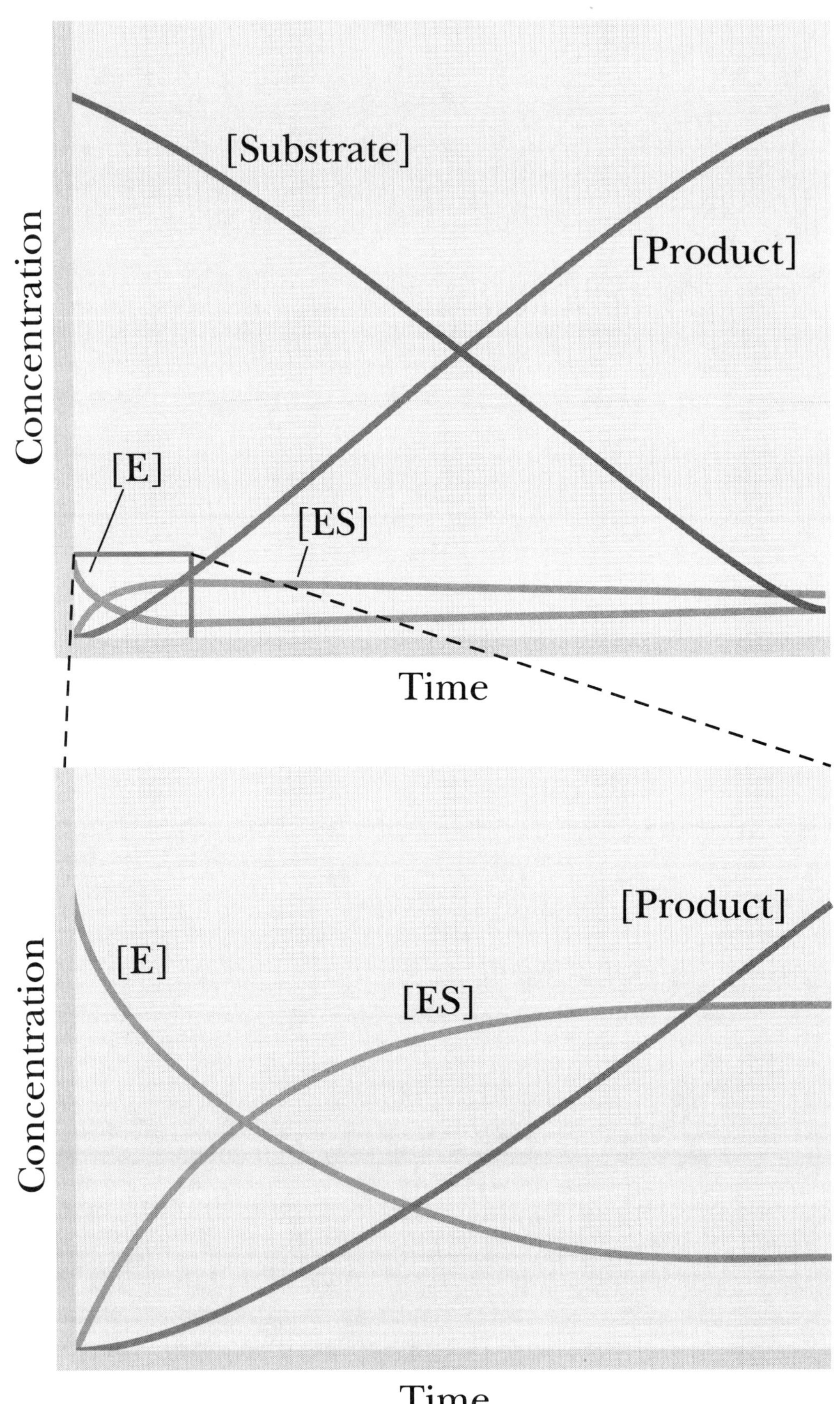

Figure 10.12   Lineweaver-Burk plot of competitive inhibition.

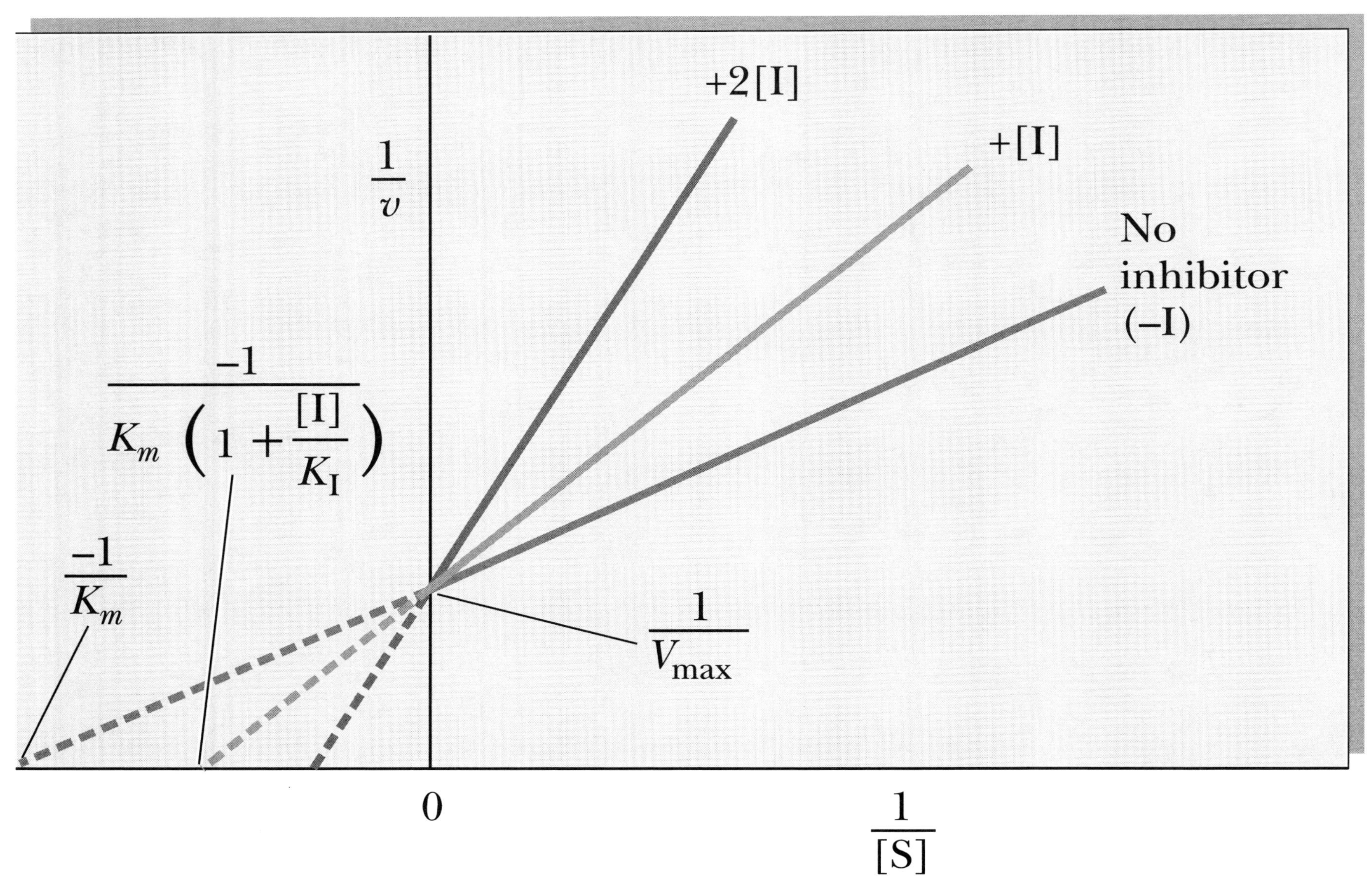

Figure 10.14   Lineweaver-Burk plot of pure noncompetitive inhibition.

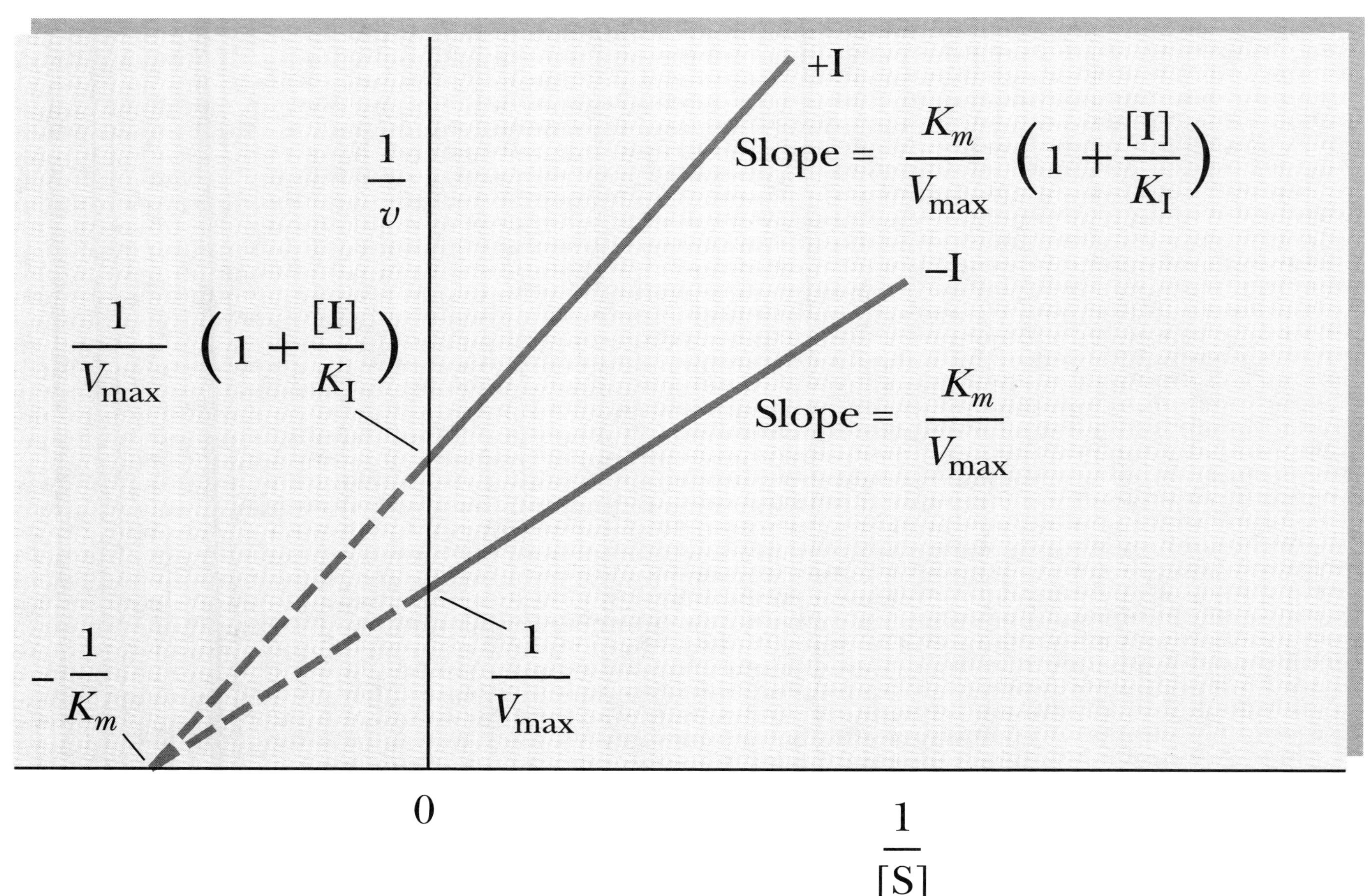

© Harcourt, Inc.

(a) $K_I < K_I'$

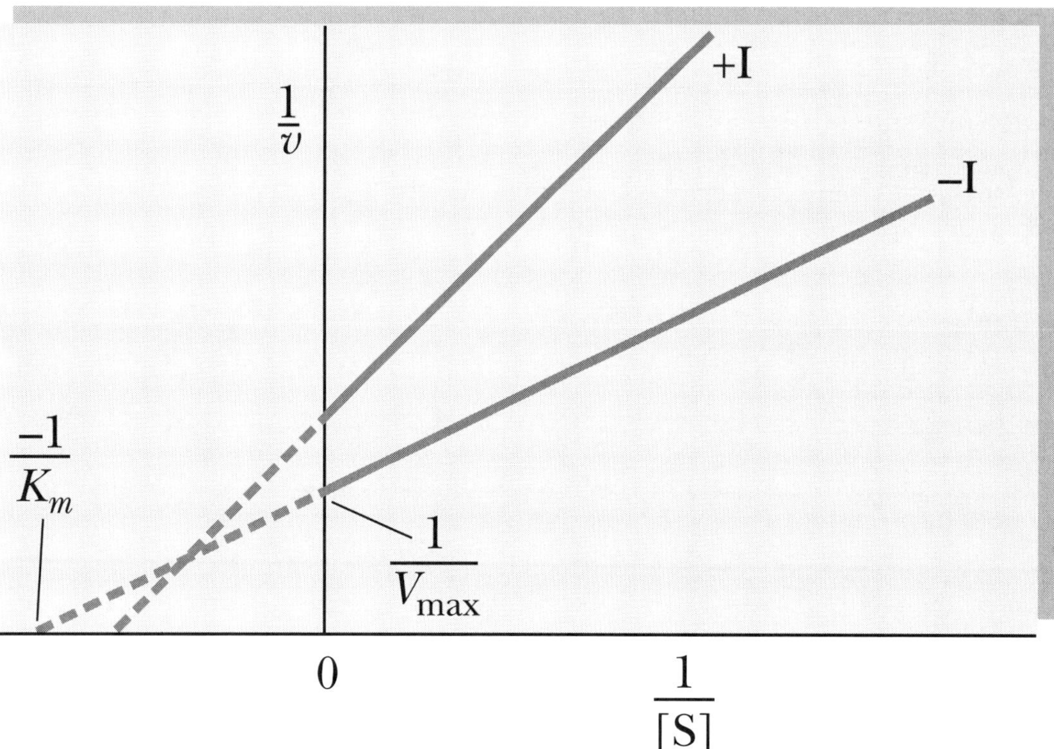

(b) $K_I' < K_I$

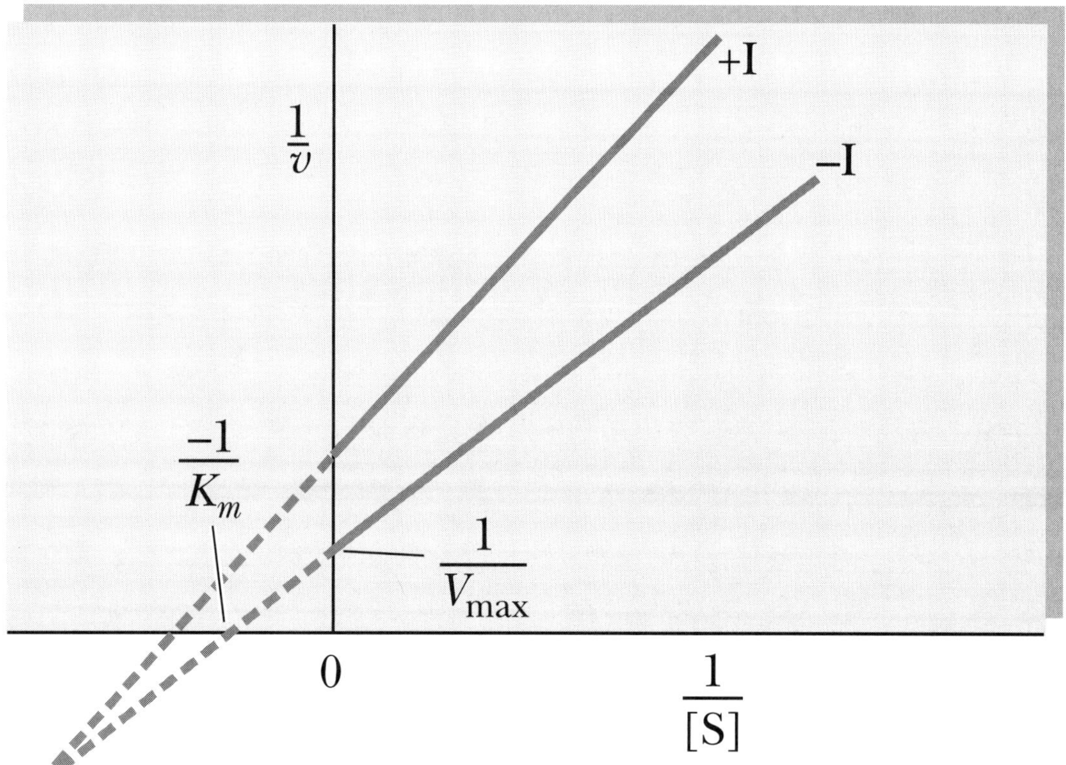

Figure 10.17   Single-displacement bisubstrate mechanism.

Double-reciprocal form of the rate equation:

$$\frac{1}{v} = \frac{1}{V_{max}} \left(K_m^A + \frac{K_S^A\, K_m^B}{[B]} \right) \left(\frac{1}{[A]} + \frac{1}{V_{max}} \left(1 + \frac{K_m^B}{[B]} \right) \right)$$

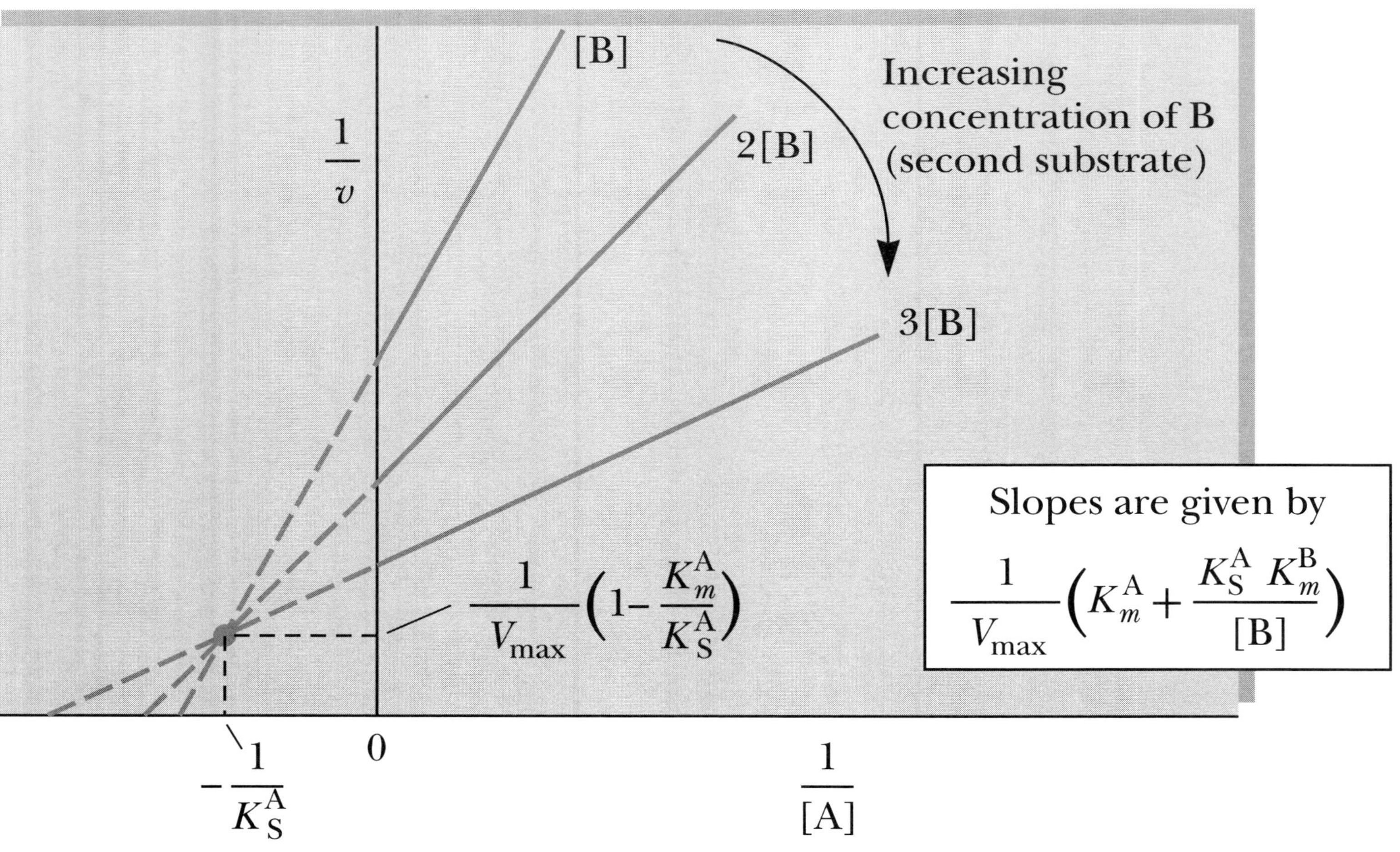

Figure 10.18   Double-displacement (ping-pong) bisubstrate mechanism.

Double-reciprocal form of the rate equation:

$$\frac{1}{v} = \frac{K_m^A}{V_{max}}\left(\frac{1}{[A]}\right) + \left(1 + \frac{K_m^B}{[B]}\right)\left(\frac{1}{V_{max}}\right)$$

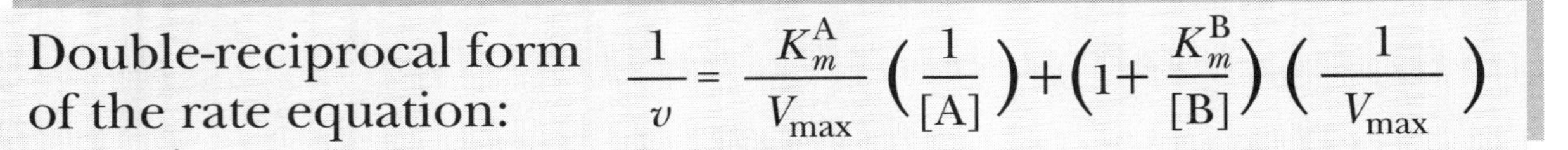

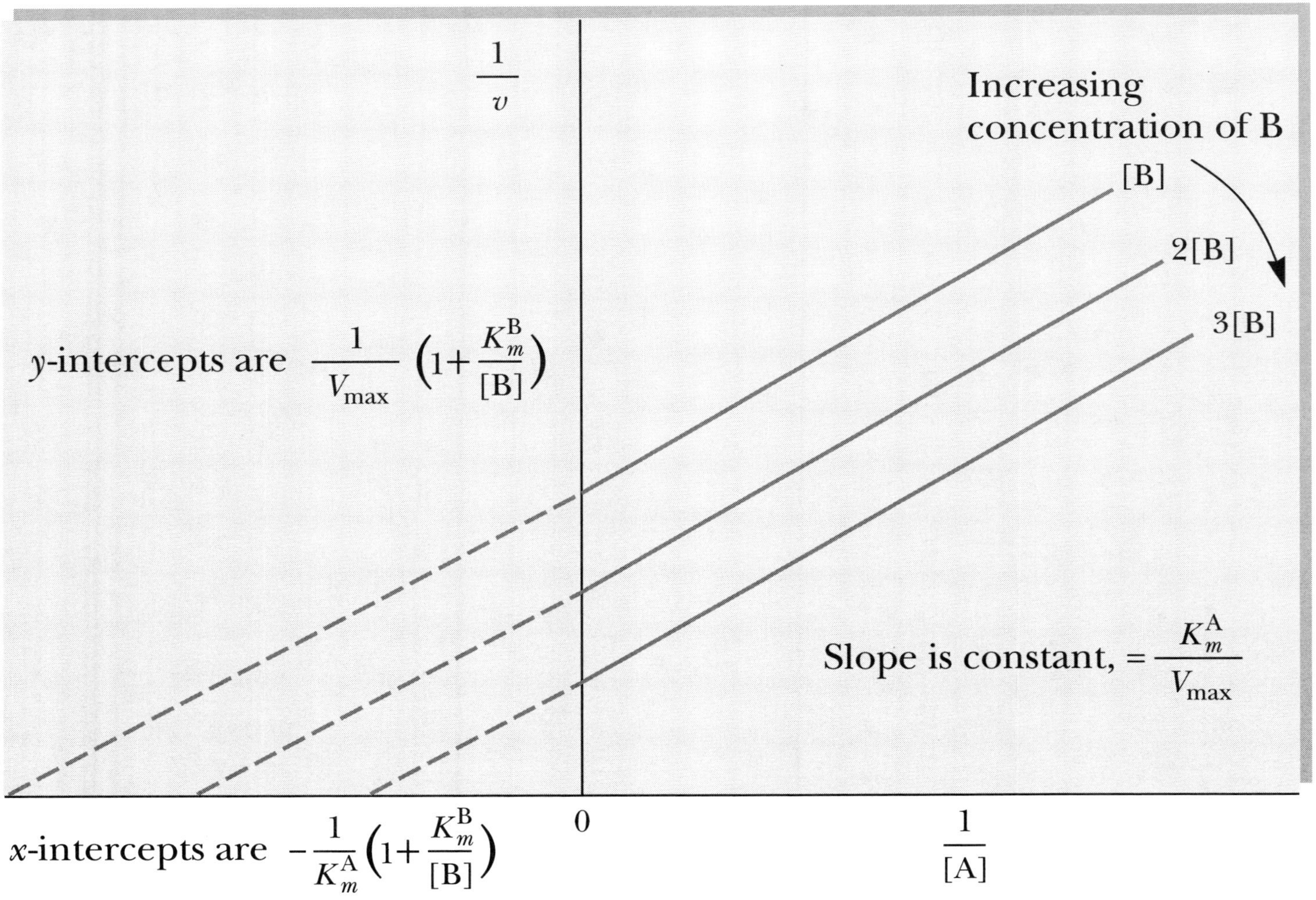

Figure 10.29 Monod-Wyman-Changeux model for allosteric transitions.

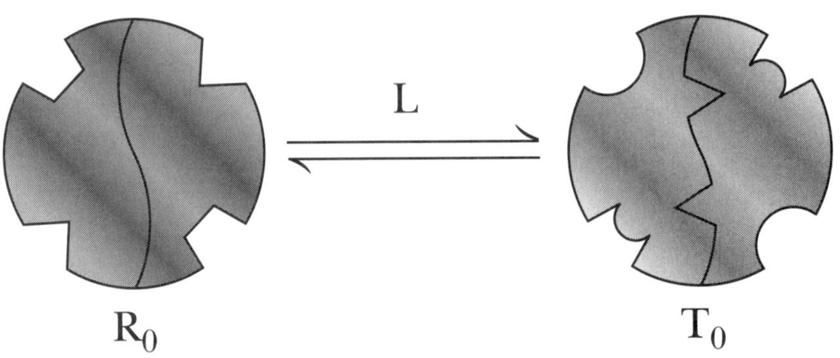

(a) A dimeric protein can exist in either of two conformational states at equilibrium.

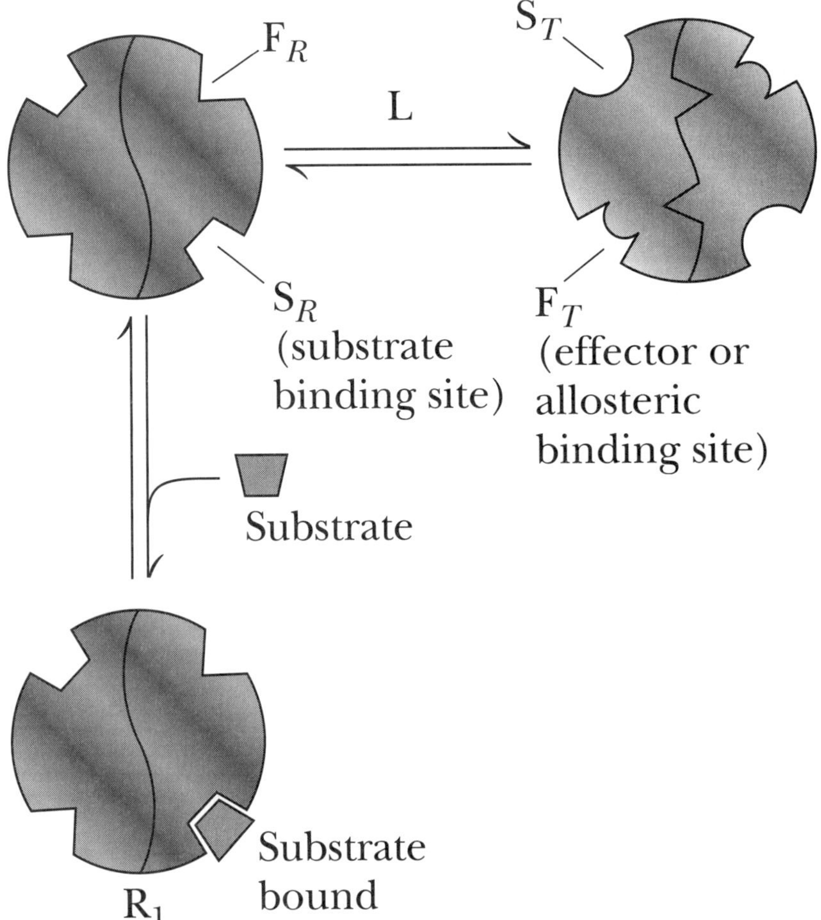

(b) Substrate binding shifts equilibrium in favor of R.

Figure 10.31   Heterotropic allosteric effects.

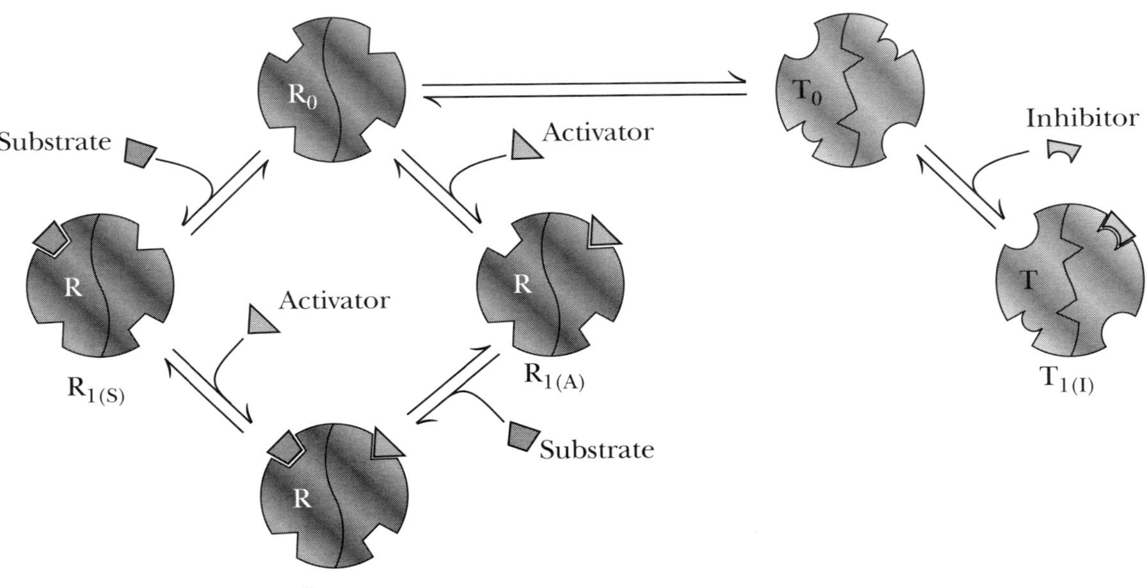

A dimeric protein which can exist in either of two states, R_0 and T_0. This protein can bind 3 ligands:

1) Substrate (S) : A positive homotropic effector that binds only to R at site S

2) Activator (A) : A positive heterotropic effector that binds only to R at site F

3) Inhibitor (I) : A negative heterotropic effector that binds only to T at site F

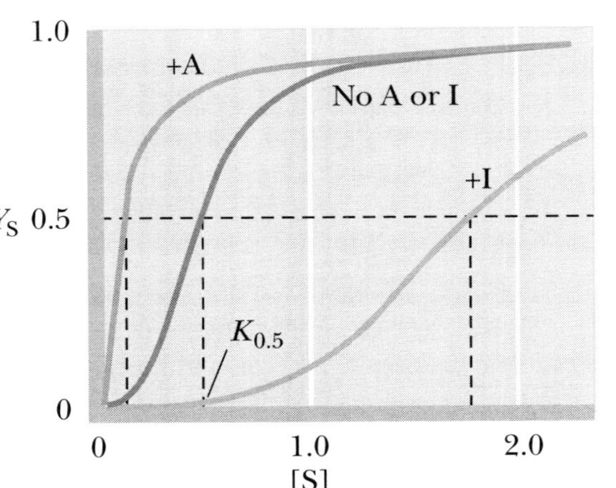

Effects of A:
$A + R_0 \longrightarrow R_{1(A)}$
Increase in number of R-conformers shifts $R_0 \rightleftharpoons T_0$ so that $T_0 \longrightarrow R_0$

1) More binding sites for S made available

2) Decrease in cooperativity of substrate saturation curve. Effector A lowers the apparent value of L.

Effects of I:
$I + T_0 \longrightarrow T_{1(I)}$
Increase in number of T-conformers (decrease in R_0 as $R_0 \longrightarrow T_0$ to restore equilibrium)

Thus, I inhibits association of S and A with R by lowering R_0 level. I increases cooperativity of substrate saturation curve. I raises the apparent value of L.

Figure 10.37 Mechanism of covalent modifications and allosteric regulation of glycogen phosphorylase.

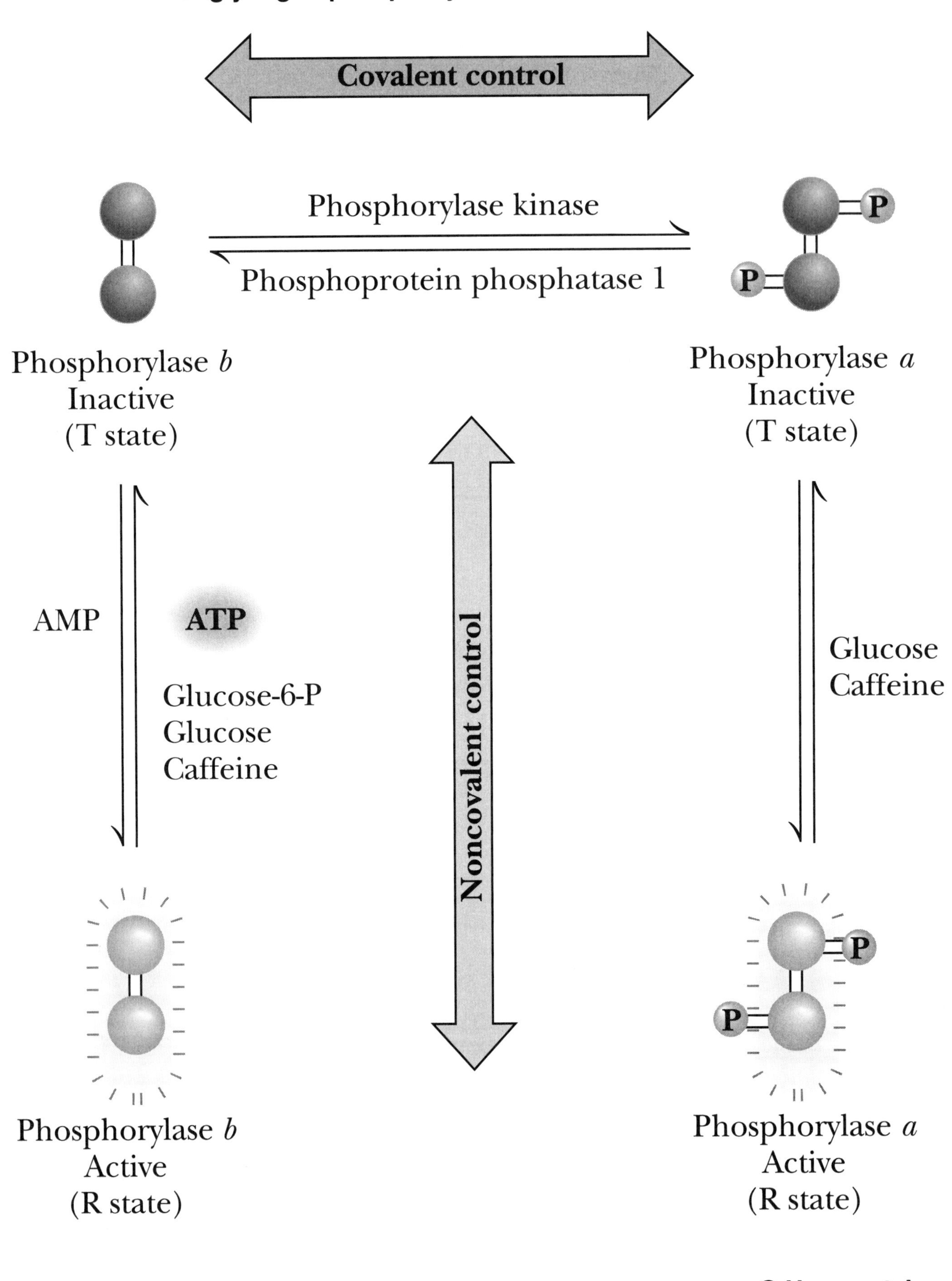

Covalent control
Phosphorylase kinase
Phosphoprotein phosphatase 1
Phosphorylase b
Inactive
(T state)
Phosphorylase a
Inactive
(T state)
AMP
ATP
Glucose-6-P
Glucose
Caffeine
Noncovalent control
Glucose
Caffeine
Phosphorylase b
Active
(R state)
Phosphorylase a
Active
(R state)

Figure 11.1   Enzymes catalyze reactions by lowering the activation energy.

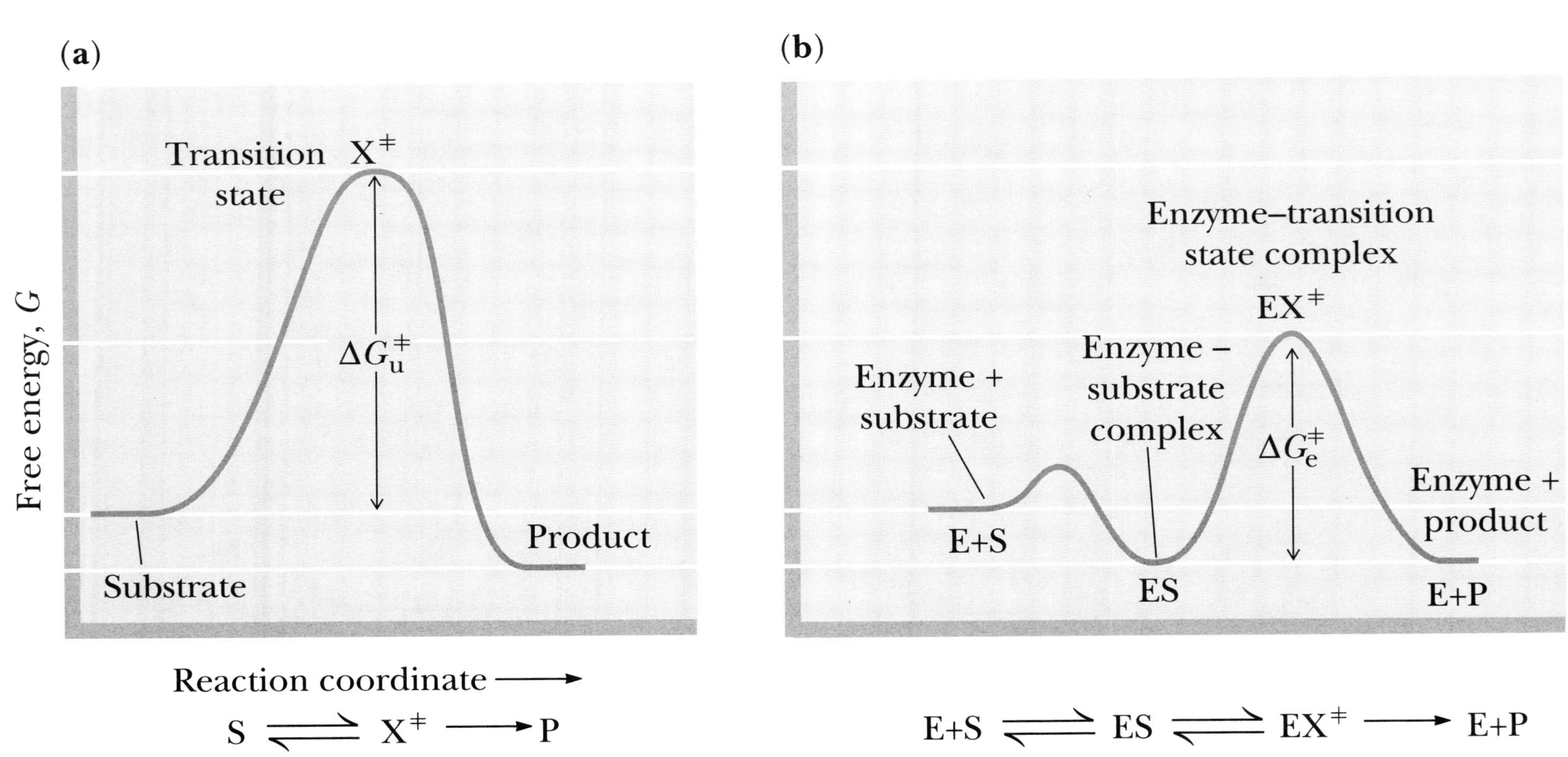

Figures 11.2, 11.3, 11.4 Entropy loss upon formation of the ES complex; electrostatic destabilization upon substrate binding.

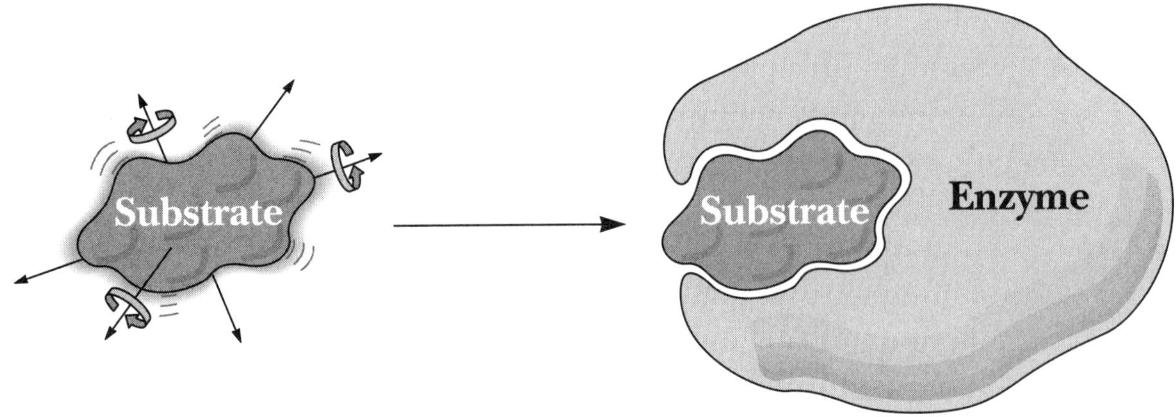

Substrate (and enzyme) are free
to undergo translational motion.
A disordered, high-entropy situation

The highly ordered,
low-entropy complex

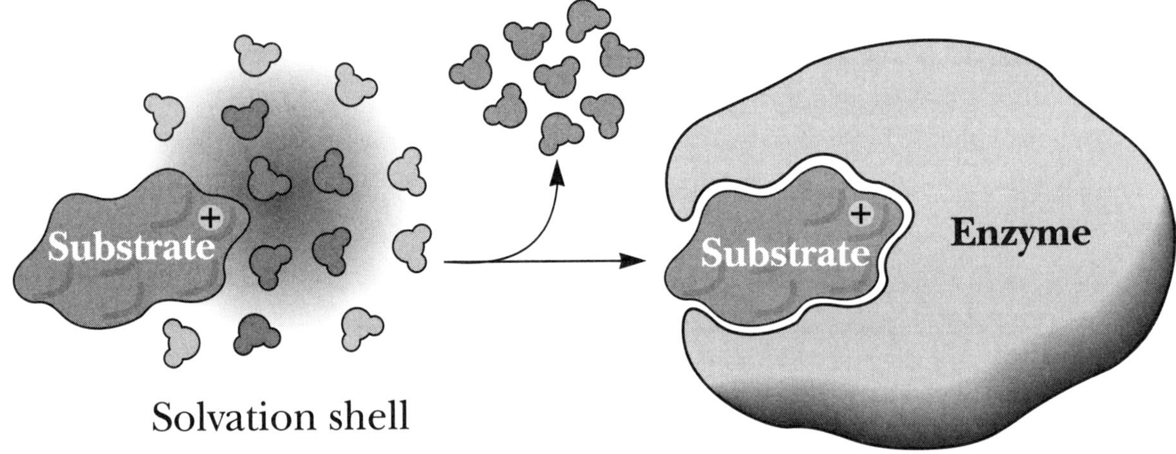

Solvation shell

Desolvated ES complex

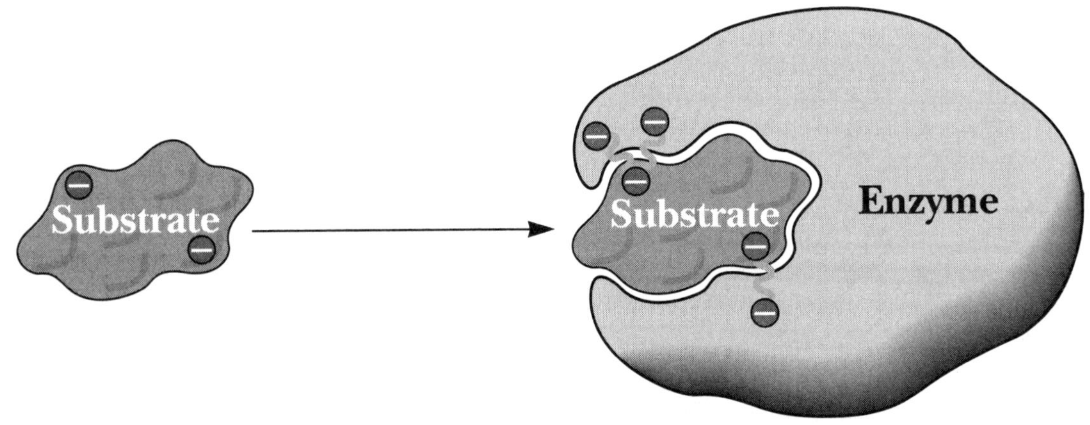

Electrostatic destabilization
in ES complex

Figure 11.21 Mechanism for the chymotrypsin reaction.

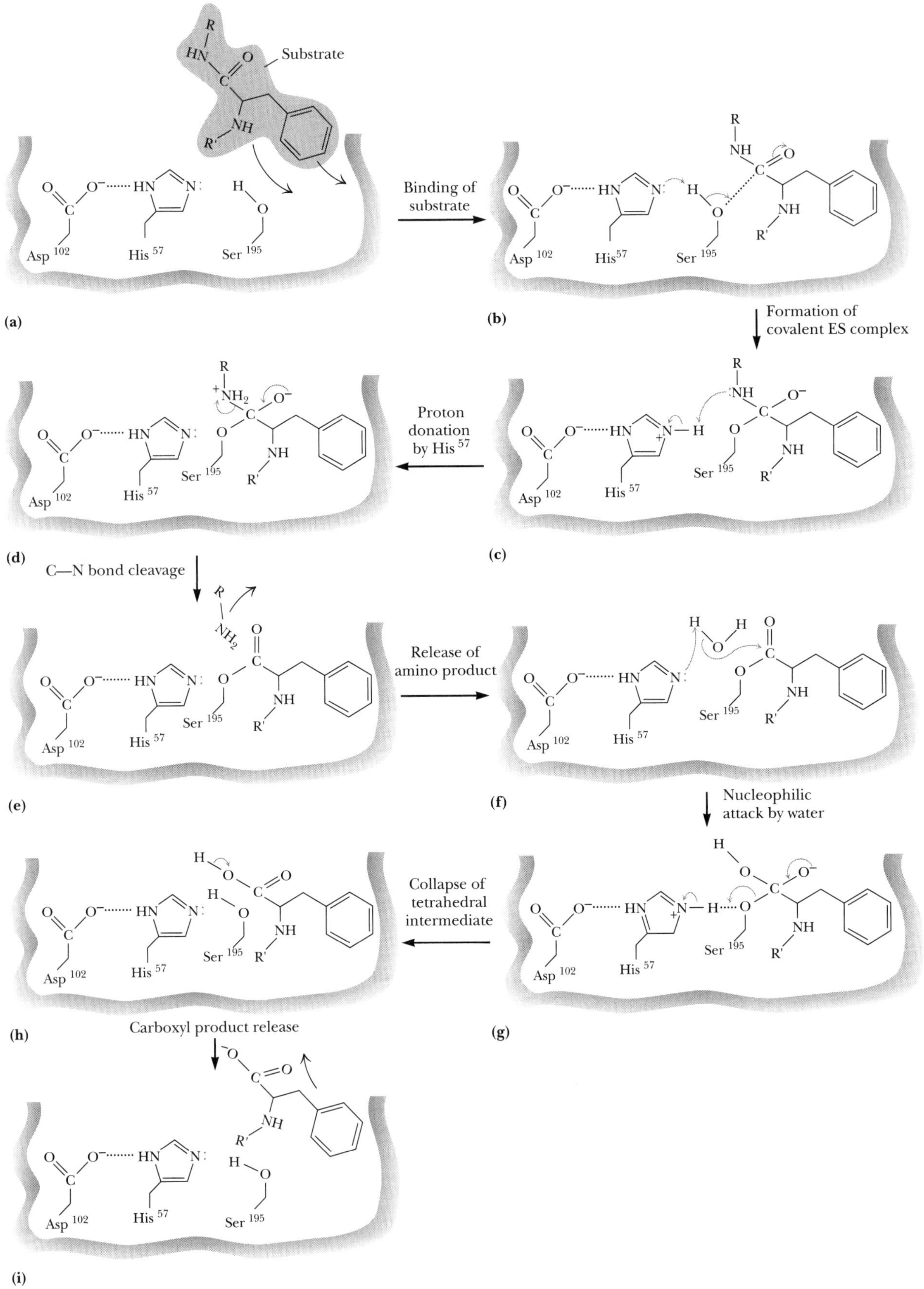

Figure 11.24   HIV mRNA provides the genetic information for synthesis of a polyprotein.

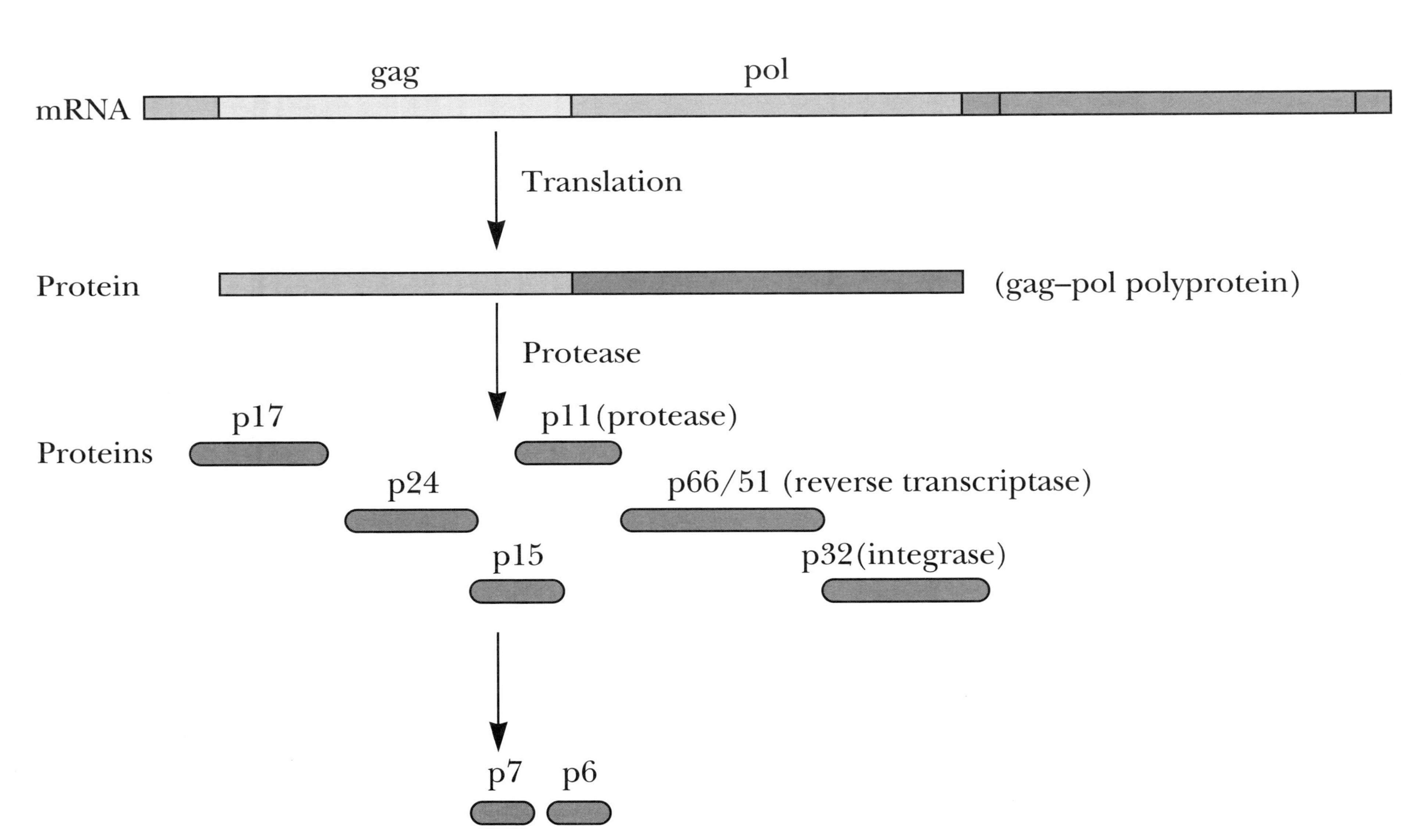

Figure 12.4 Lipoprotein components are synthesized in the ER of liver cells.

Endoplasmic reticulum
1 Synthesis of apoproteins, phosphatidylcholine, triacylglycerol, cholesterol, cholesterol esters occurs in the endoplasmic reticulum
2 Assembly of components into prelipoprotein particles in the ER, then transfer to Golgi
3 Golgi processes particles with additional phospholipids and perhaps also cholesterol and cholesterol esters added
Secretory vesicle
Golgi
4 Formation of secretory vesicle containing lipoprotein particles
VLDL
Liver cell
5 The VLDL is released into the circulation
Extracellular space

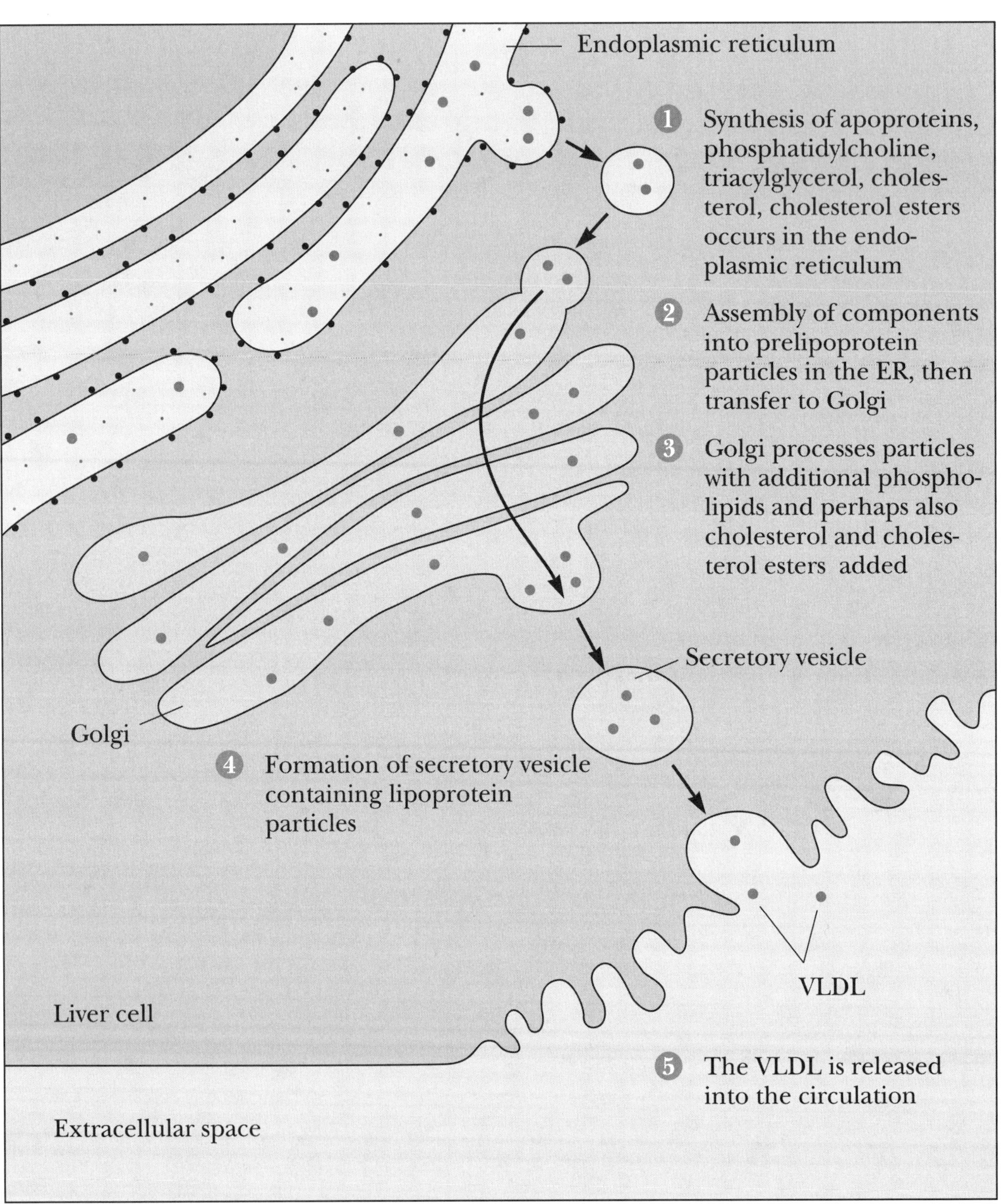

Figure 12.5   Endocytosis and degradation of lipoprotein particles.

Apoprotein B
LDL
LDL
LDL receptor
Recycling
vesicle
Vesicle loses
its coating
and forms
endosome
LDL receptors
bud off and form
a small recycling
vesicle
Endosome formation
may or may not include
fusion with another
vesicle
Lysosome
Synthesis of
LDL receptors
Remaining vesicle
fuses with lysosome
Protease
Inhibits
Synthesis of
cholesterol
Apoprotein B is degraded
by lysosomal protease and
released as amino acids
HMG-CoA
reductase
Free
cholesterol
released
Amino acids
Cholesterol esters in
core are hybridized by
ACAT and stored in cell
Free
cholesterol
Oversupply of
cholesterol
ACAT
Activates

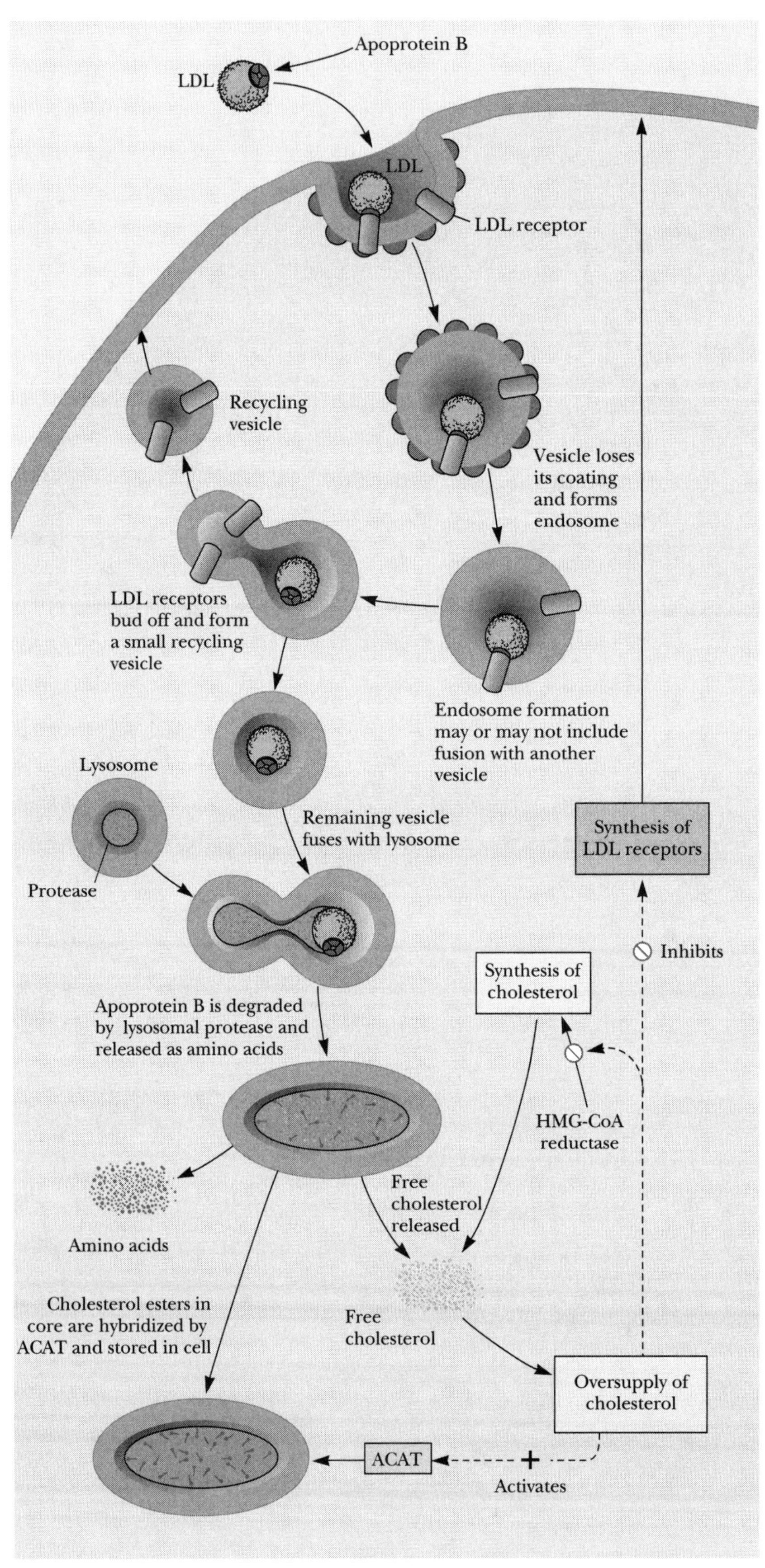

Figure 12.6 The organization of the IgG molecule.

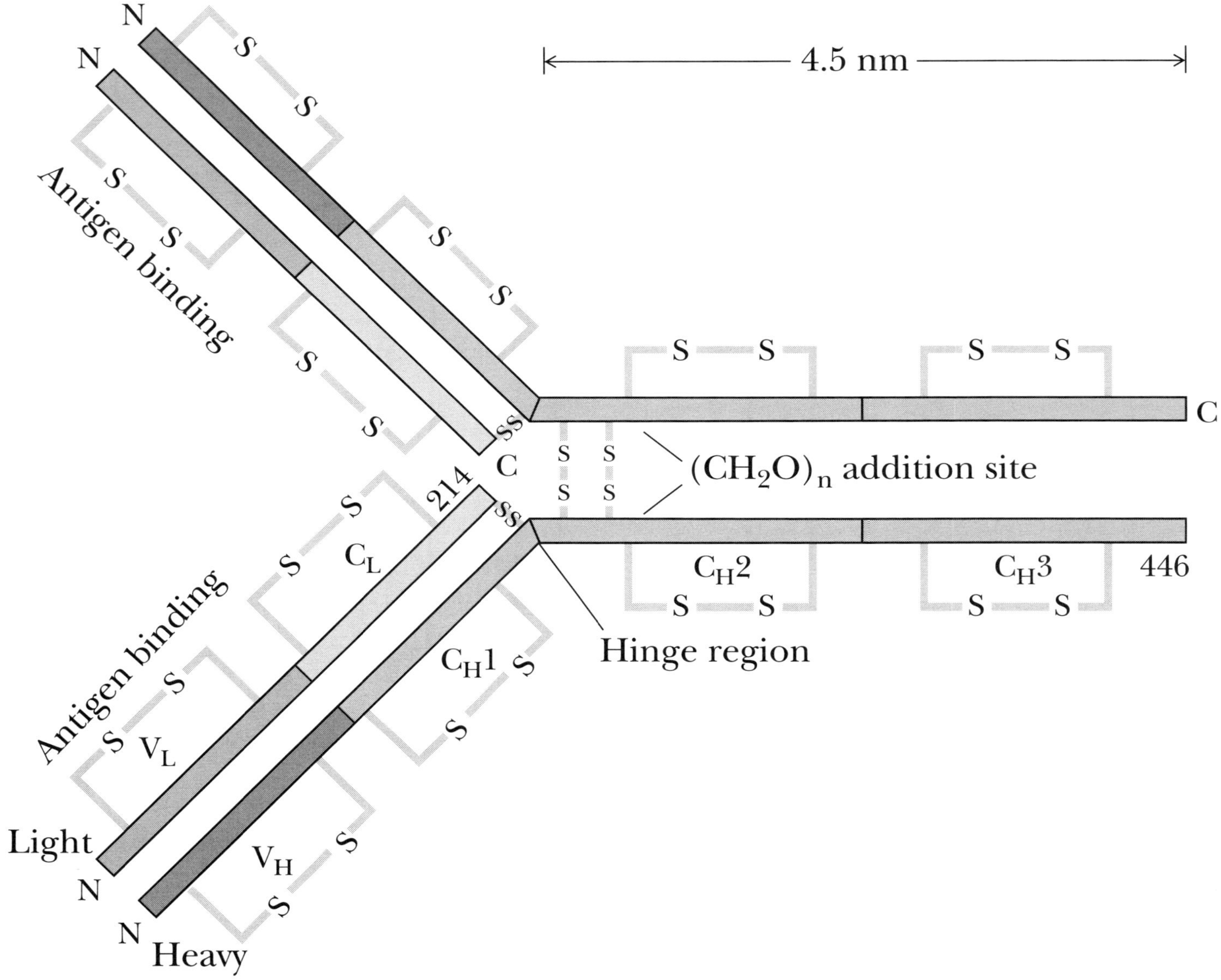

Figure 12.7 "Collapsed β-barrel domain" known as the immunoglobulin fold.

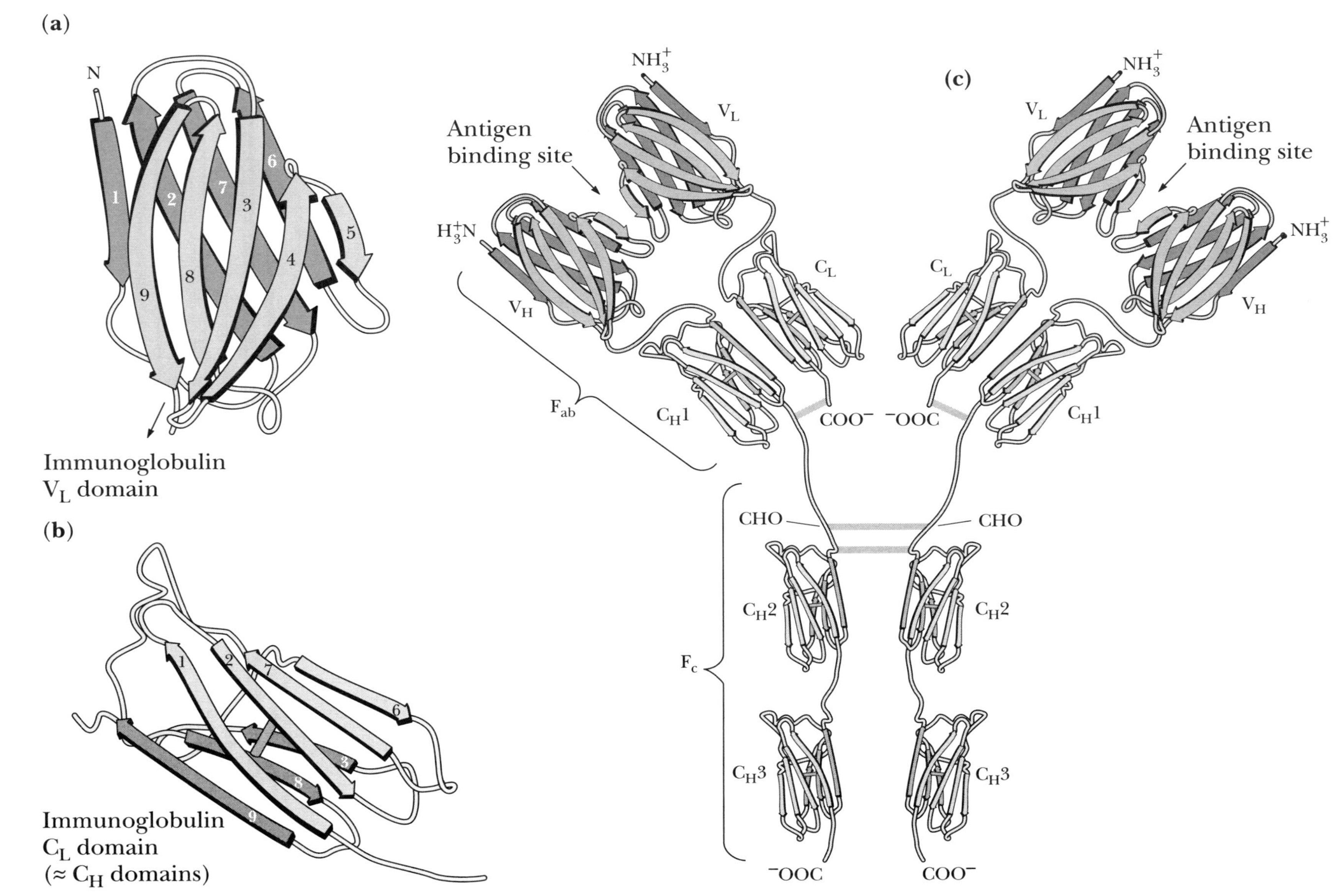

Figure 12.13   Oxygen-binding curves for hemoglobin and myoglobin.

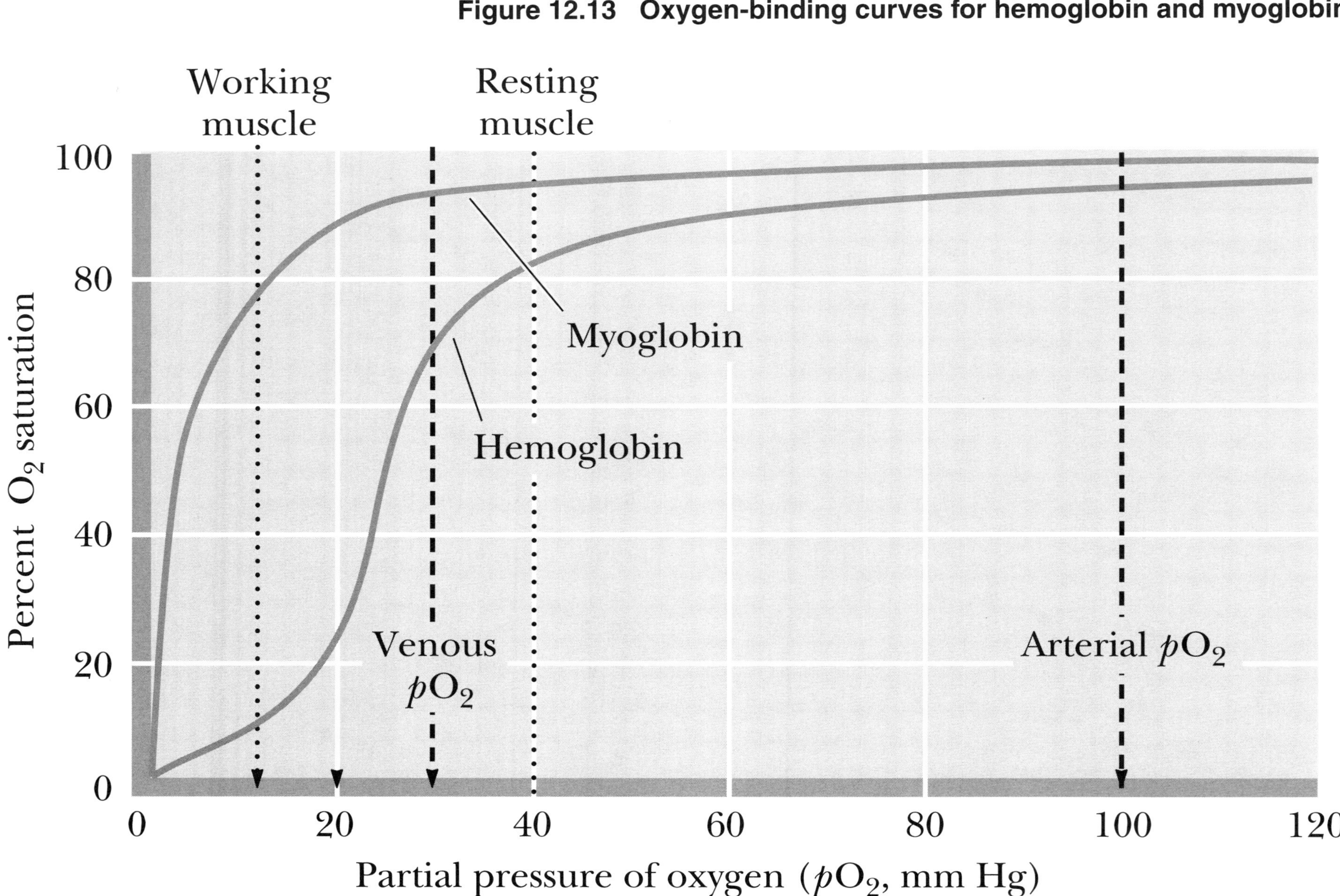

Figure 12.23 Oxygen saturation curves for myoglobin and hemoglobin at various pH values.

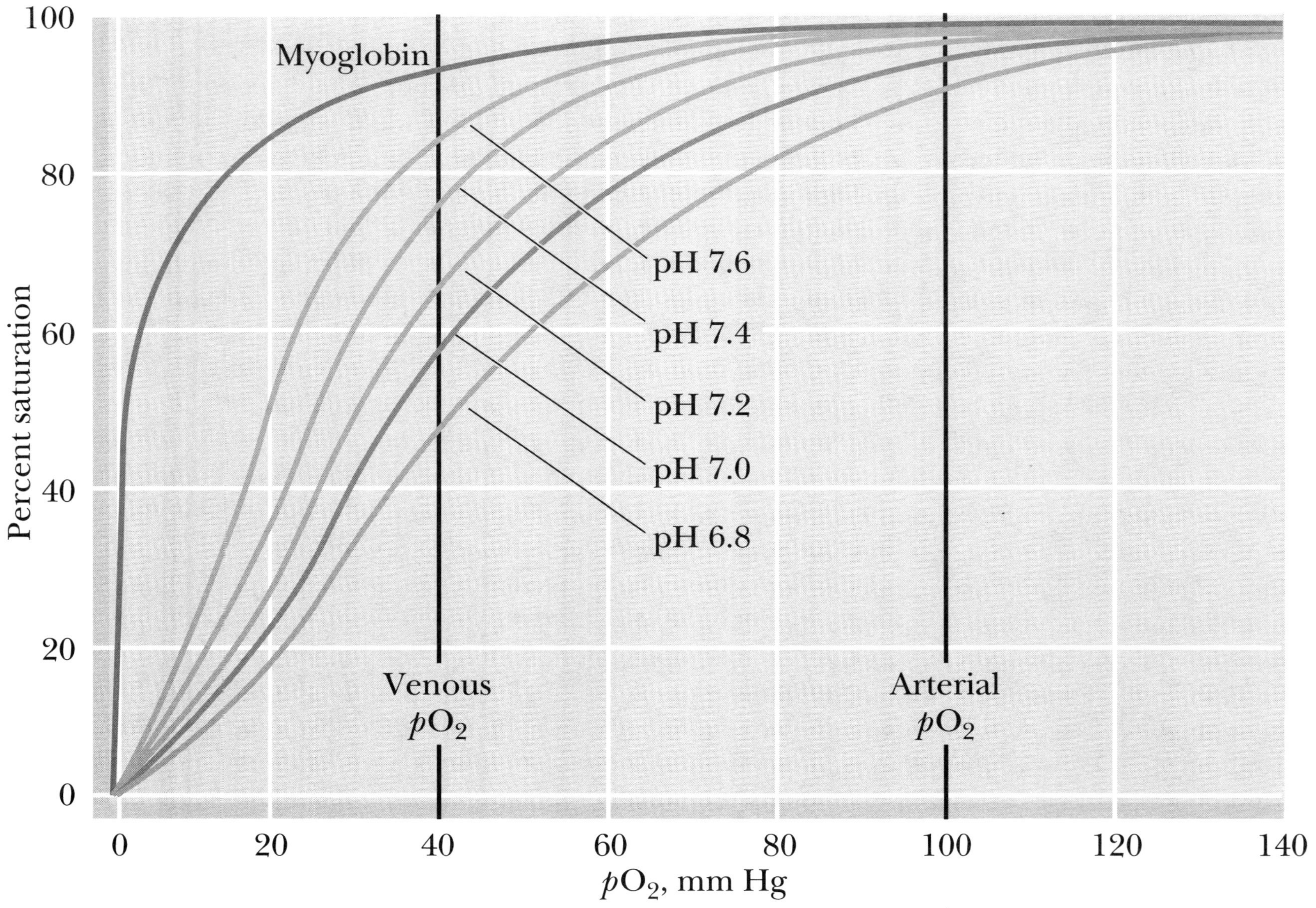

Figure 13.2 The structure of the tubulin $\alpha\beta$ heterodimer.

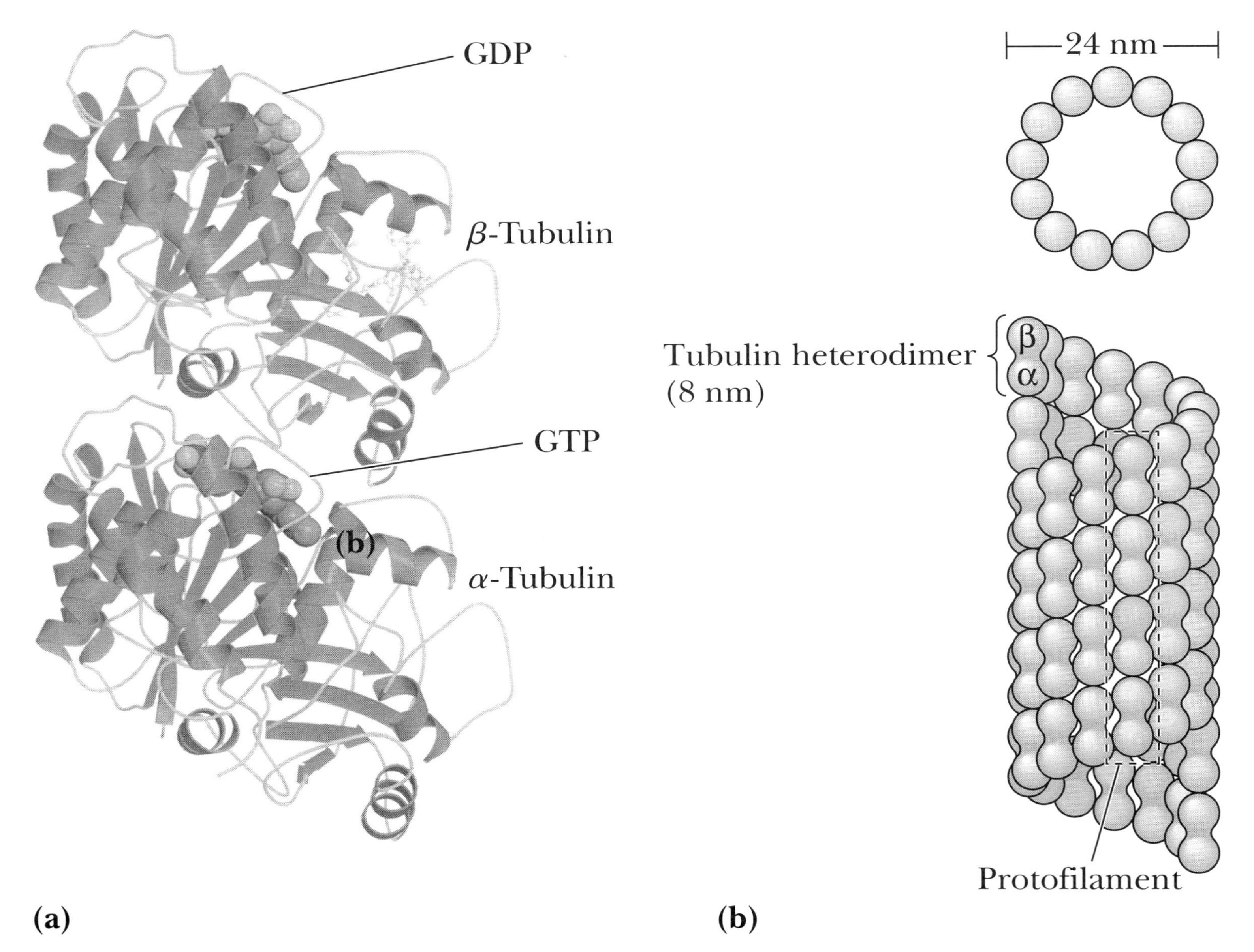

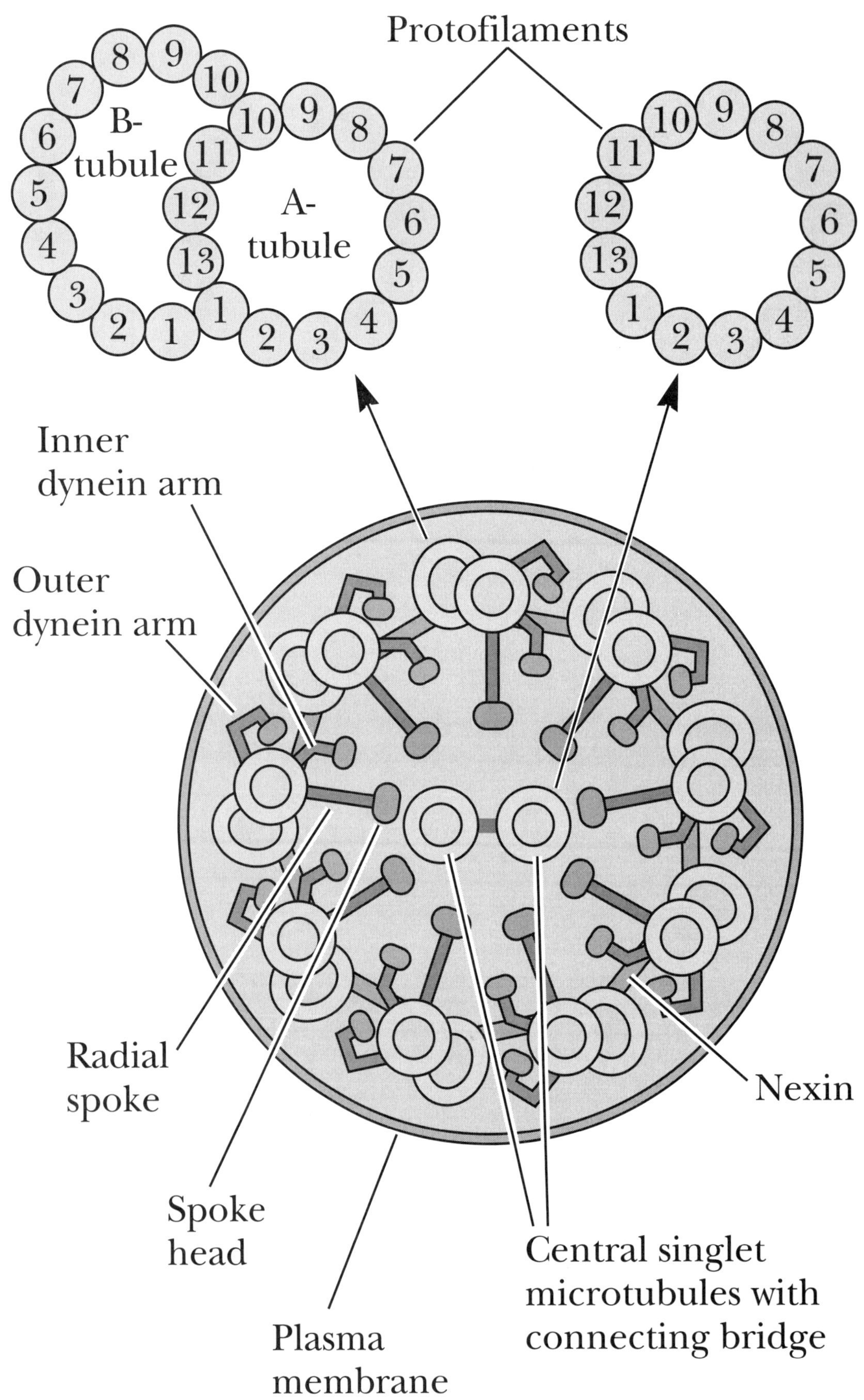

Protofilaments
B-tubule
A-tubule
Inner dynein arm
Outer dynein arm
Radial spoke
Spoke head
Plasma membrane
Central singlet microtubules with connecting bridge
Nexin

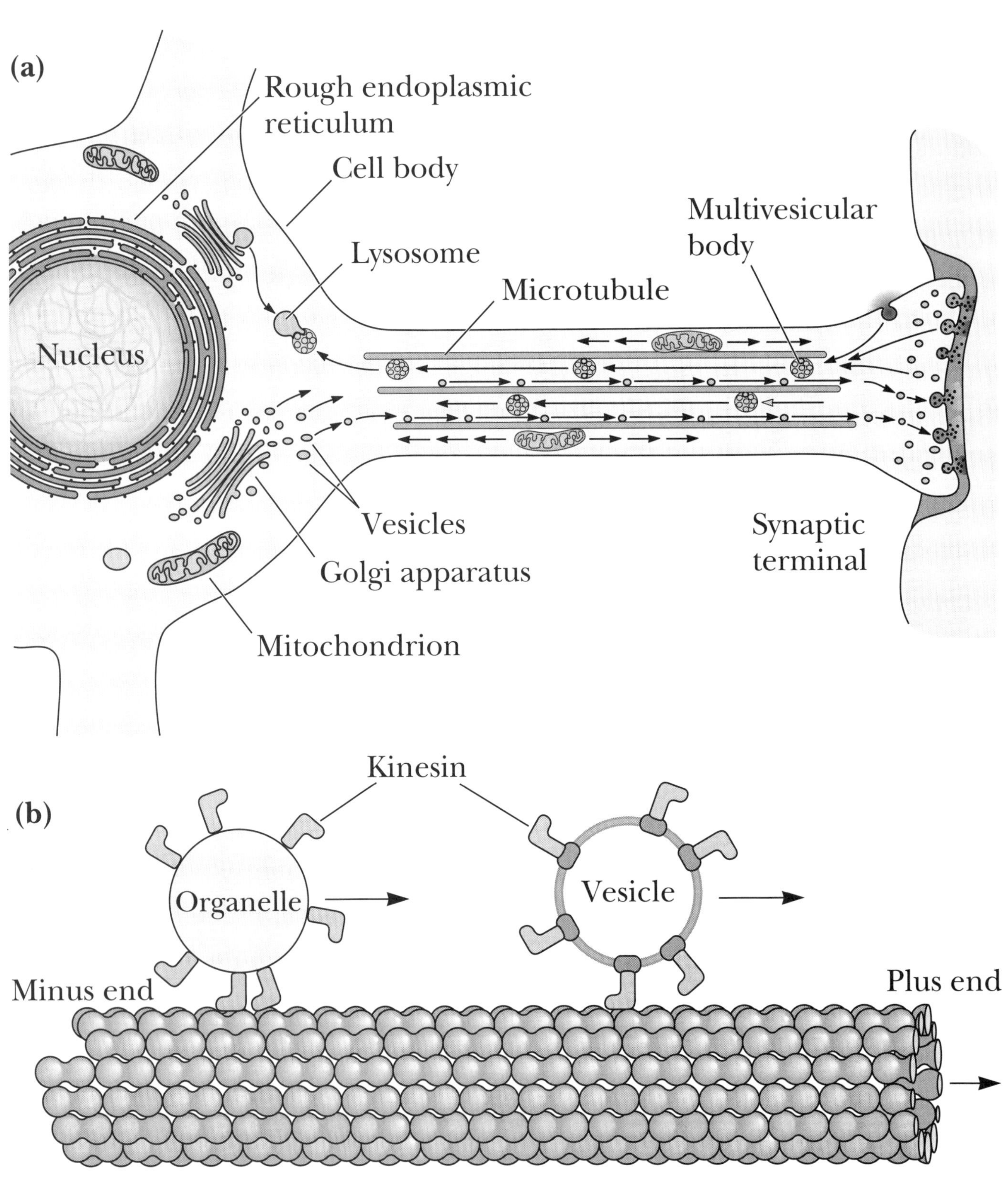
(a)
Rough endoplasmic reticulum
Cell body
Lysosome
Microtubule
Multivesicular body
Nucleus
Vesicles
Golgi apparatus
Synaptic terminal
Mitochondrion
(b)
Kinesin
Organelle
Vesicle
Minus end
Plus end

Figure 13.11   Skeletal muscle with t-tubules shown.

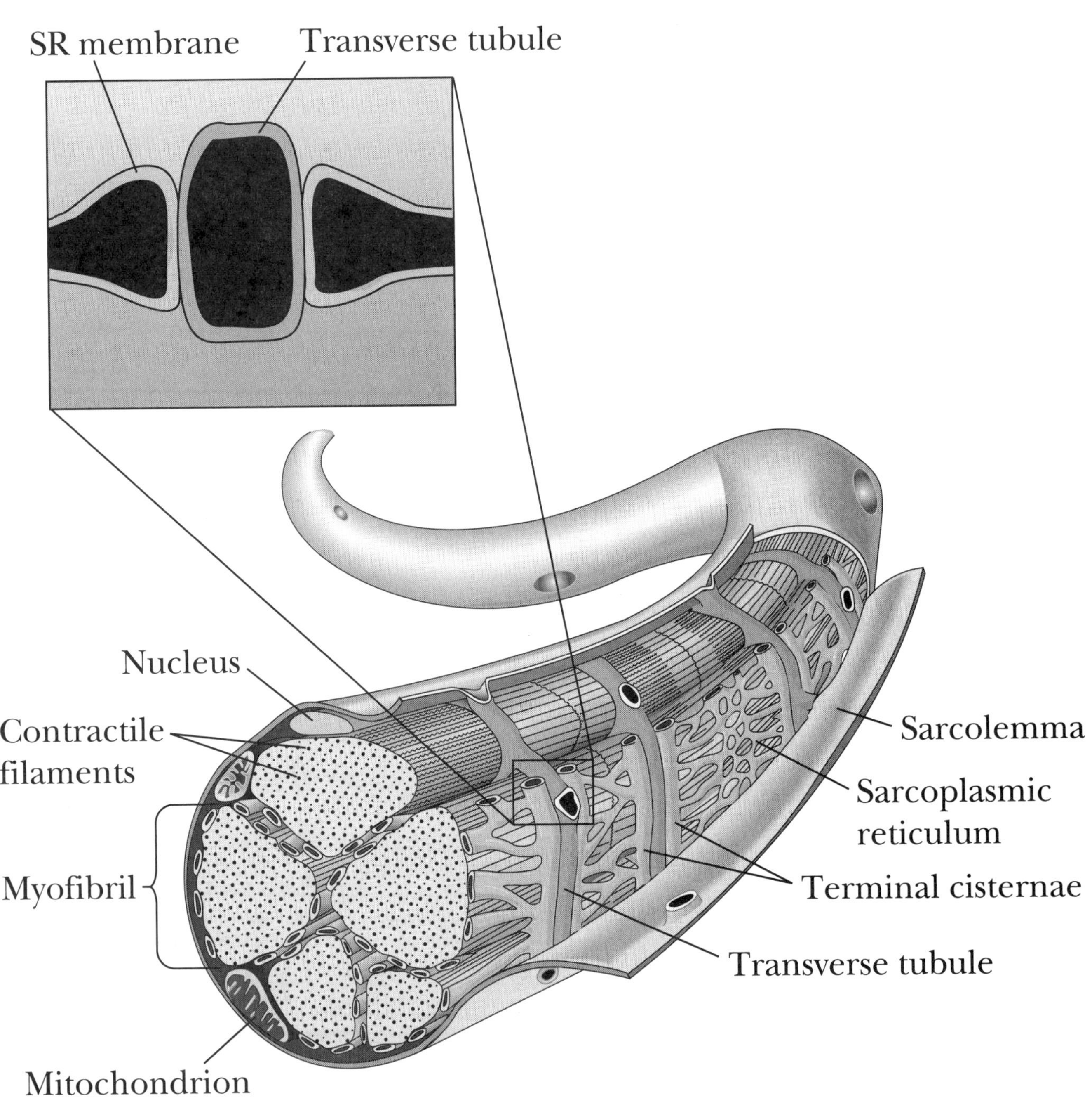

Figure 13.19 The sliding filament model of skeletal muscle contraction.

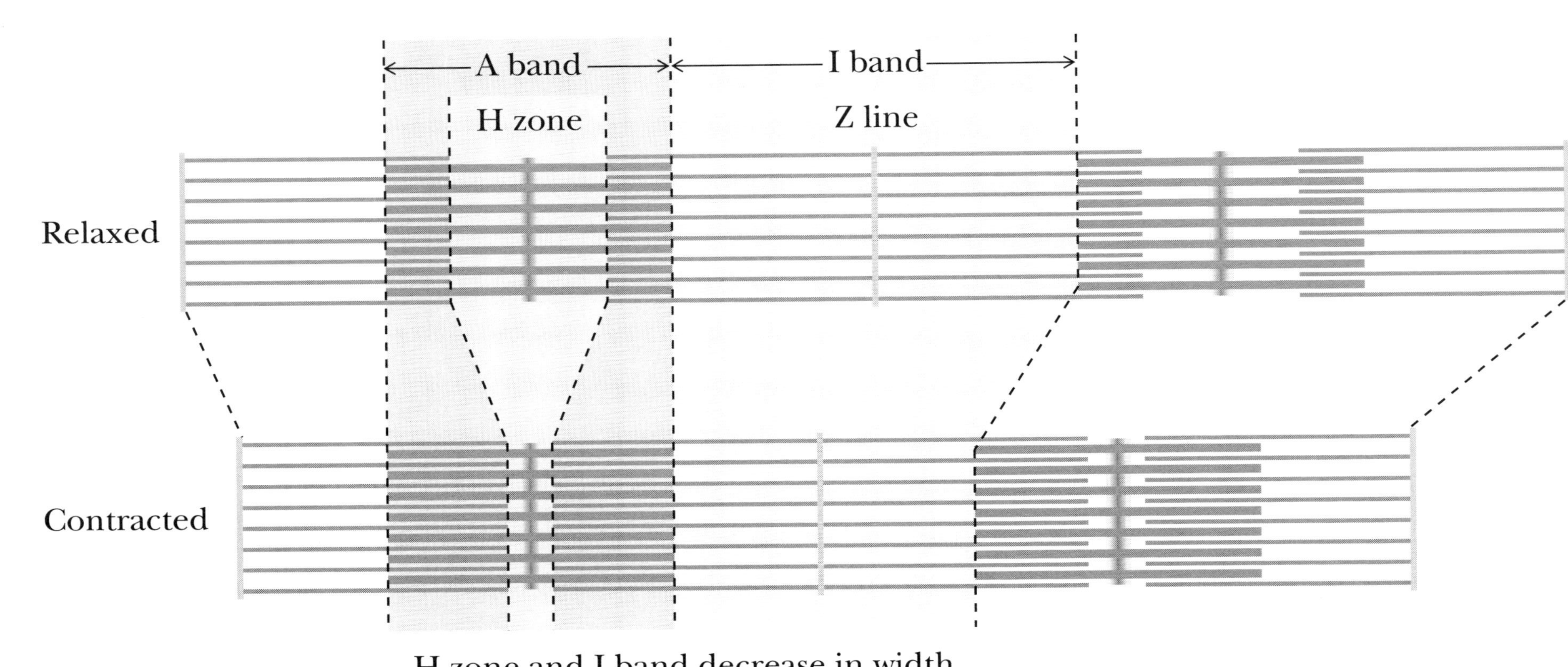

Figure 13.20 The mechanism of skeletal muscle contraction.

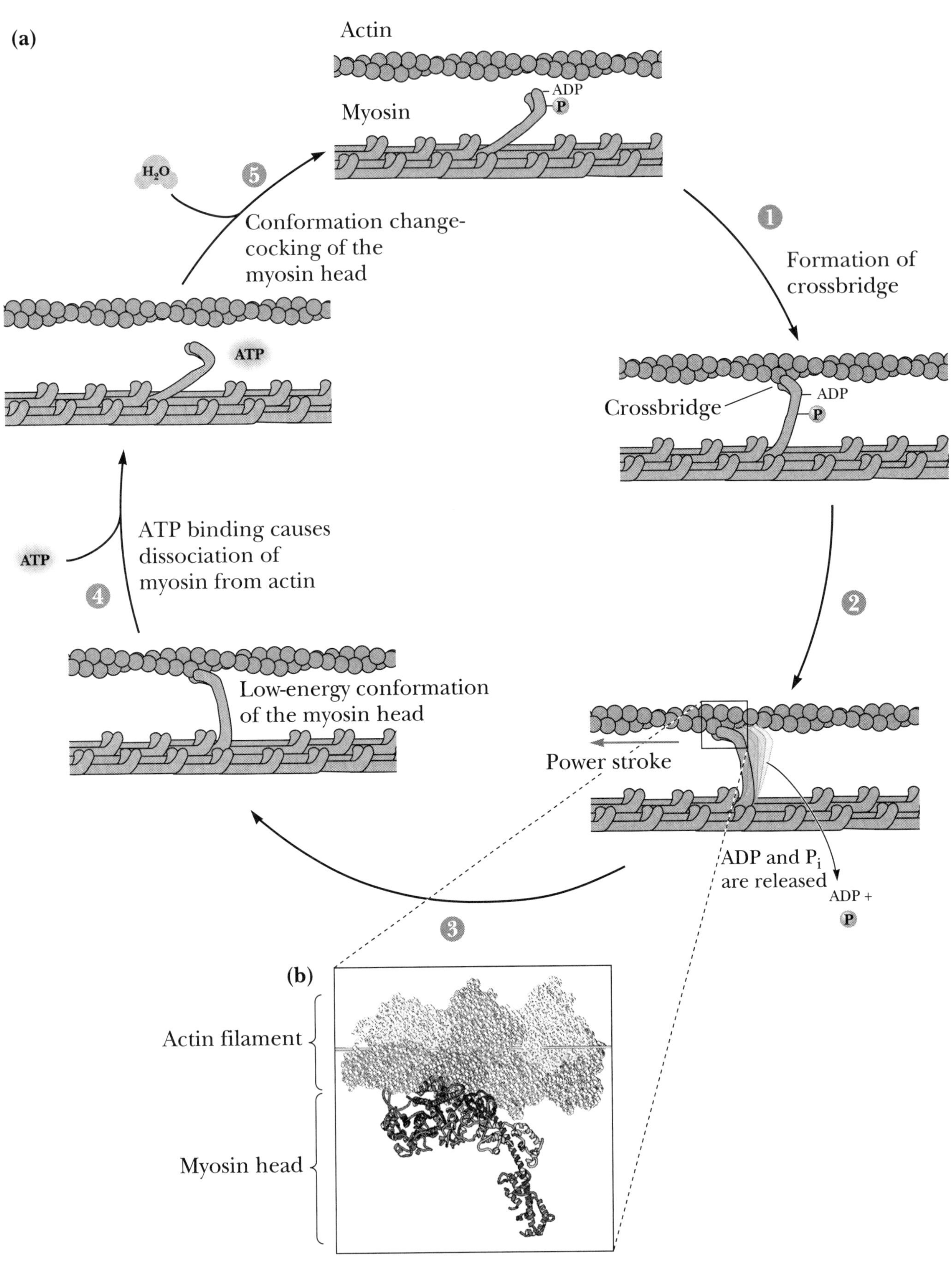

Figure 13.25   A model of flagellar motor assembly of *E. coli*.

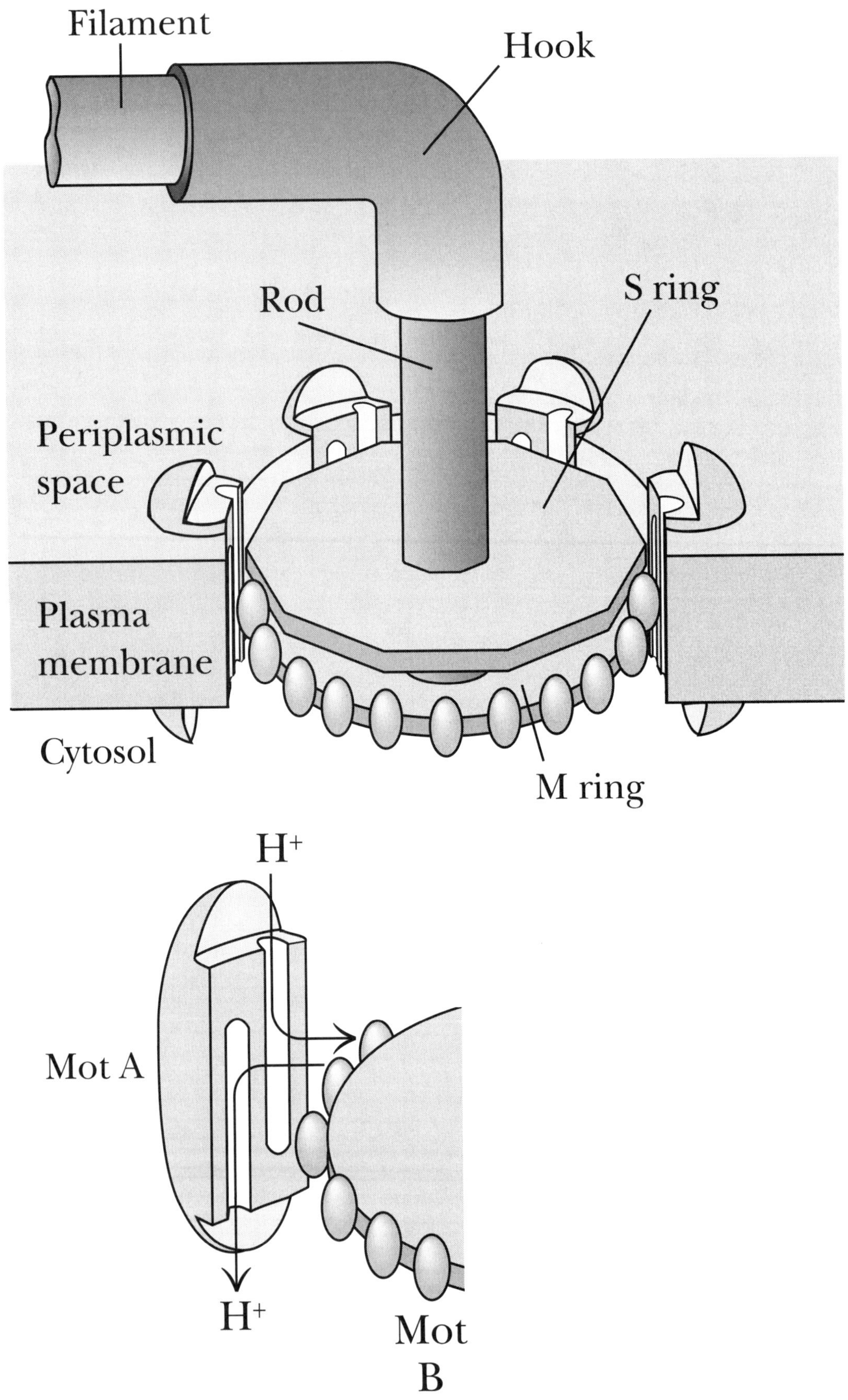

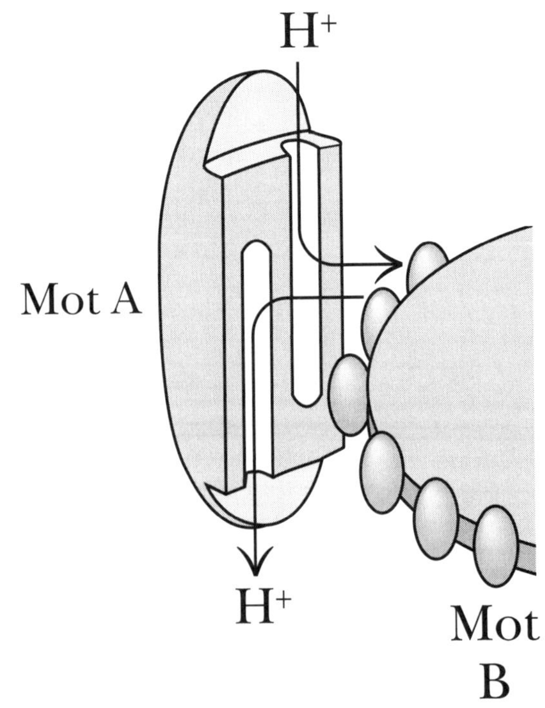

Figure 14.3 Energy relationships between the pathways of catabolism and anabolism.

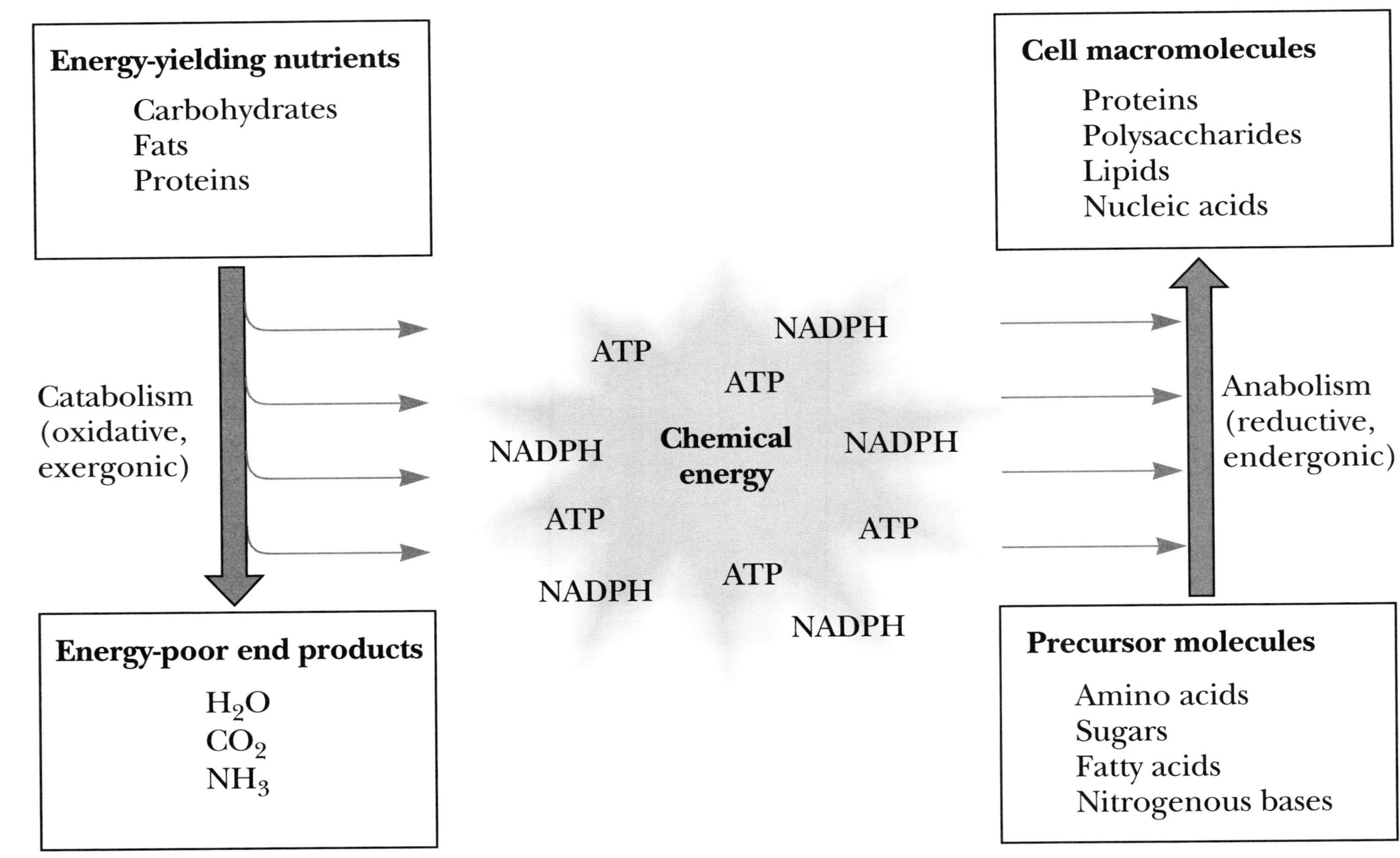

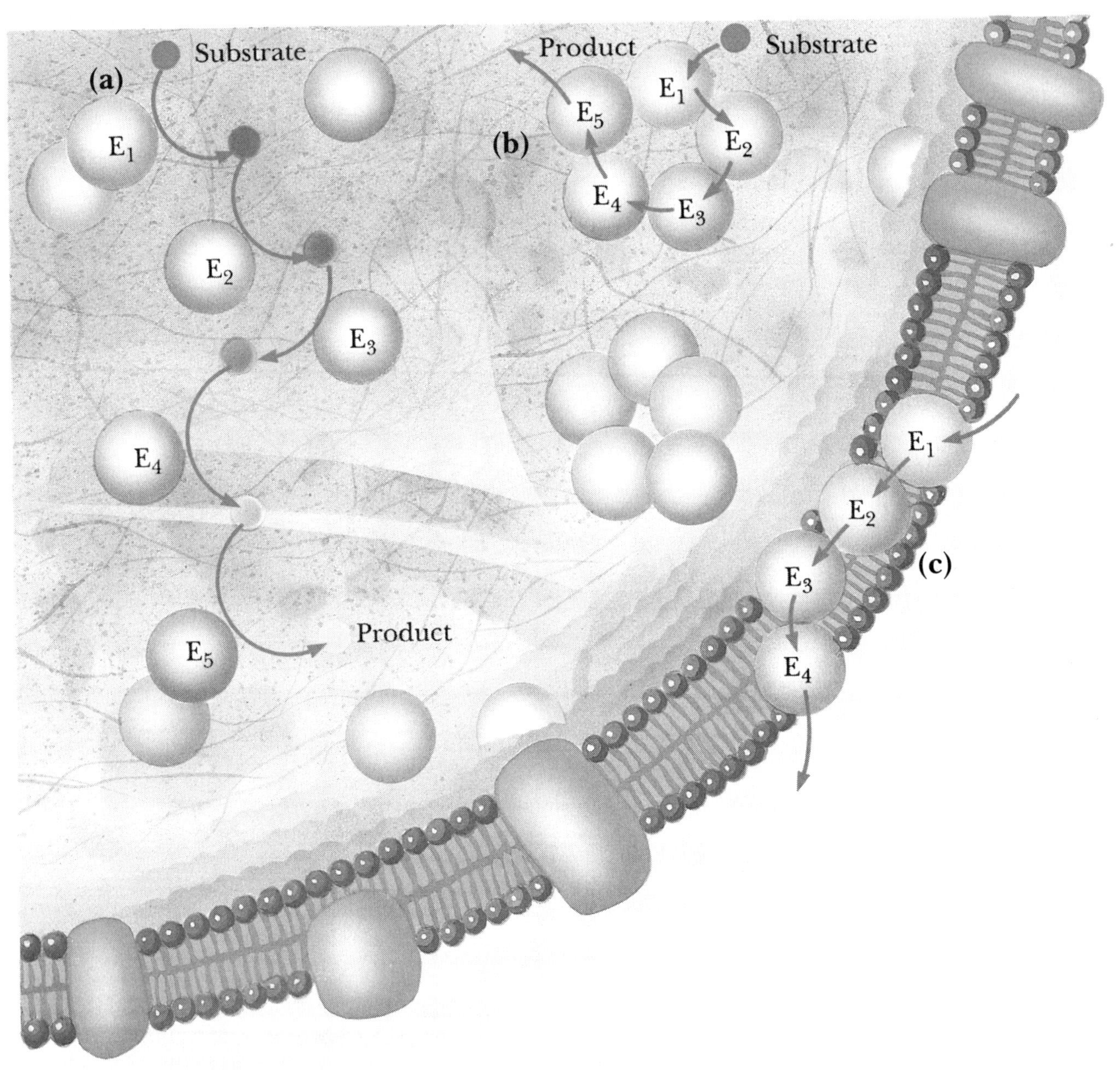
(a)
Substrate
E₁
E₂
E₃
E₄
E₅
Product
(b)
Product
Substrate
E₁
E₂
E₃
E₄
E₅
(c)
E₁
E₂
E₃
E₄
Substrate

Figure 14.5 The three stages of catabolism.

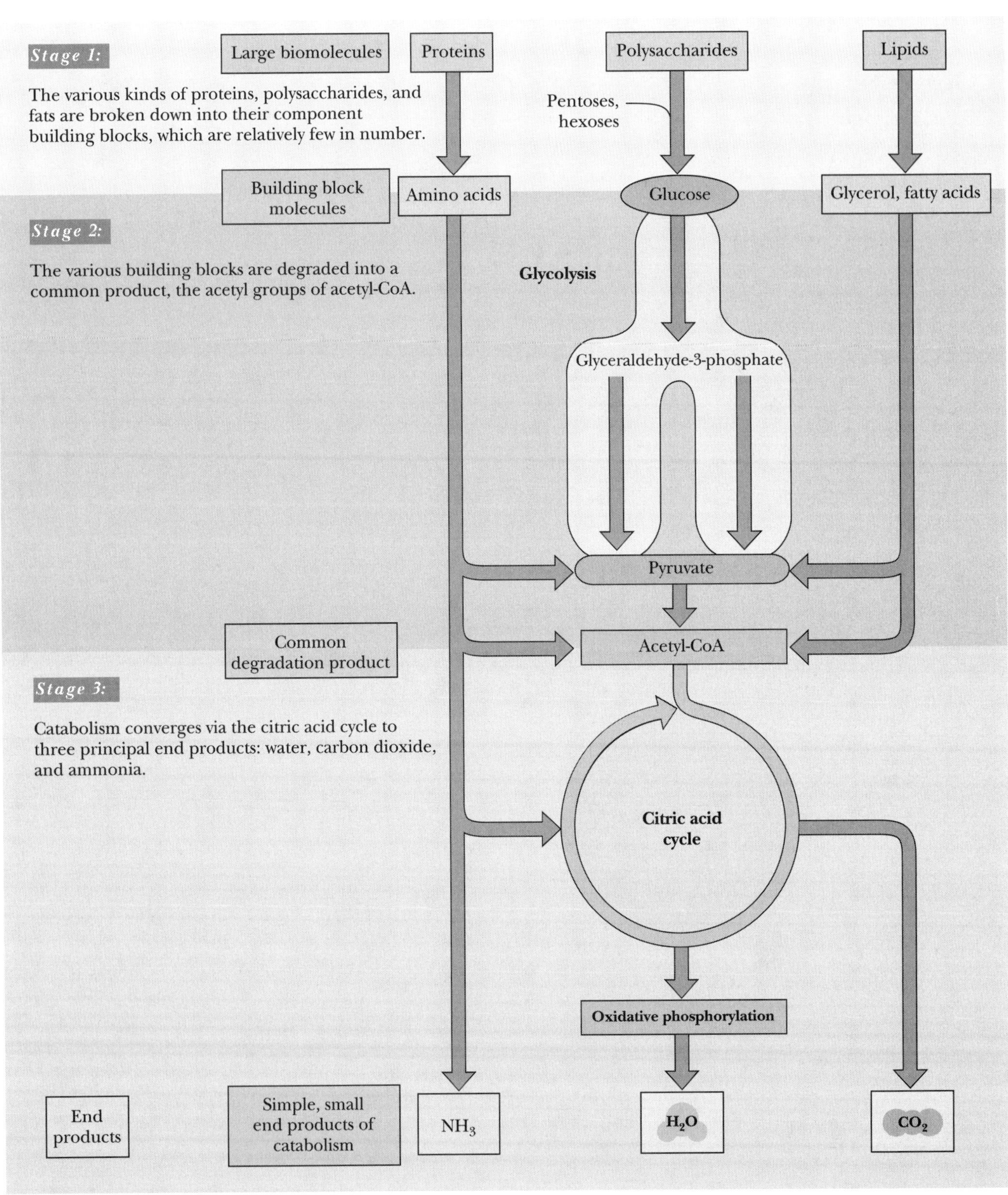

Figure 14.7   The ATP cycle in cells.

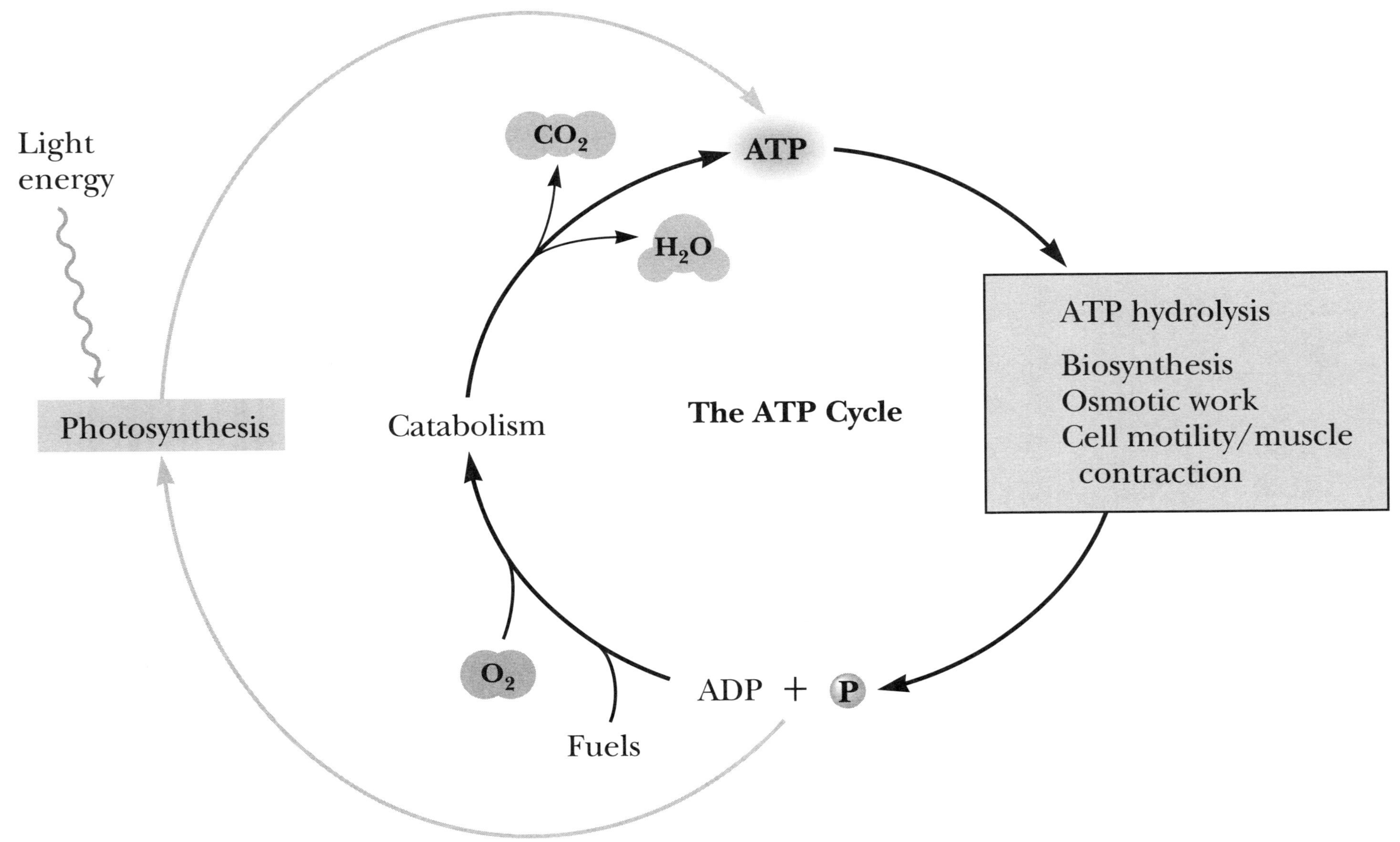

Figure 14.21 (part 1)   Seven classes of reactions catalyzed by pyrioxal-5-phosphate.

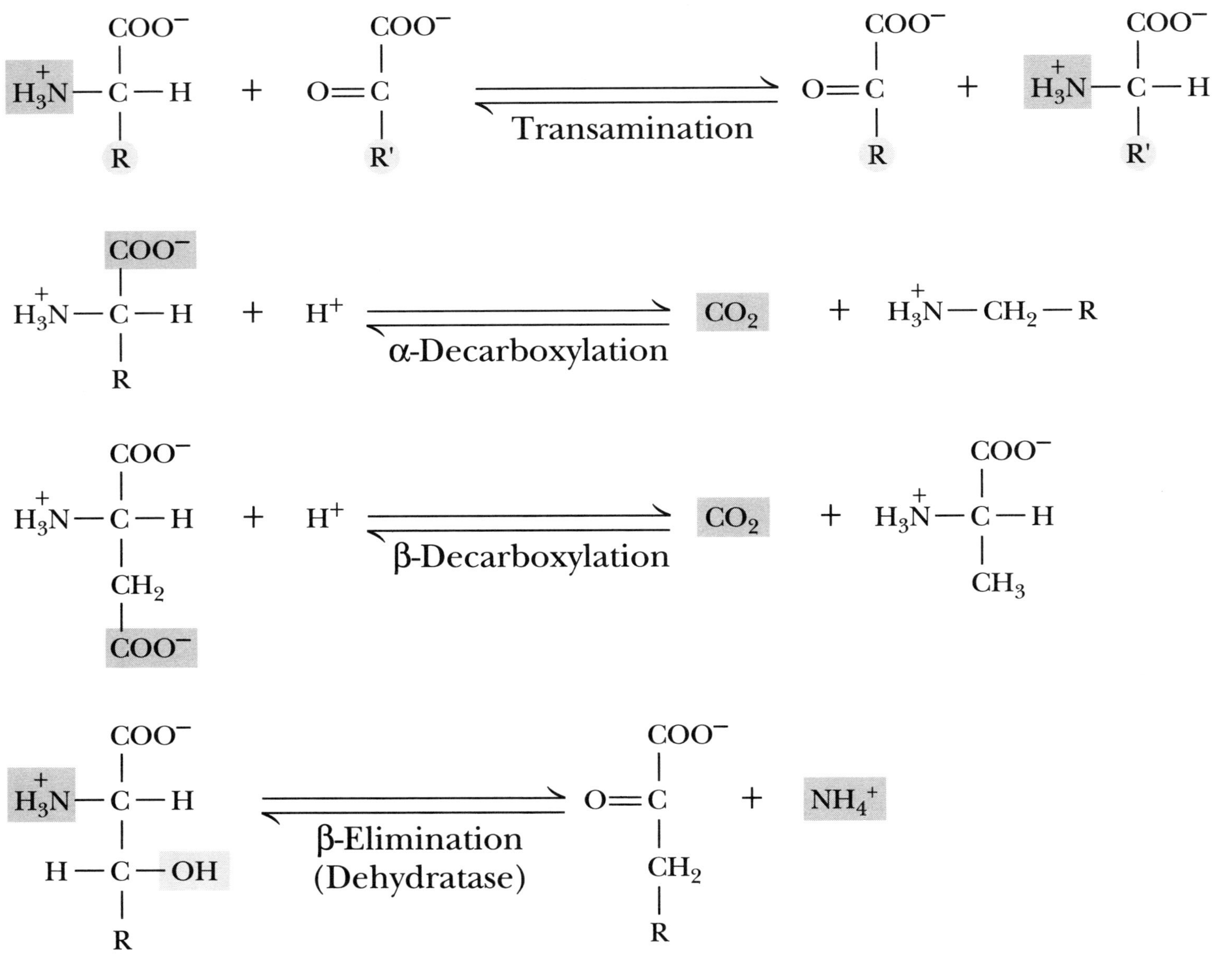

$$\overset{+}{H_3N}-\underset{R}{\overset{COO^-}{\underset{|}{\overset{|}{C}}}}-H \ + \ O{=}\underset{R'}{\overset{COO^-}{\underset{|}{\overset{|}{C}}}} \ \underset{\text{Transamination}}{\rightleftharpoons} \ O{=}\underset{R}{\overset{COO^-}{\underset{|}{\overset{|}{C}}}} \ + \ \overset{+}{H_3N}-\underset{R'}{\overset{COO^-}{\underset{|}{\overset{|}{C}}}}-H$$

$$\overset{+}{H_3N}-\underset{R}{\overset{COO^-}{\underset{|}{\overset{|}{C}}}}-H \ + \ H^+ \ \underset{\alpha\text{-Decarboxylation}}{\rightleftharpoons} \ CO_2 \ + \ \overset{+}{H_3N}-CH_2-R$$

$$\overset{+}{H_3N}-\underset{CH_2-COO^-}{\overset{COO^-}{\underset{|}{\overset{|}{C}}}}-H \ + \ H^+ \ \underset{\beta\text{-Decarboxylation}}{\rightleftharpoons} \ CO_2 \ + \ \overset{+}{H_3N}-\underset{CH_3}{\overset{COO^-}{\underset{|}{\overset{|}{C}}}}-H$$

$$\overset{+}{H_3N}-\underset{H-C-OH,\,R}{\overset{COO^-}{\underset{|}{\overset{|}{C}}}}-H \ \underset{\substack{\beta\text{-Elimination}\\(\text{Dehydratase})}}{\rightleftharpoons} \ O{=}\underset{CH_2-R}{\overset{COO^-}{\underset{|}{\overset{|}{C}}}} \ + \ NH_4^+$$

Figure 14.21 (part 2)   Seven classes of reactions catalyzed by pyrioxal-5-phosphate.

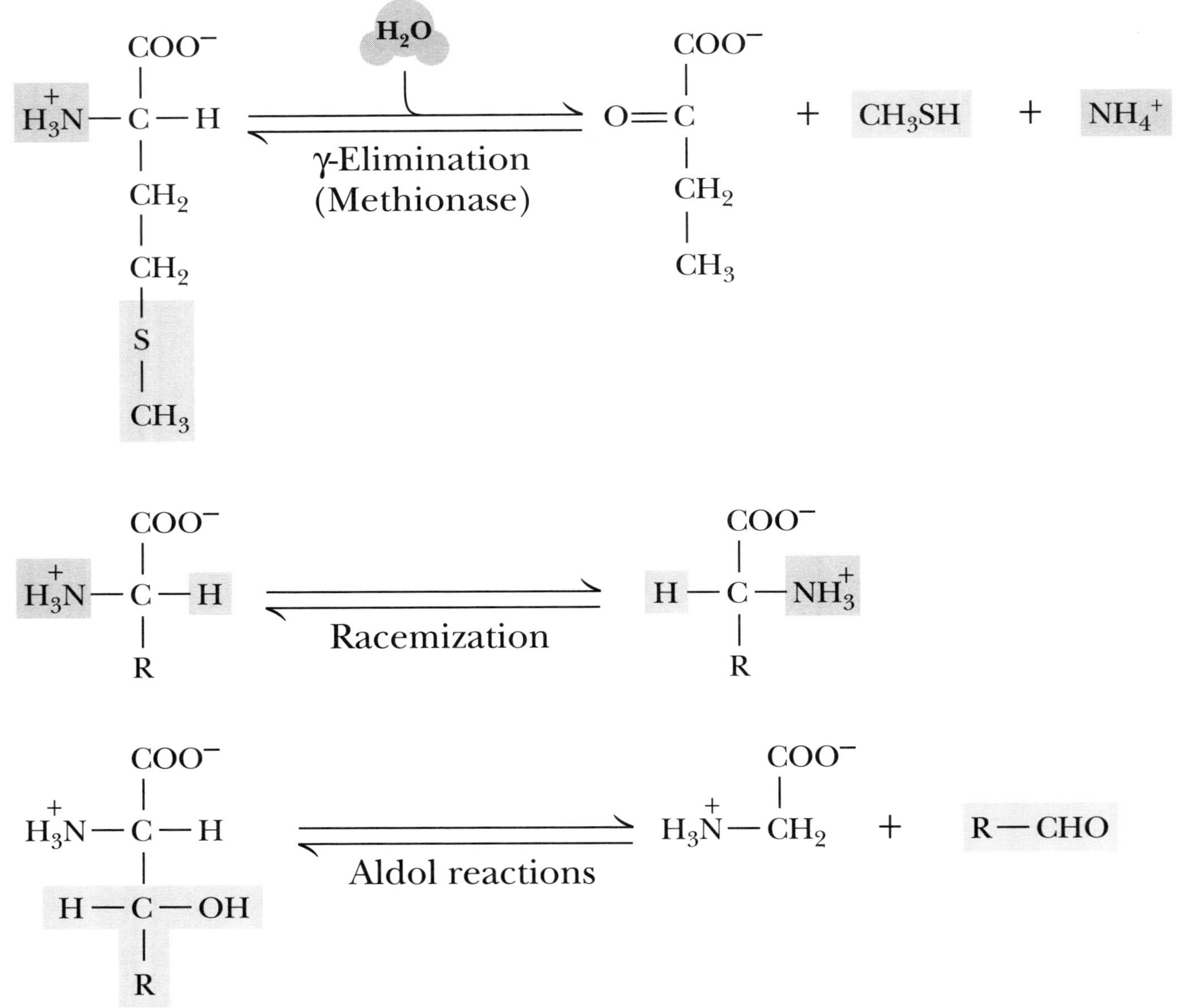

© Harcourt, Inc.

Figure 14.22 Pyridoxal-5-phosphate forms stable Schiff-base adducts with amino acids.

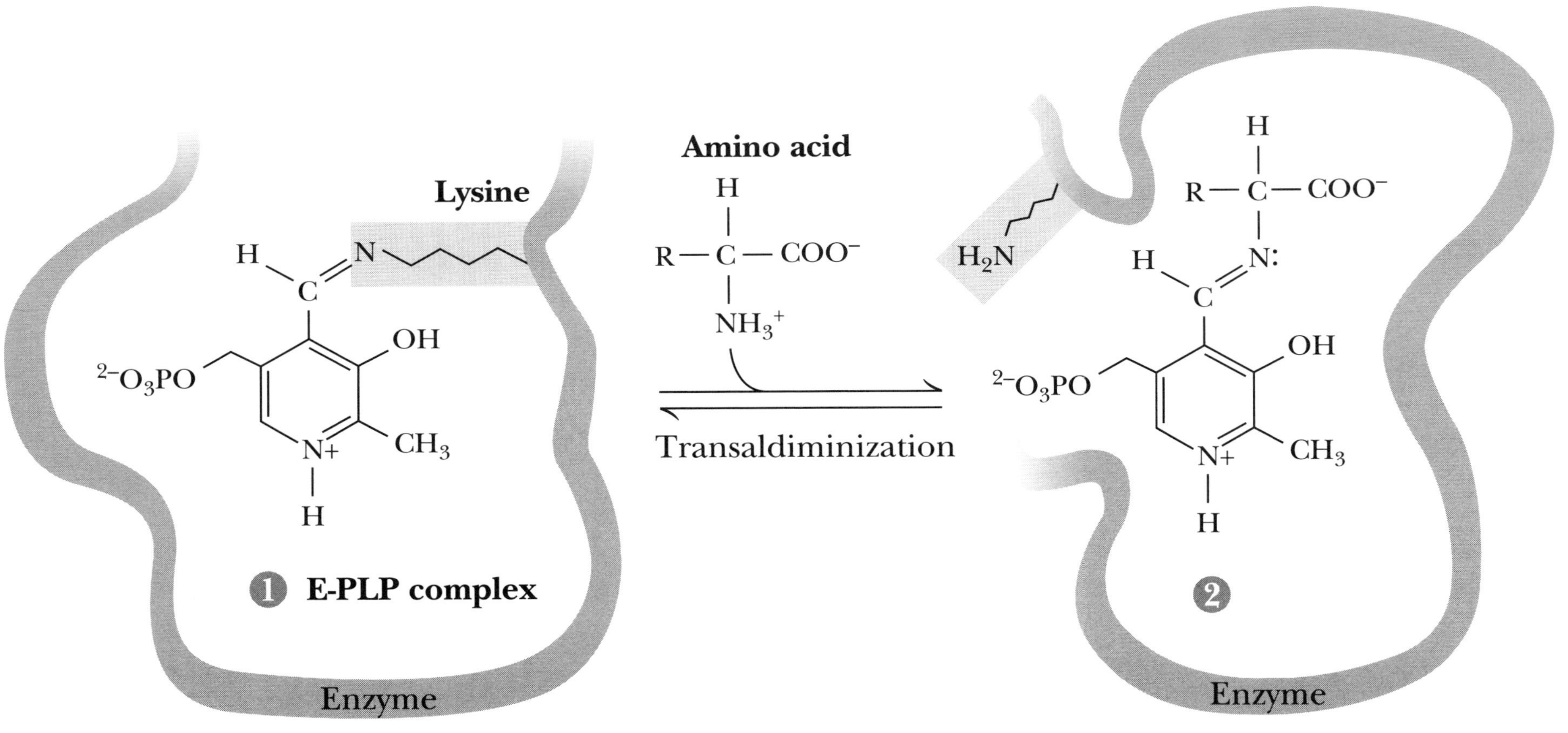

Figure 14.32a Biosynthesis of cholecalciferol and ergocalciferol.

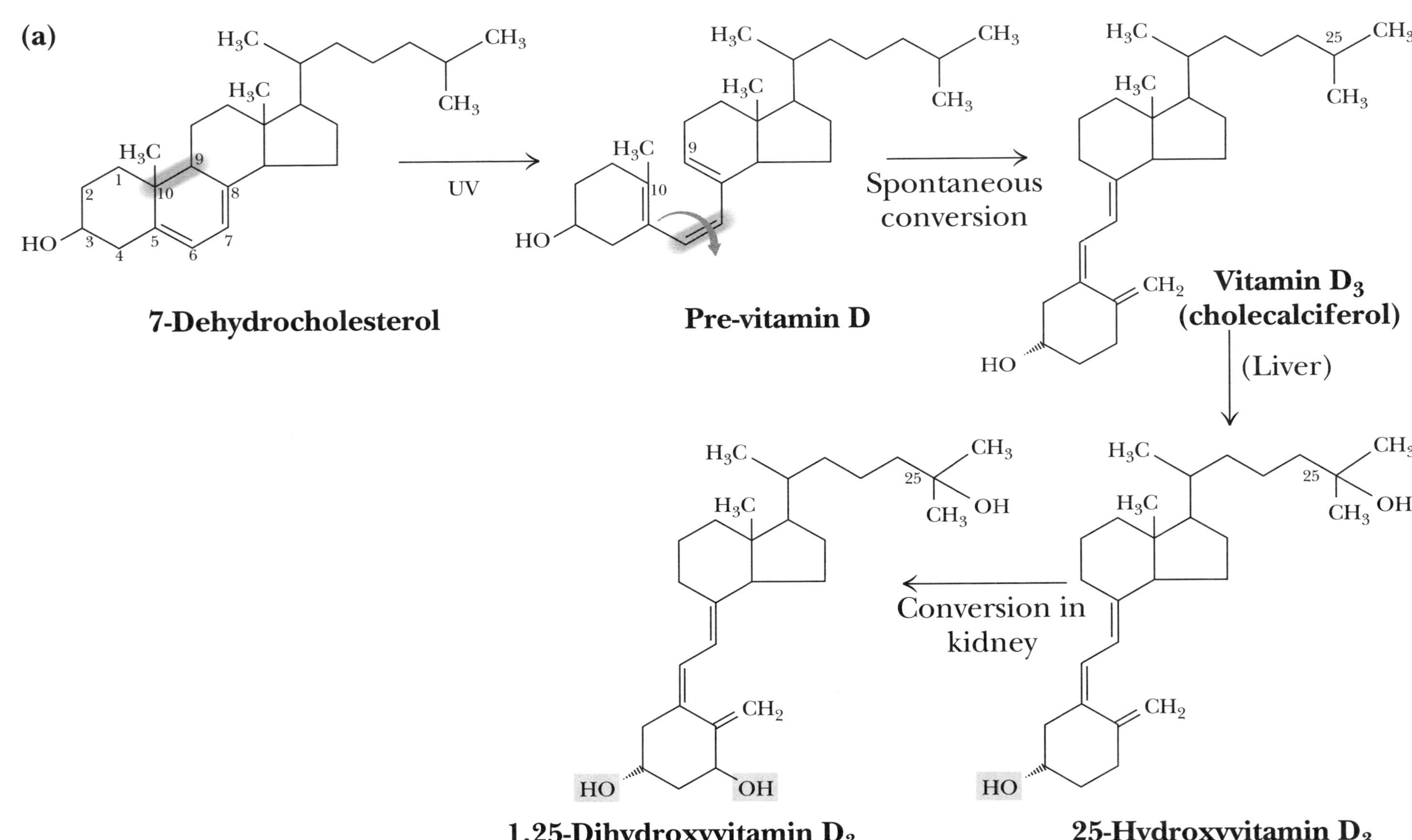

© Harcourt, Inc.

Figure 14.32b Biosynthesis of cholecalciferol and ergocalciferol.

(b)

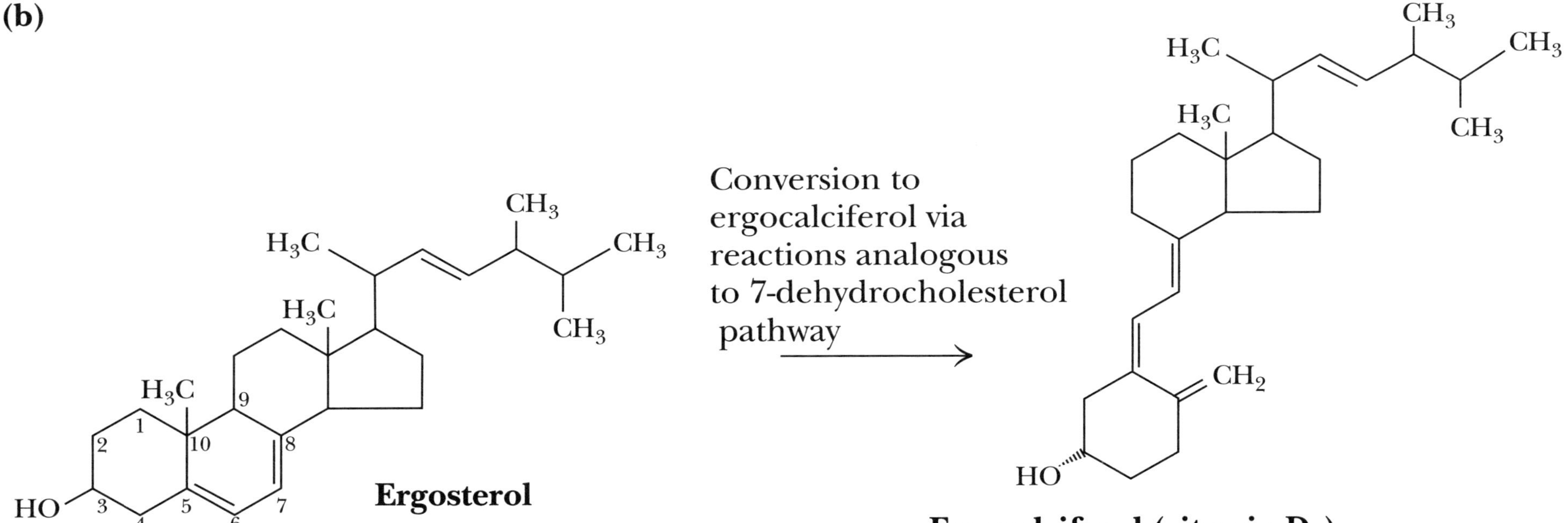

Ergosterol

Ergocalciferol (vitamin D_2)

Figure 15.1 The glycolytic pathway.

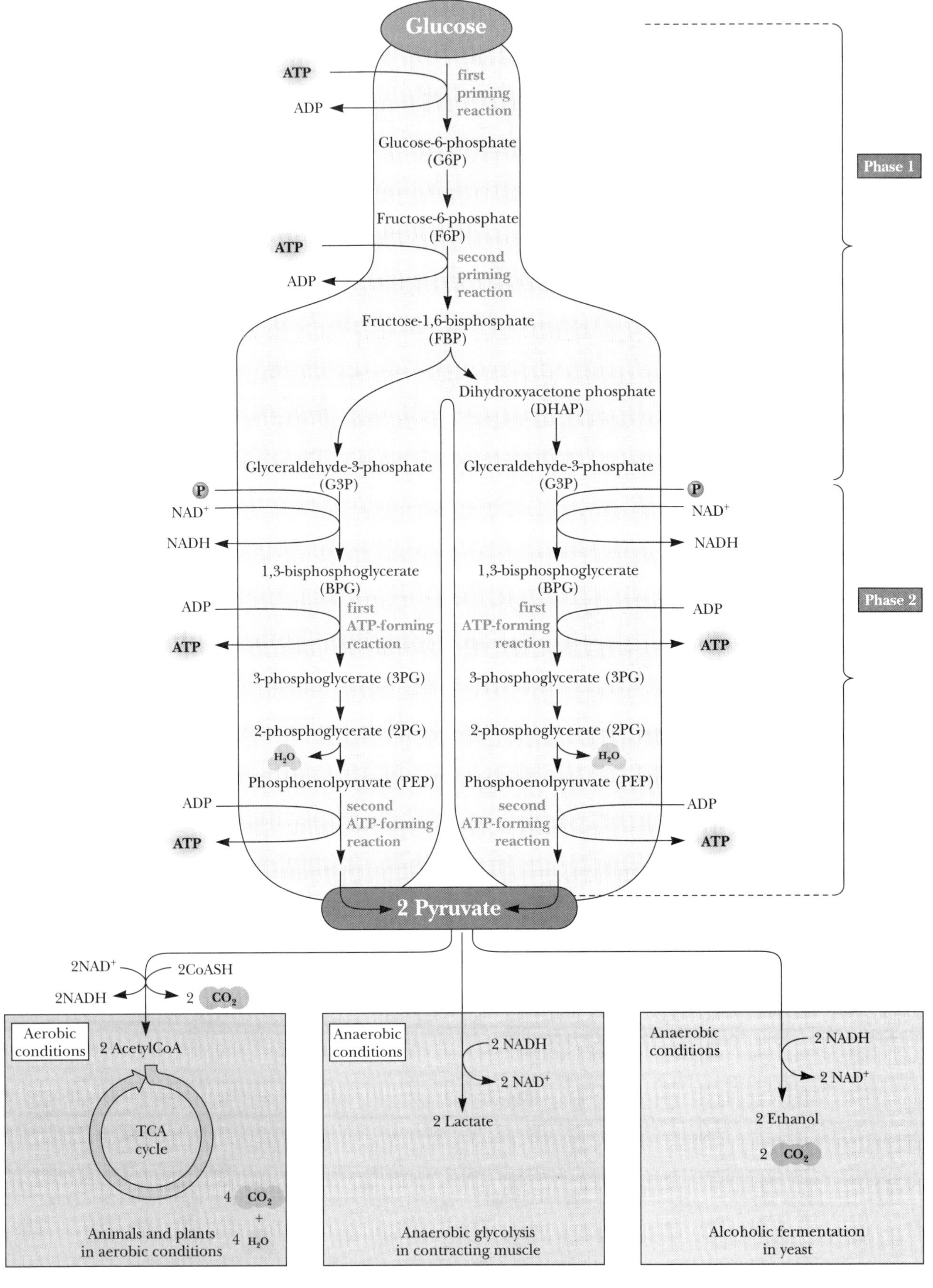

Figure 15.2 The first phase of glycolysis.

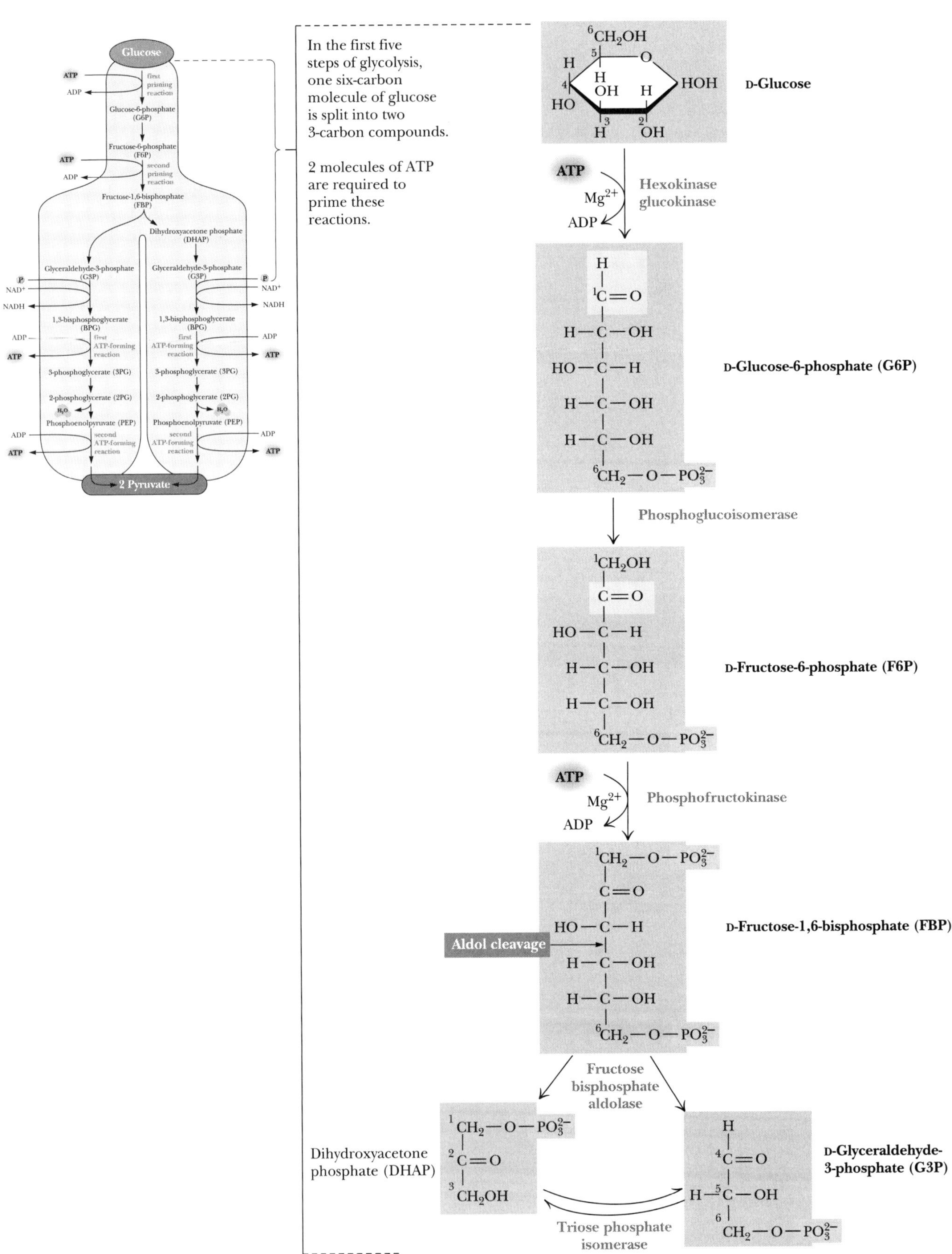

Table 15.1   Reactions and thermodynamics of glycolysis.

Table 15.1   Reactions and Thermodynamics of Glycolysis

Reaction	Enzyme	$\Delta G^{\circ\prime}$ (kJ/mol)	ΔG (kJ/mol)
α-D-Glucose + ATP^{4-} $\rightleftharpoons$ glucose-6-phosphate^{2-} + ADP^{3-} + H^+	Hexokinase/Glucokinase	-16.7	$-33.9*$
Glucose-6-phosphate^{2-} $\rightleftharpoons$ fructose-6-phosphate^{2-}	Phosphoglucoisomerase	$+1.67$	-2.92
Fructose-6-phosphate^{2-} + ATP^{4-} $\rightleftharpoons$ fructose-1,6-bisphosphate^{4-} + ADP^{3-} + H^+	Phosphofructokinase	-14.2	-18.8
Fructose-1,6-bisphosphate^{4-} $\rightleftharpoons$ dihydroxyacetone-P^{2-} + glyceraldehyde-3-P^{2-}	Fructose bisphosphate aldolase	$+23.9$	-0.23
Dihydroxyacetone-P^{2-} $\rightleftharpoons$ glyceraldehyde-3-P^{2-}	Triose phosphate isomerase	$+7.56$	$+2.41$
Glyceraldehyde-3-P^{2-} + P_i^{2-} + NAD^+ $\rightleftharpoons$ 1,3-bisphosphoglycerate^{4-} + NADH + H^+	Glyceraldehyde-3-P dehydrogenase	$+6.30$	-1.29
1,3-Bisphosphoglycerate^{4-} + ADP^{3-} $\rightleftharpoons$ 3-P-glycerate^{3-} + ATP^{4-}	Phosphoglycerate kinase	-18.9	$+0.1$
3-Phosphoglycerate^{3-} $\rightleftharpoons$ 2-phosphoglycerate^{3-}	Phosphoglycerate mutase	$+4.4$	$+0.83$
2-Phosphoglycerate^{3-} $\rightleftharpoons$ phosphoenolpyruvate^{3-} + H_2O	Enolase	$+1.8$	$+1.1$
Phosphoenolpyruvate^{3-} + ADP^{3-} + H^+ $\rightleftharpoons$ pyruvate$^-$ + ATP^{4-}	Pyruvate kinase	-31.7	-23.0
Pyruvate$^-$ + NADH + H^+ $\rightleftharpoons$ lactate$^-$ + NAD^+	Lactate dehydrogenase	-25.2	-14.8

*ΔG values calculated for 310 K (37°C) using the data in Table 15.2 for metabolite concentrations in erythrocytes. $\Delta G^{\circ\prime}$ values are assumed to be the same at 25°C and 37°C.

Figure 15.10 The second phase of glycolysis.

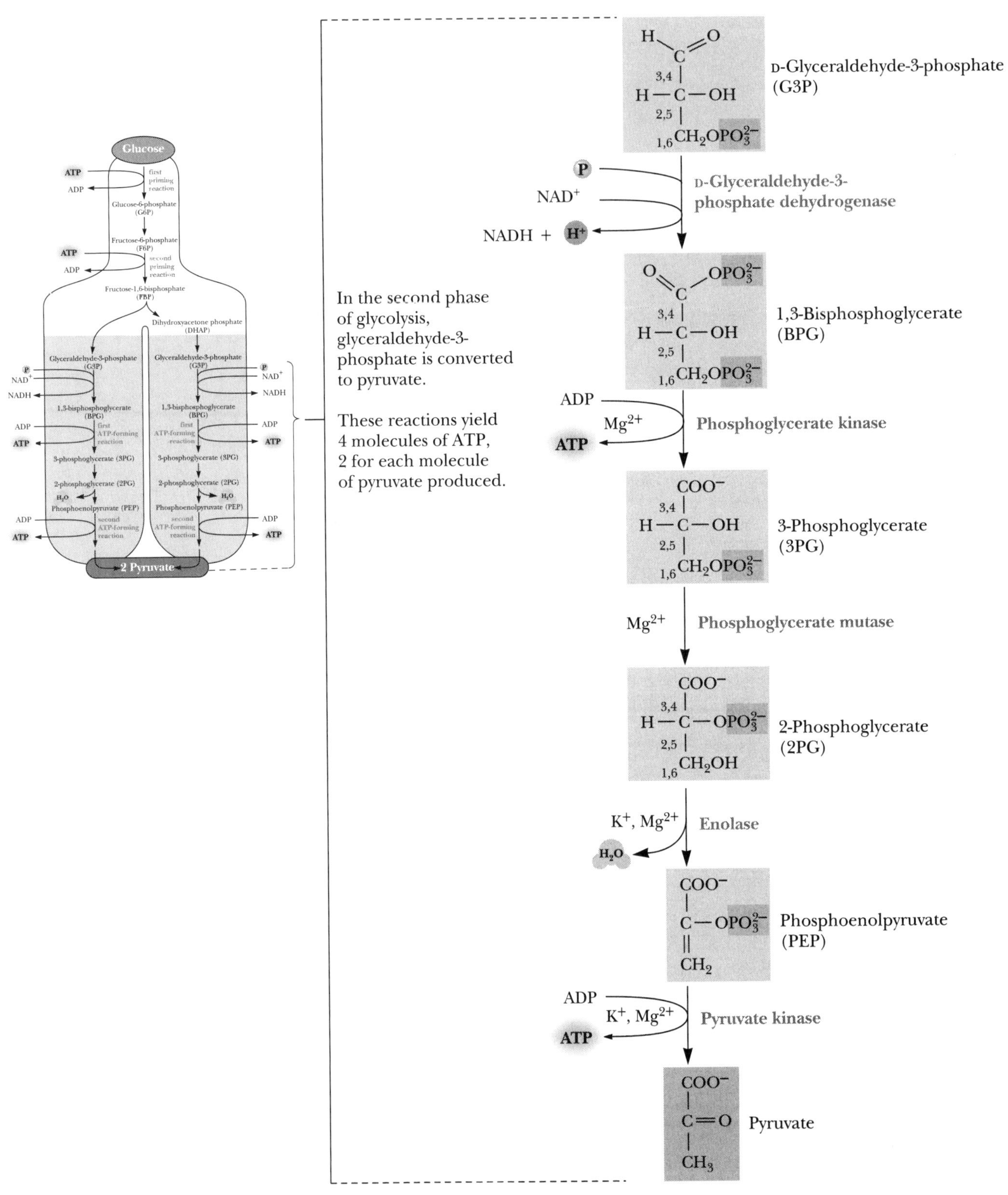

Figure 15.20   A mechanism for the pyruvate kinase reaction.

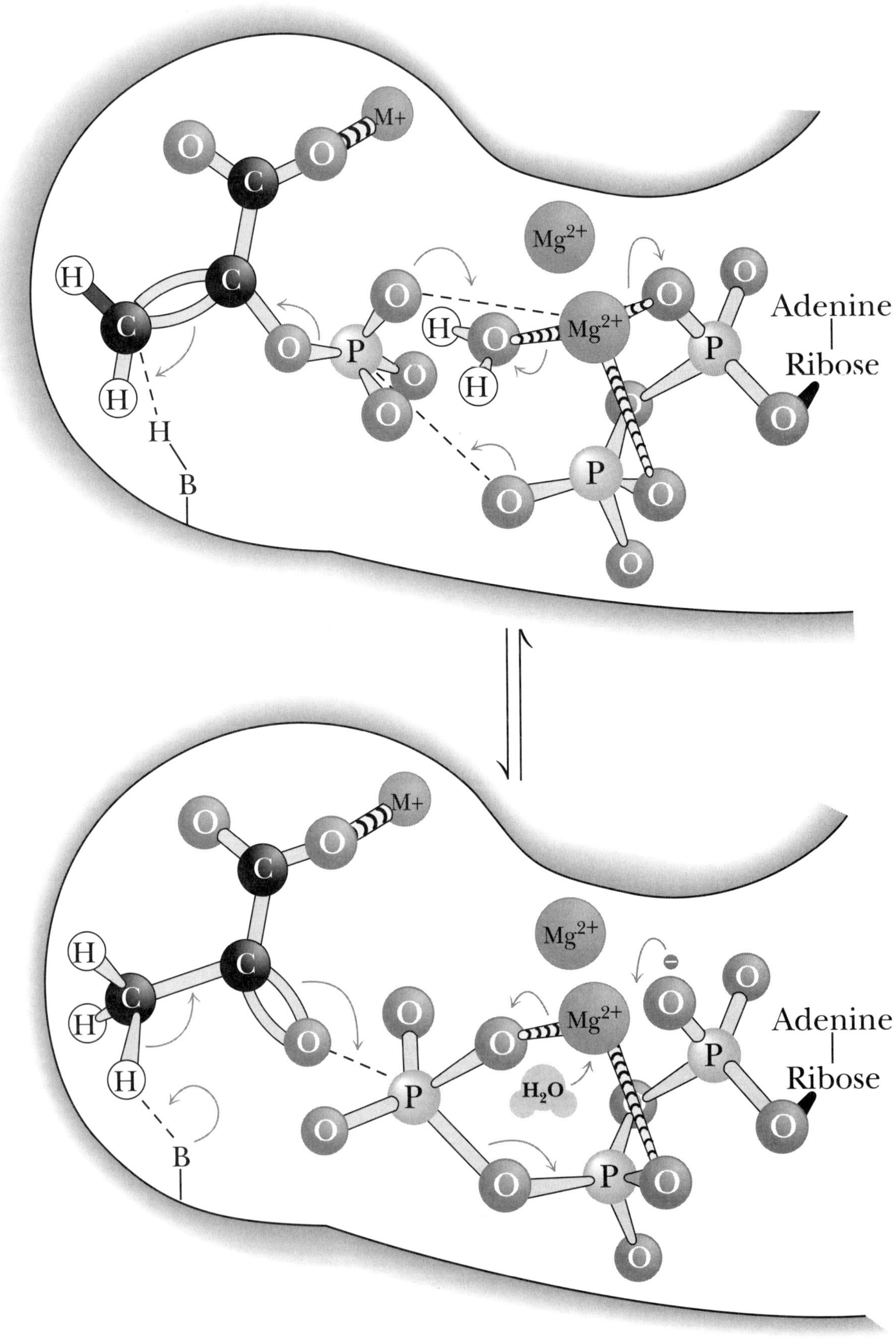

Figure 15.22   A comparison of free energy changes for the reactions of glycolysis.

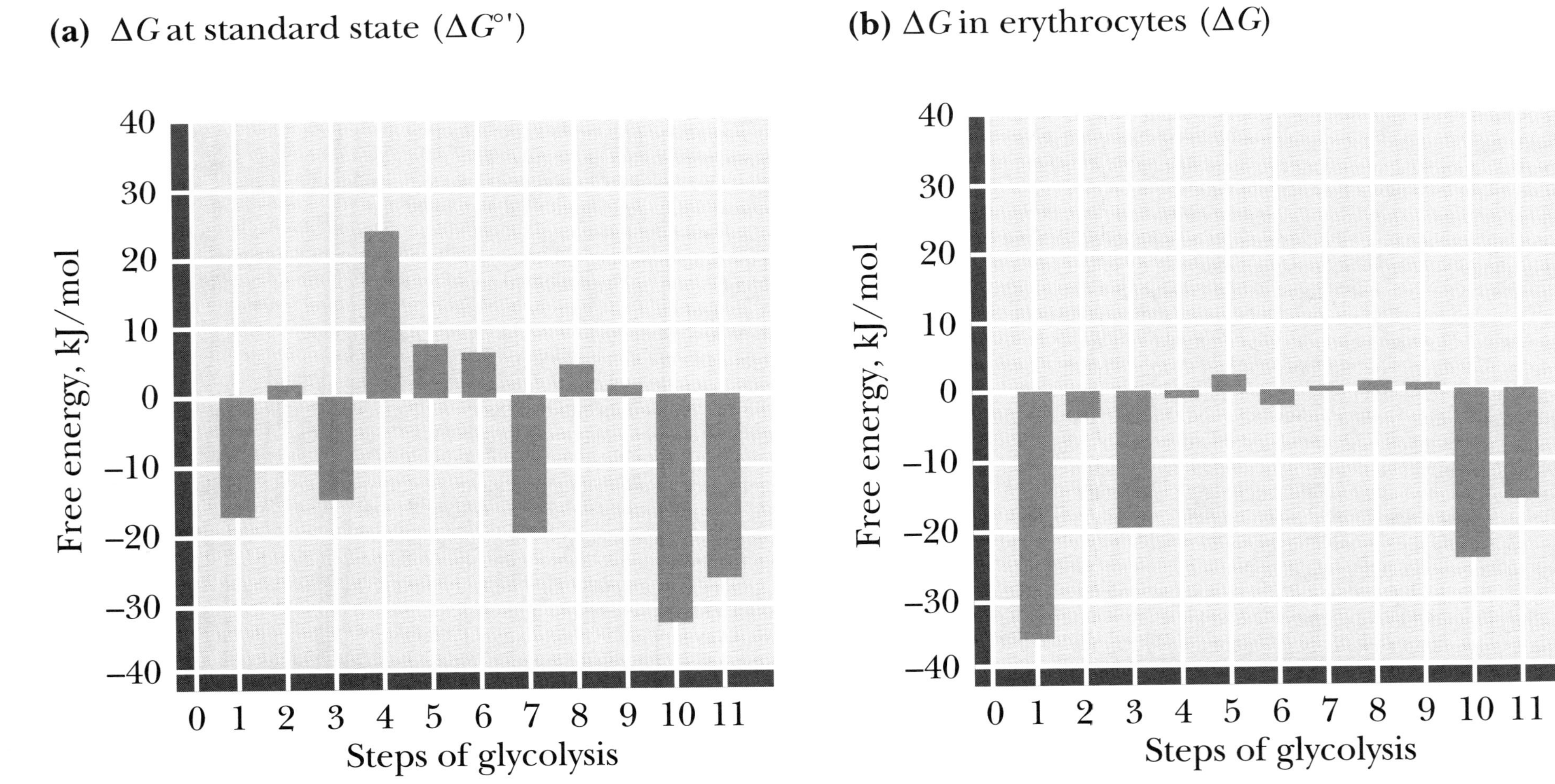

Figure 15.23 Simple metabolites can enter the glycolytic pathway.

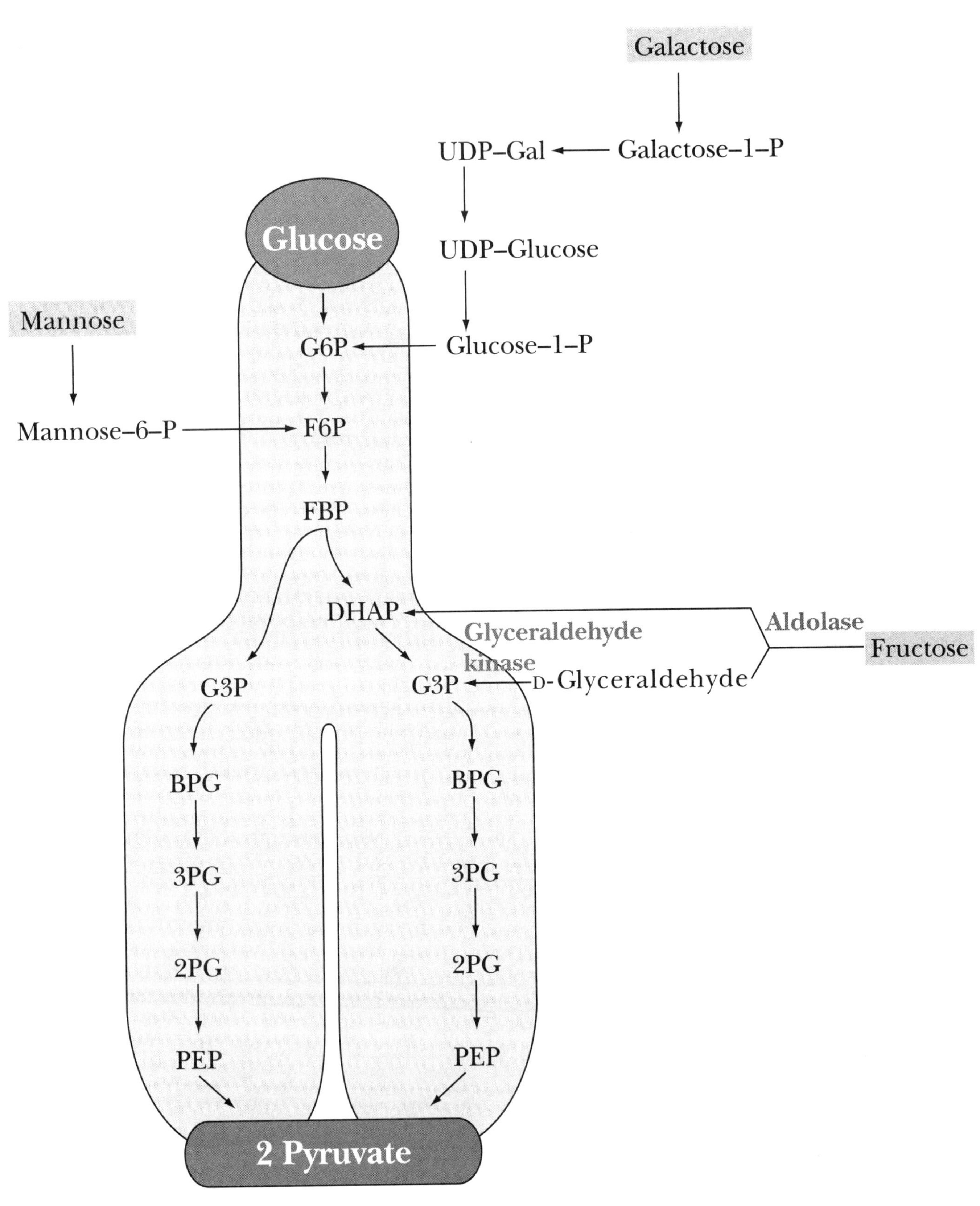

© Harcourt, Inc.

Figure 16.1 Oxidative phosphorylation.

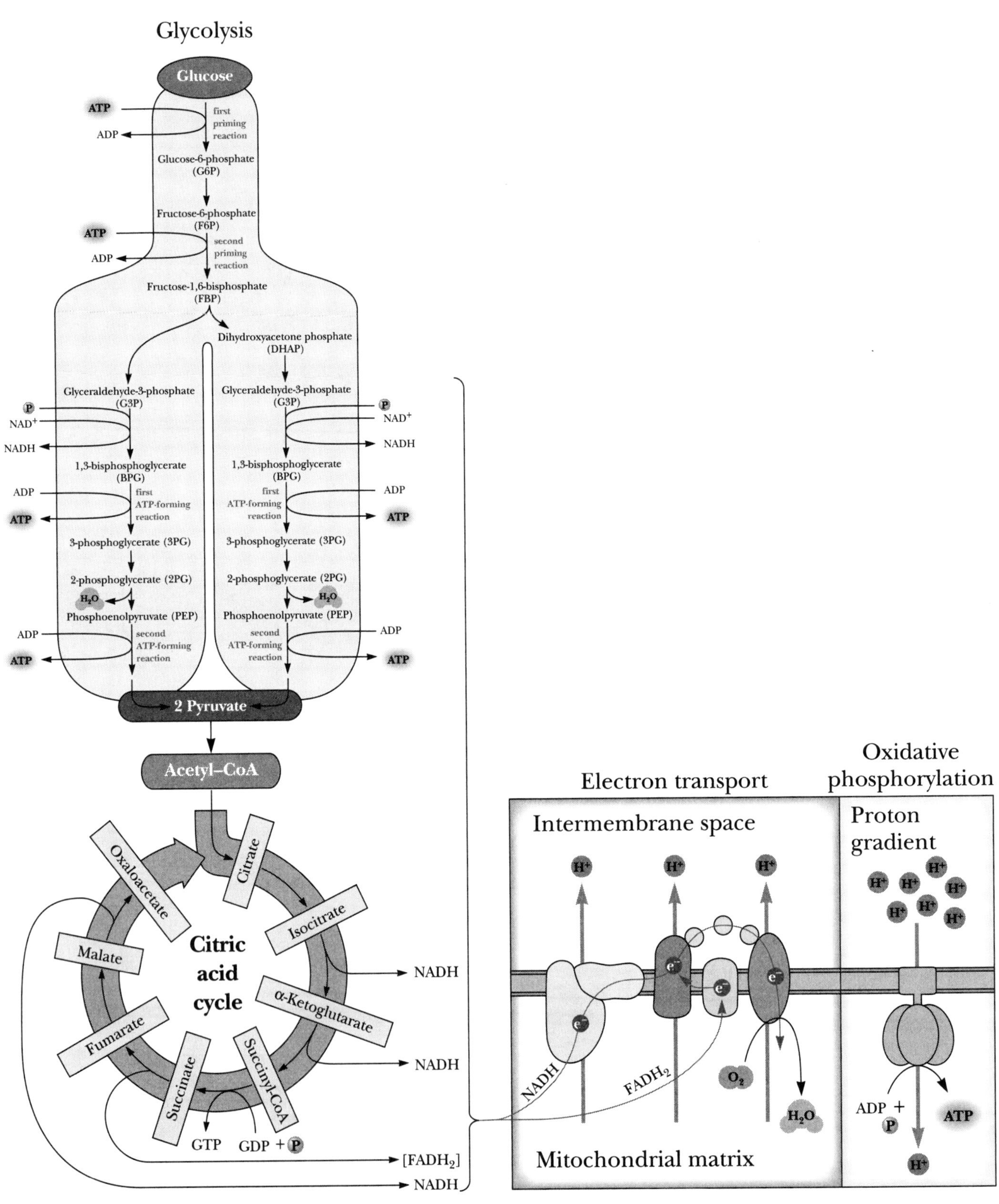

Figure 16.2 The tricarboxylic acid cycle.

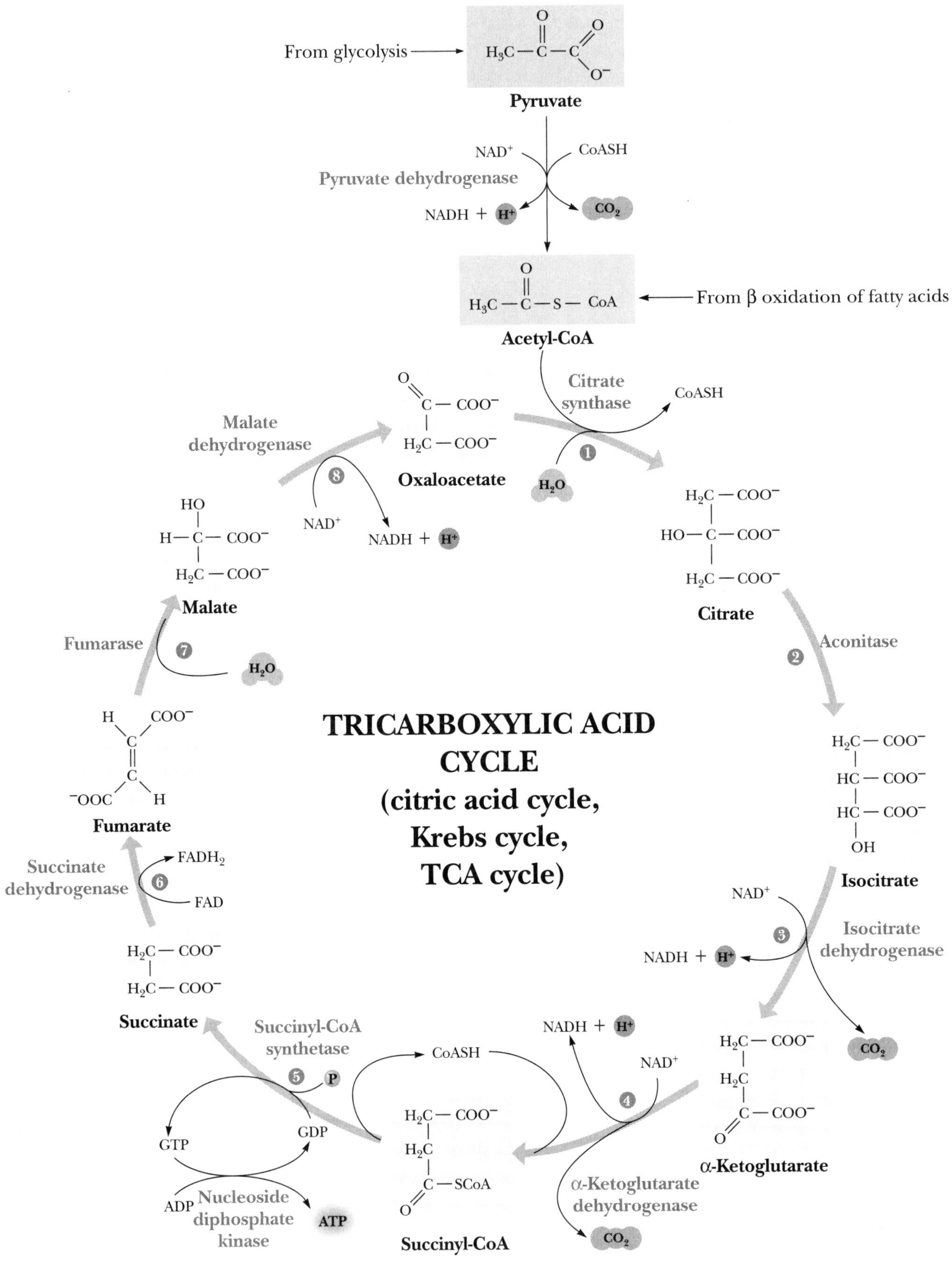

Table 16.1   The enzymes and reactions of the TCA cycle.

Table 16.1   The Enzymes and Reactions of the TCA Cycle

Reaction	Enzyme	$\Delta G^{\circ\prime}$ (kJ/mol)	ΔG (kJ/mol)
1. Acetyl-CoA + oxaloacetate + H_2O $\rightleftharpoons$ CoASH + citrate	Citrate synthase	-31.4	-53.9
2. Citrate $\rightleftharpoons$ isocitrate	Aconitase	$+6.7$	$+0.8$
3. Isocitrate + NAD^+ $\rightleftharpoons$ α-ketoglutarate + NADH + CO_2 + H^+	Isocitrate dehydrogenase	-8.4	-17.5
4. α-Ketoglutarate + CoASH + NAD^+ $\rightleftharpoons$ succinyl-CoA + NADH + CO_2 + H^+	α-Ketoglutarate dehydrogenase complex	-30	-43.9
5. Succinyl-CoA + GDP + P_i $\rightleftharpoons$ succinate + GTP + CoASH	Succinyl-CoA synthetase	-3.3	≈ 0
6. Succinate + [FAD] $\rightleftharpoons$ fumarate + [$FADH_2$]	Succinate dehydrogenase	$+0.4$	≈ 0
7. Fumarate + H_2O $\rightleftharpoons$ L-malate	Fumarase	-3.8	≈ 0
8. L-Malate + NAD^+ $\rightleftharpoons$ oxaloacetate + NADH + H^+	Malate dehydrogenase	$+29.7$	≈ 0

Net for reactions 1–8:

Actelyl-CoA + 3 NAD^+ + [FAD] + GDP + P_i + 2 H_2O $\rightleftharpoons$ CoASH + 3 NADH + [$FADH_2$] + GTP + 2 CO_2 + 3 H^+ -40 $\approx(-115)$

Simple combustion of acetate: Acetate + 2 O_2 + H^+ $\rightleftharpoons$ 2 CO_2 + 2 H_2O -849

ΔG values from Newsholme, E. A., and Leech, A. R., 1983. *Biochemistry for the Medical Sciences.* New York: Wiley & Sons.

Figure 16.14   Intermediates of the TCA cycle.

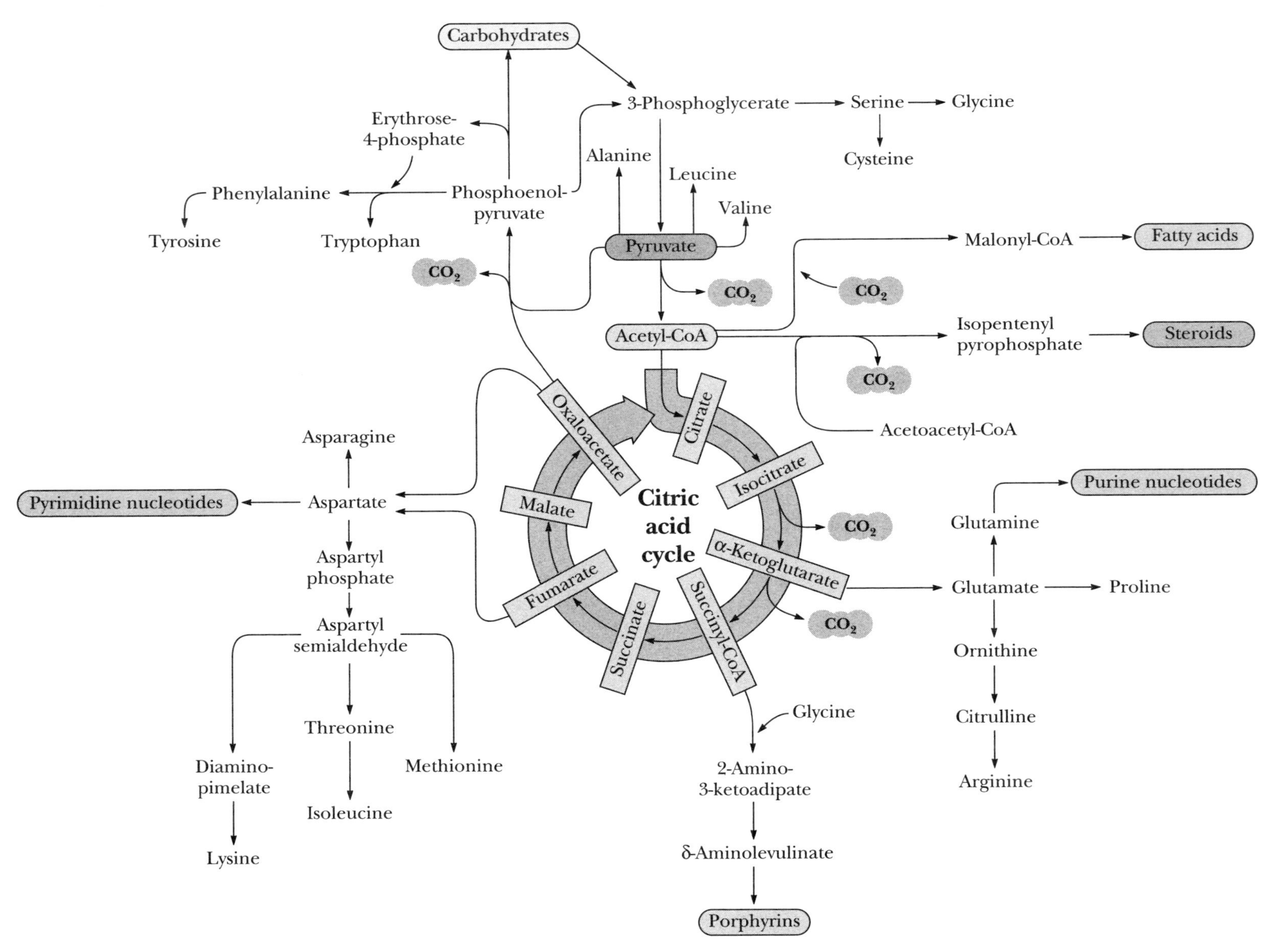

© Harcourt, Inc.

Figure 16.15 Export of citrate from mitochondria and cytosolic breakdown produces oxaloacetate and acetyl-CoA.

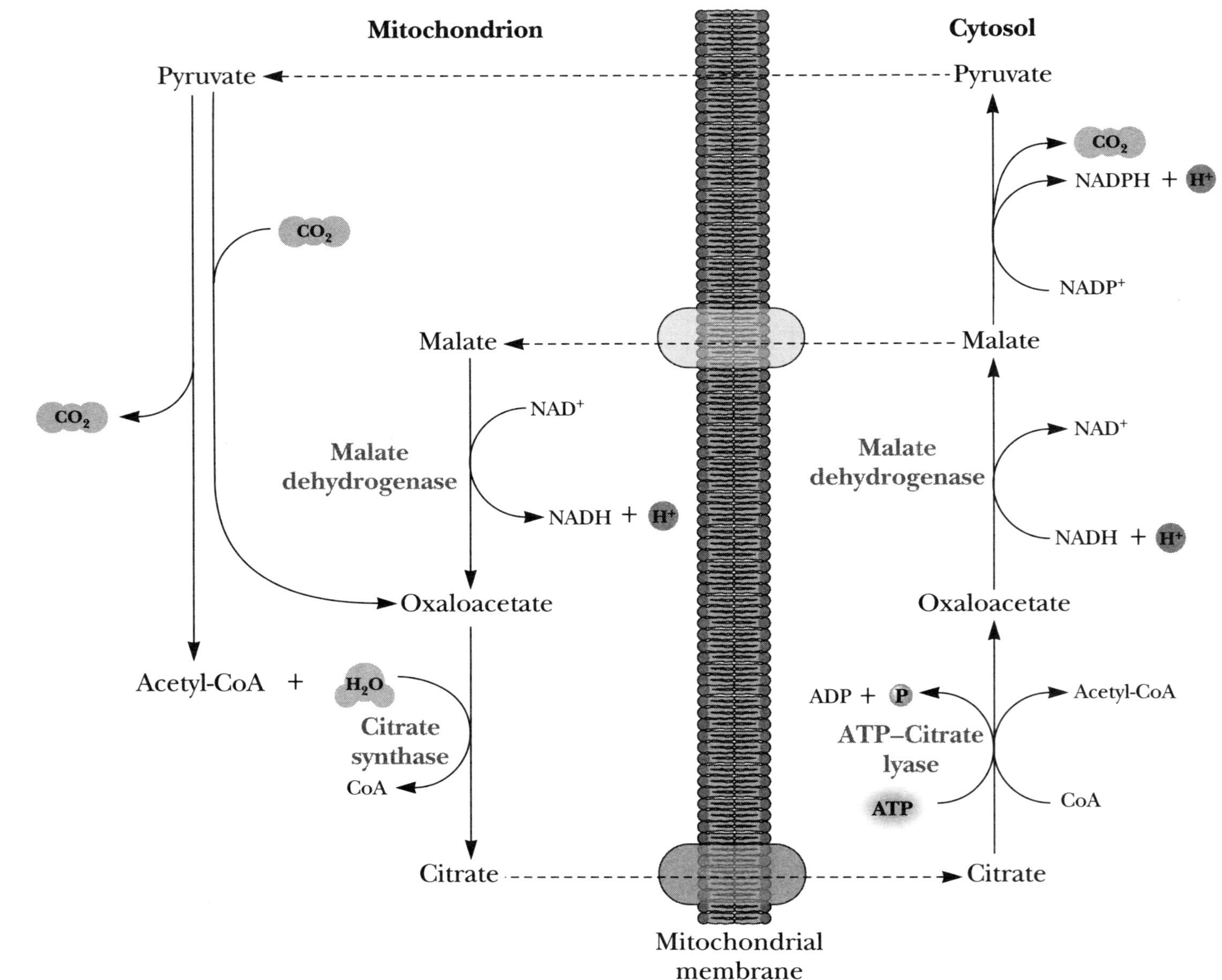

© Harcourt, Inc.

Figure 16.18   Regulation of the TCA cycle.

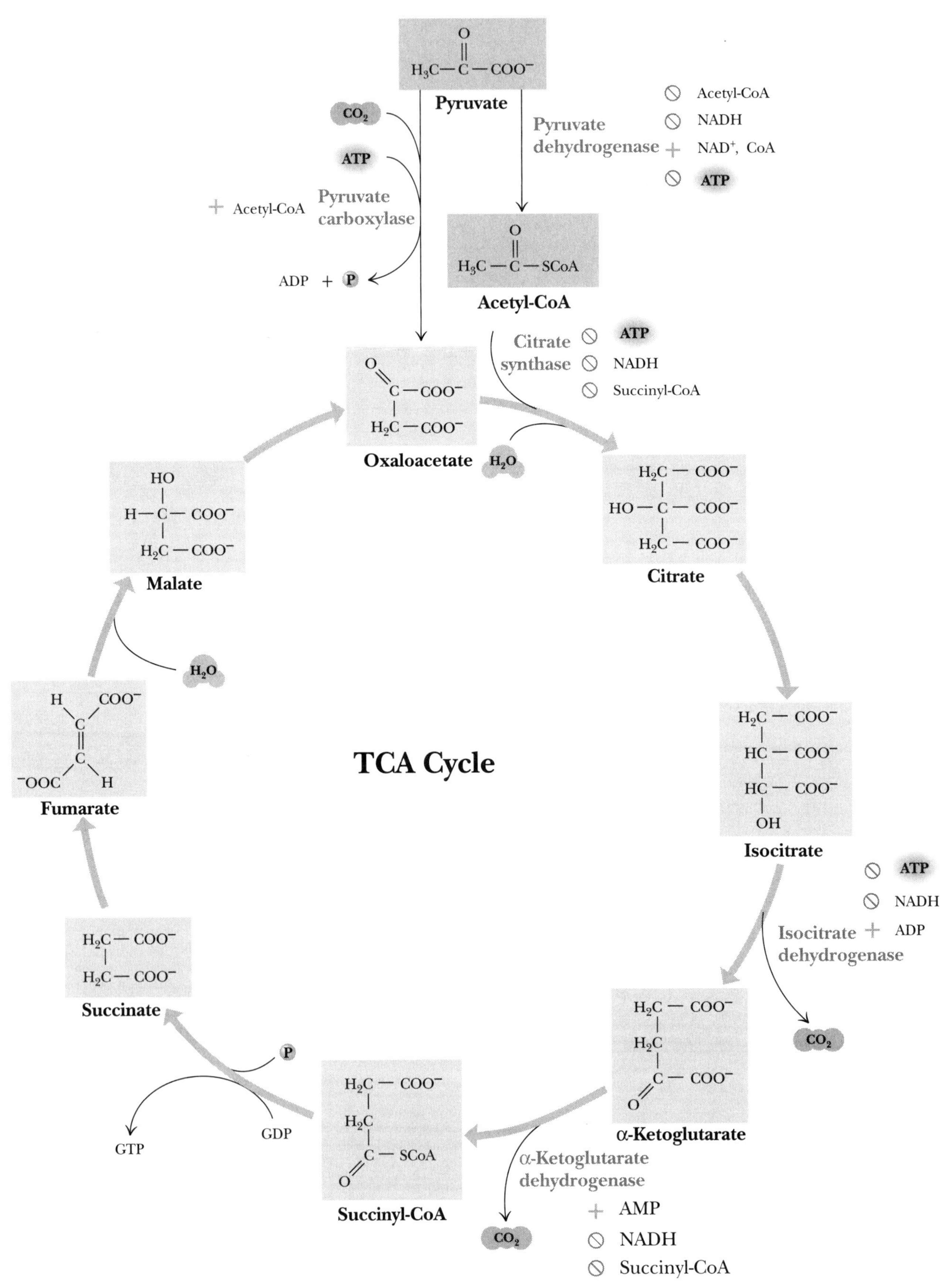

Figure 16.19   Regulation of the pyruvate dehydrogenase reaction.

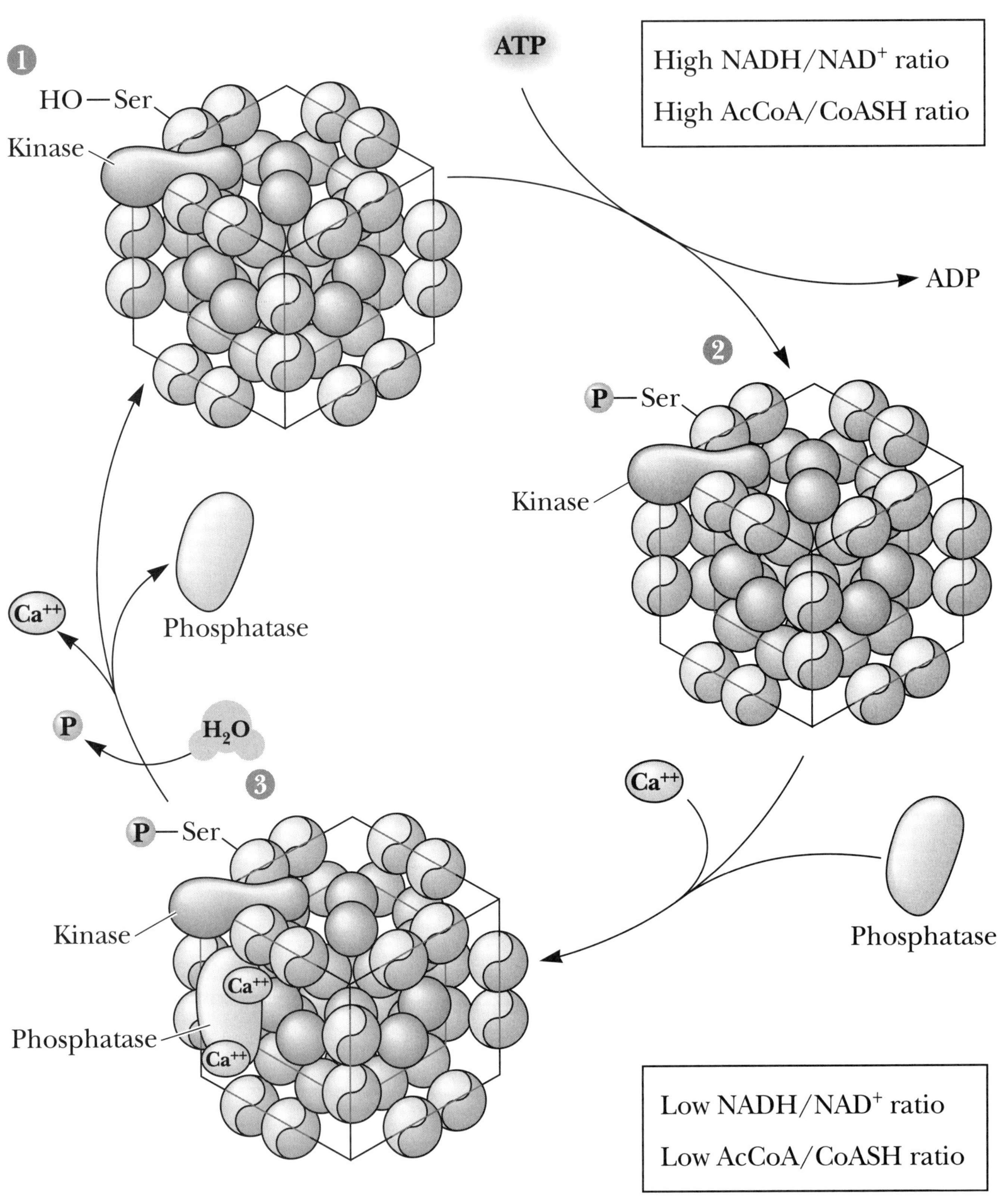

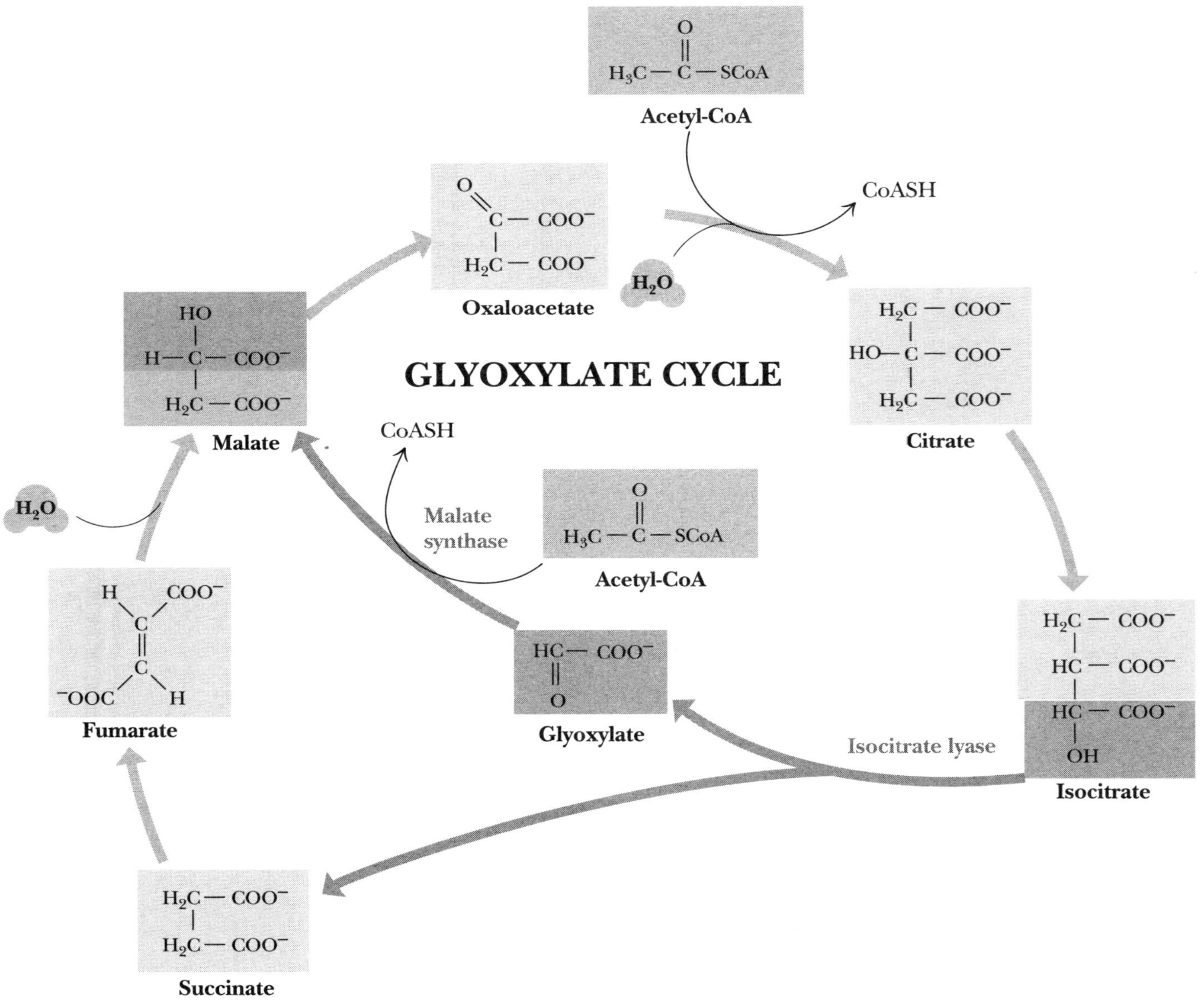
Acetyl-CoA
CoASH
Oxaloacetate
GLYOXYLATE CYCLE
Citrate
Malate
Malate synthase
Acetyl-CoA
Glyoxylate
Fumarate
Isocitrate lyase
Isocitrate
Succinate
H2O

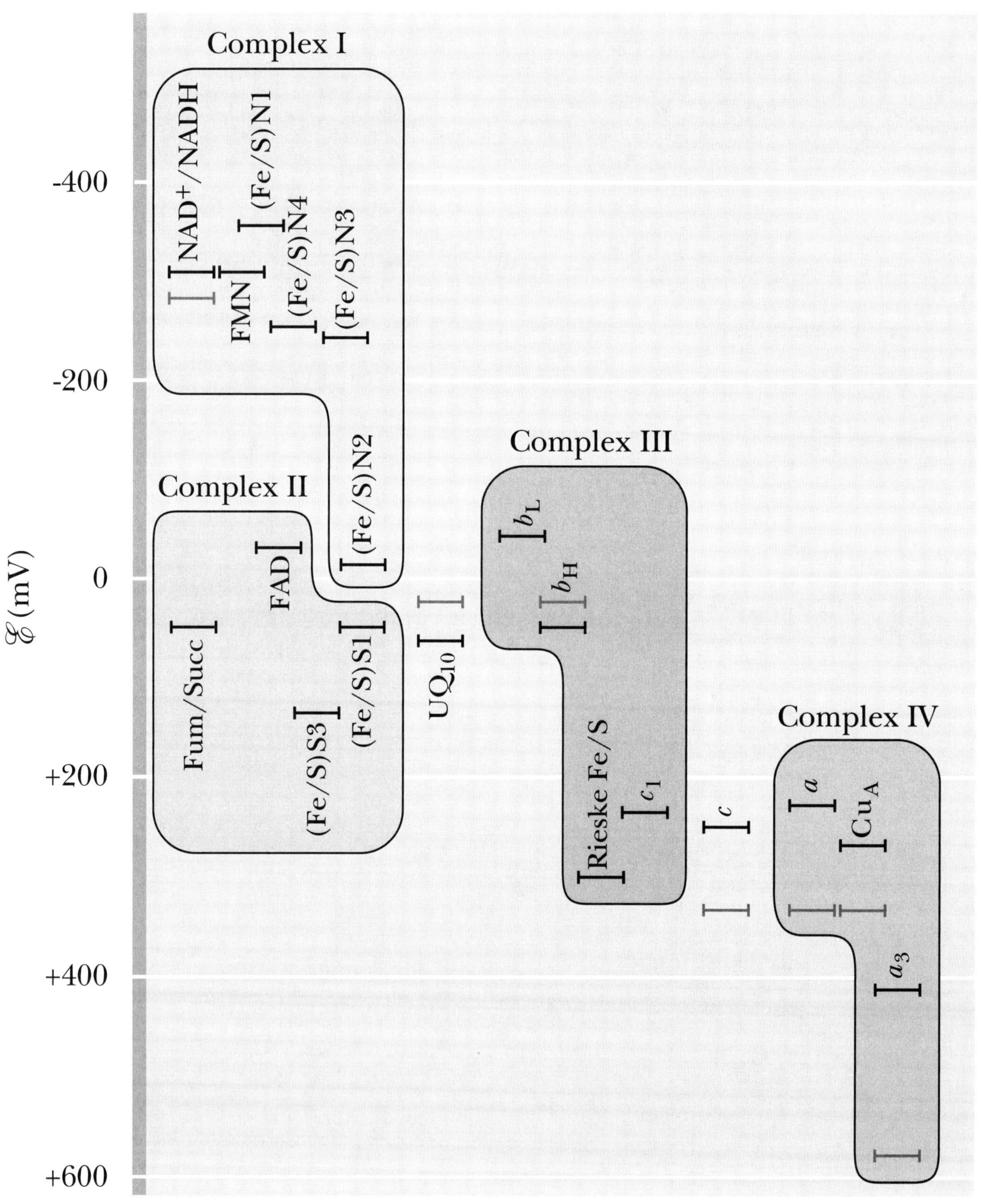
Complex I
Complex II
Complex III
Complex IV
NAD+/NADH
FMN
(Fe/S)N1
(Fe/S)N4
(Fe/S)N3
(Fe/S)N2
Fum/Succ
FAD
(Fe/S)S3
(Fe/S)S1
UQ10
bL
bH
Rieske Fe/S
c1
c
a
CuA
a3
-400
-200
0
+200
+400
+600

Figure 17.4   The complexes and pathways in the mitochondrial electron transport chain.

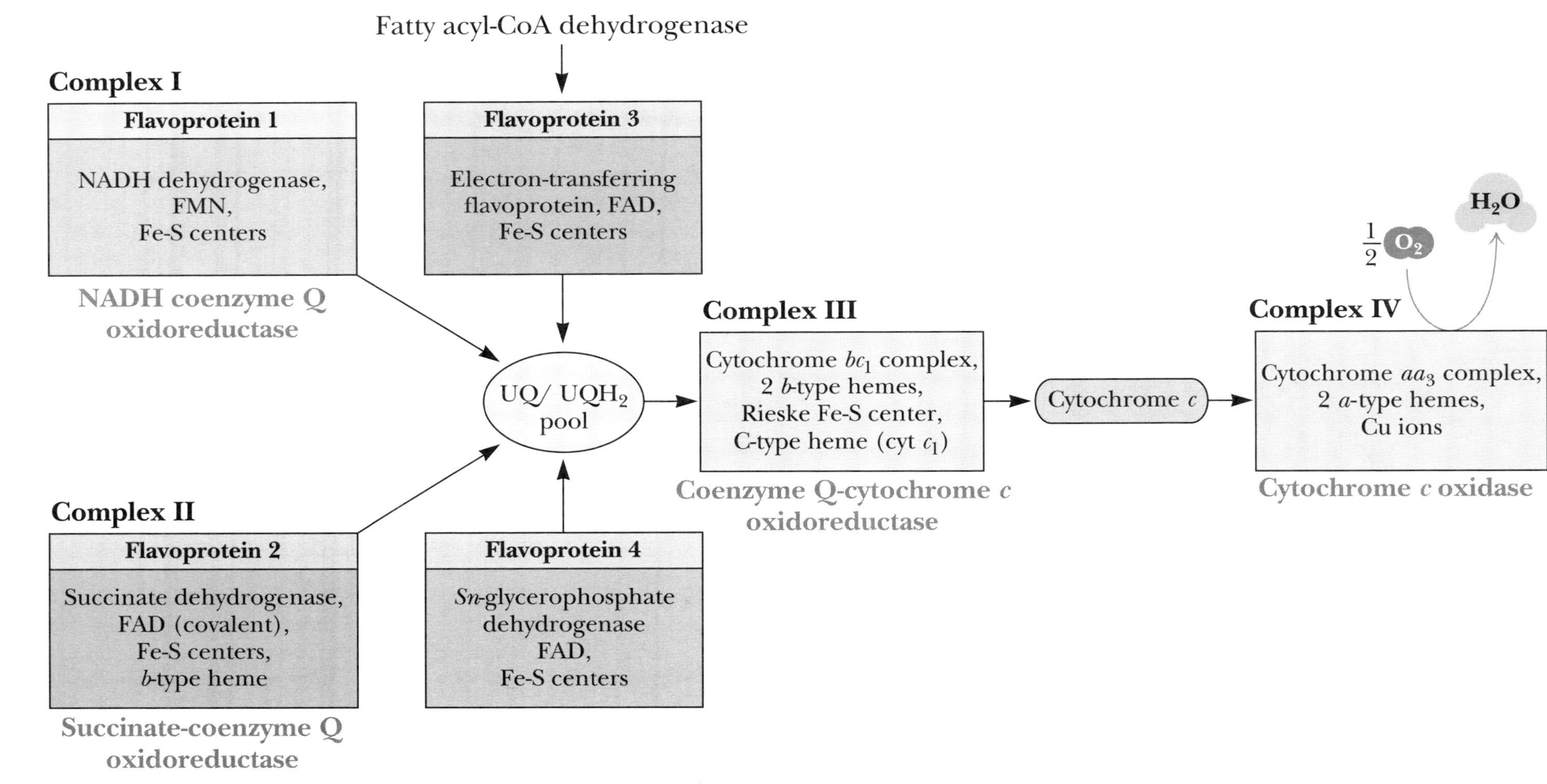

Figure 17.6   Structural model and electron transport pathway for Complex I.

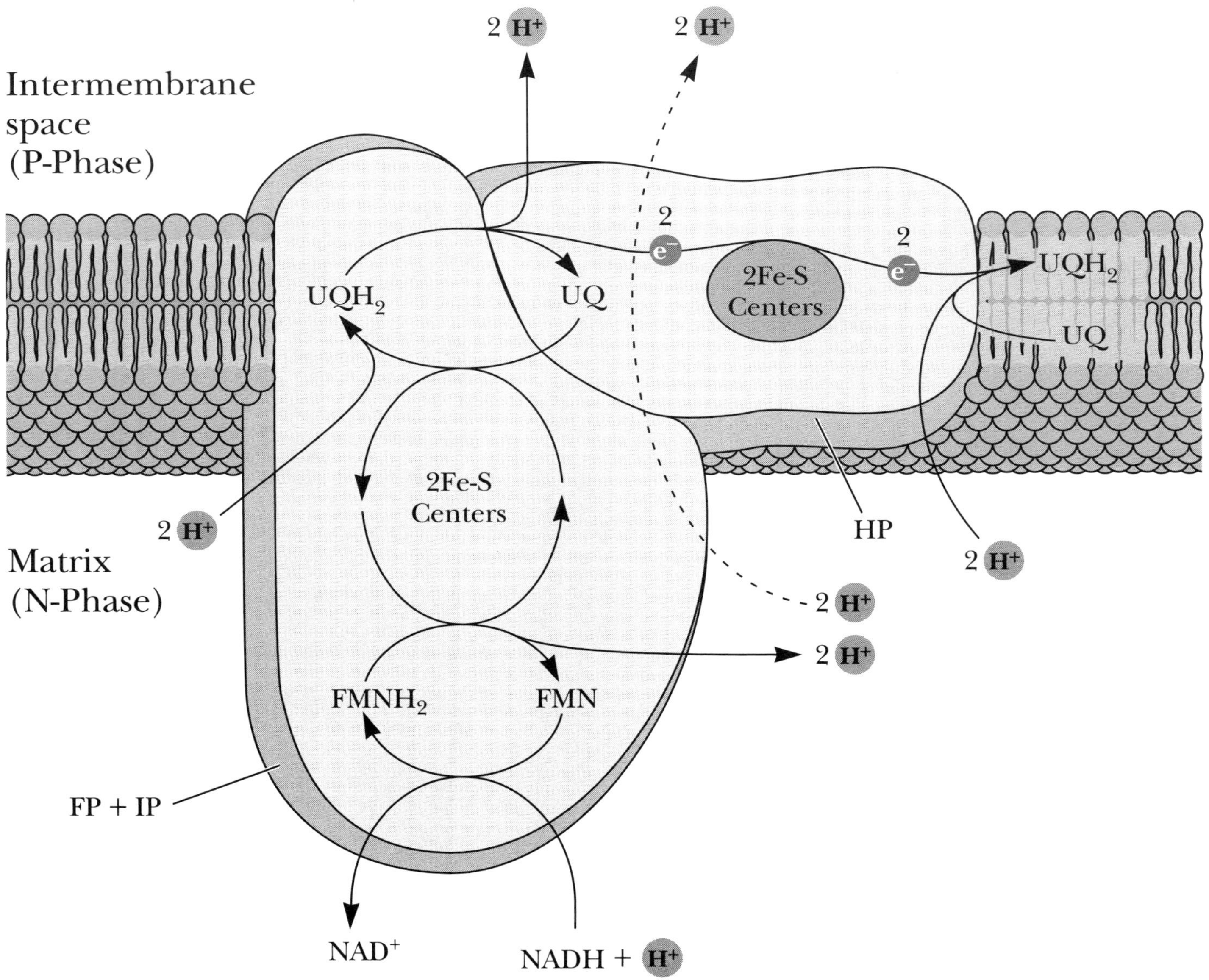

Figure 17.8   A scheme for electron flow in Complex II.

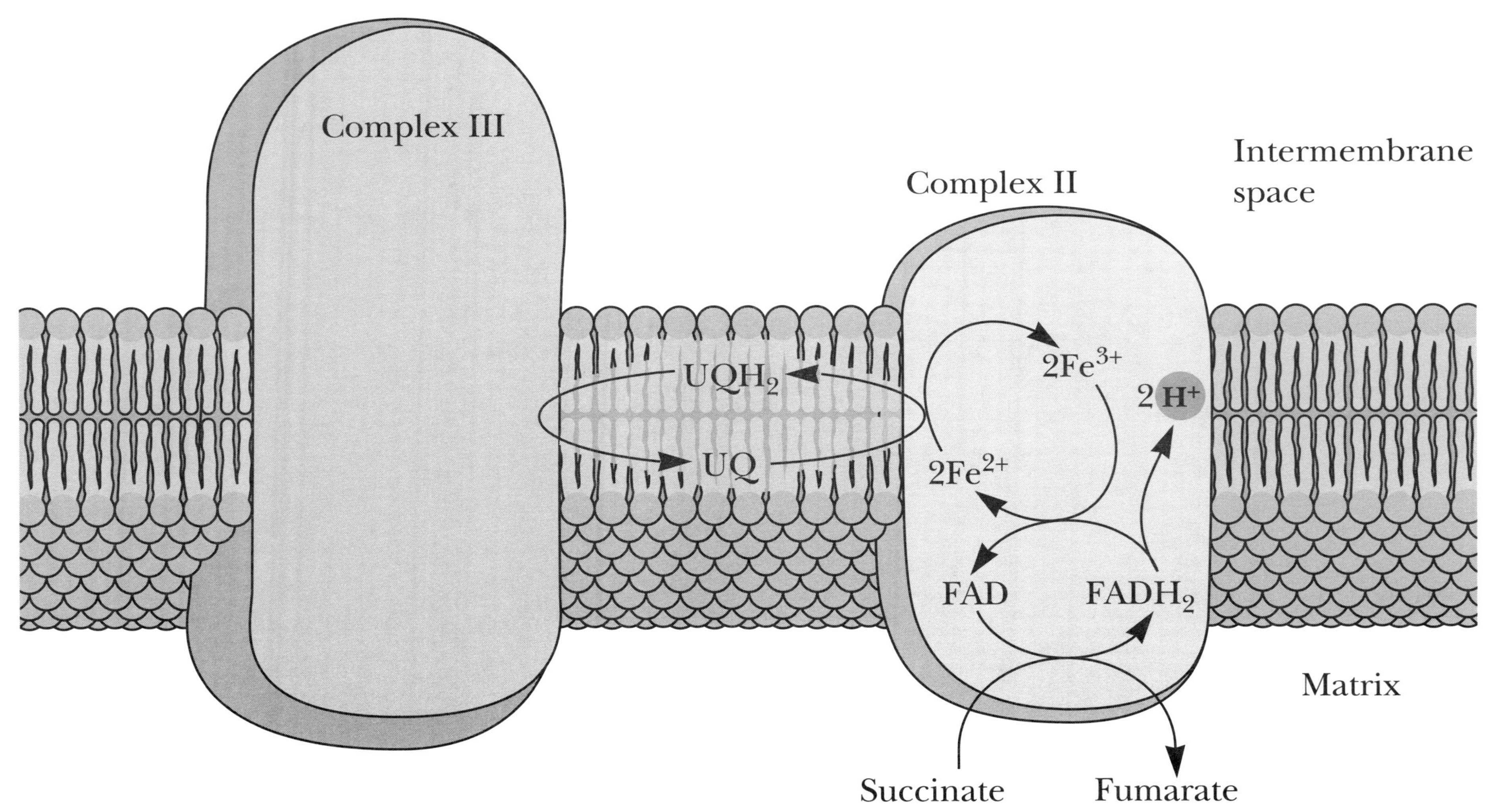

Figure 17.12a   The Q cycle in mitochondria.

(a) First half of Q cycle

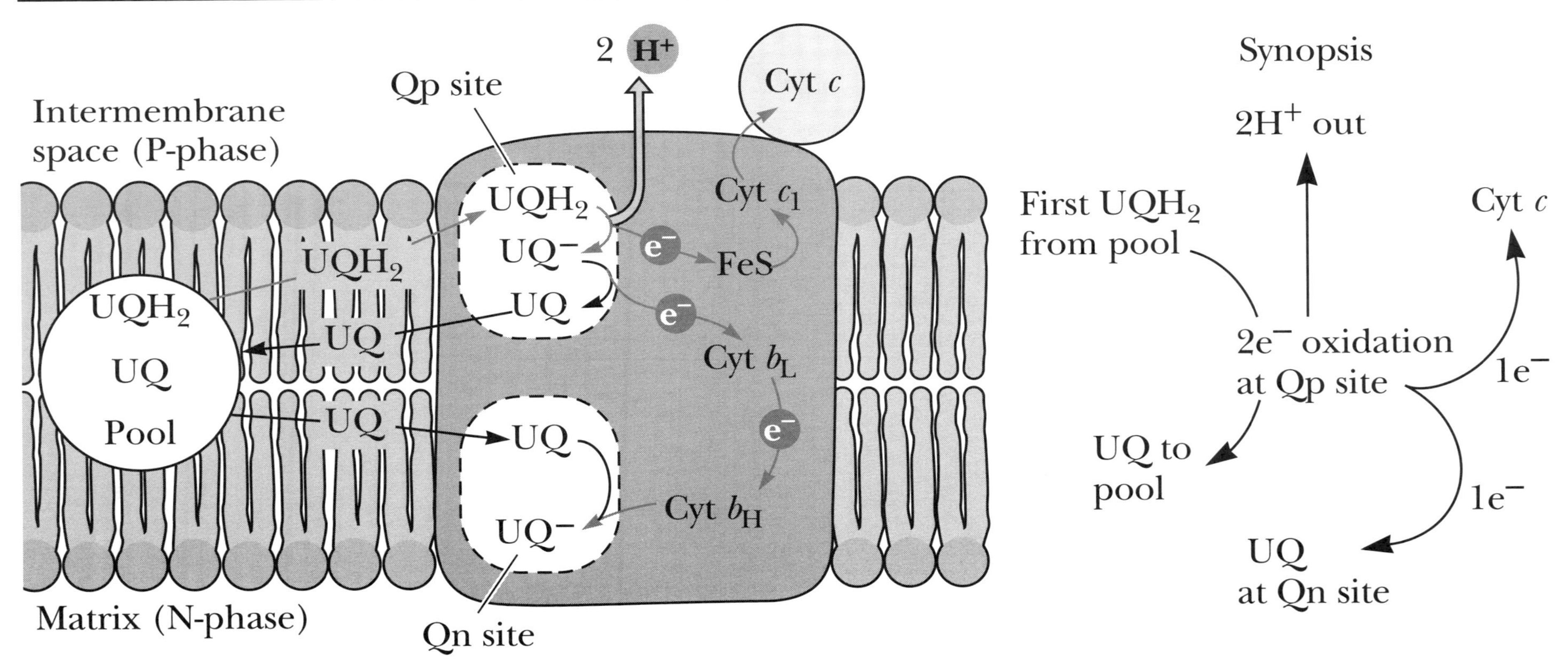

Figure 17.12b   The Q cycle in mitochondria.

(b) Second half of Q cycle

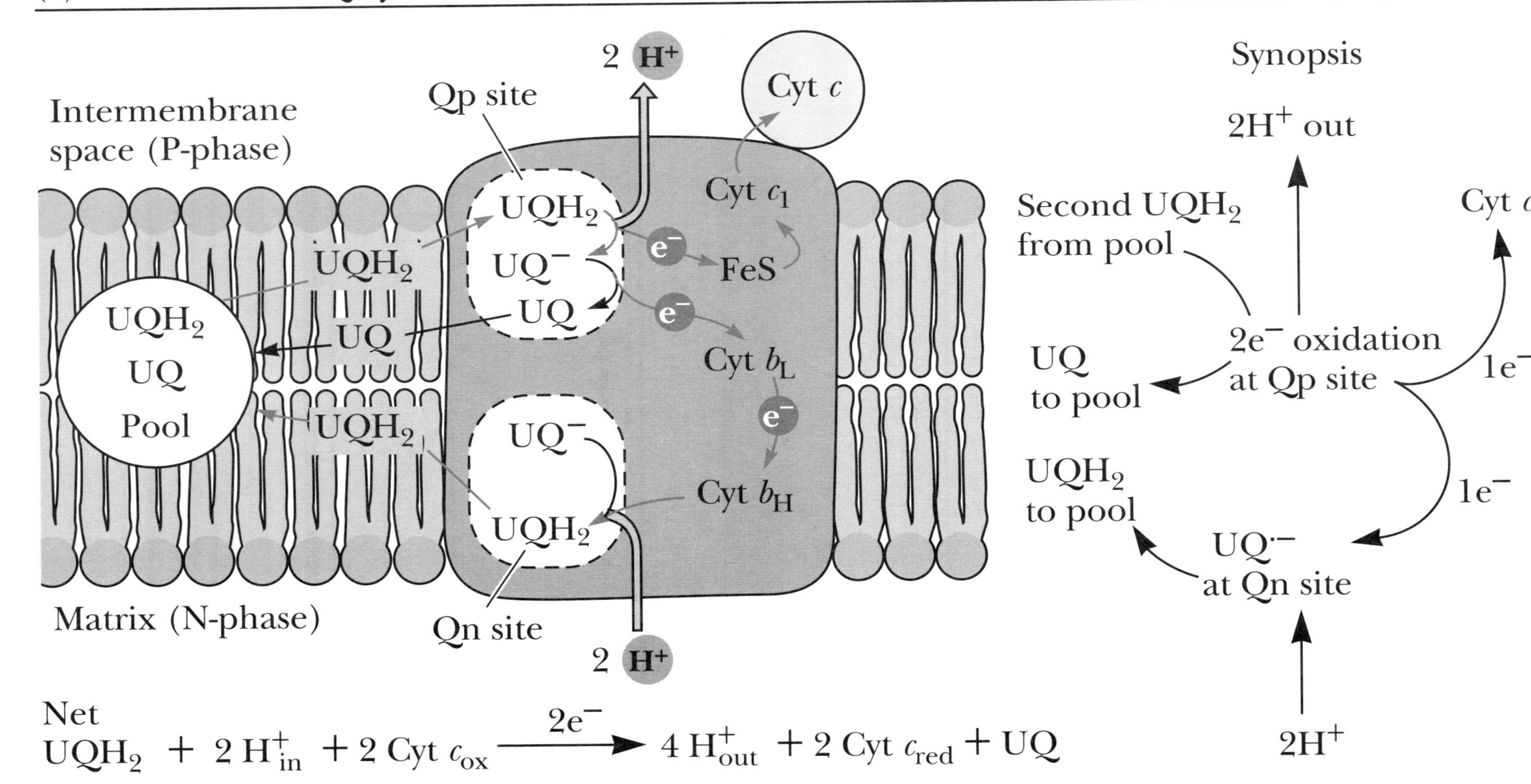

Figure 17.17   A model for the electron transport pathway in the mitochondrial inner membrane.

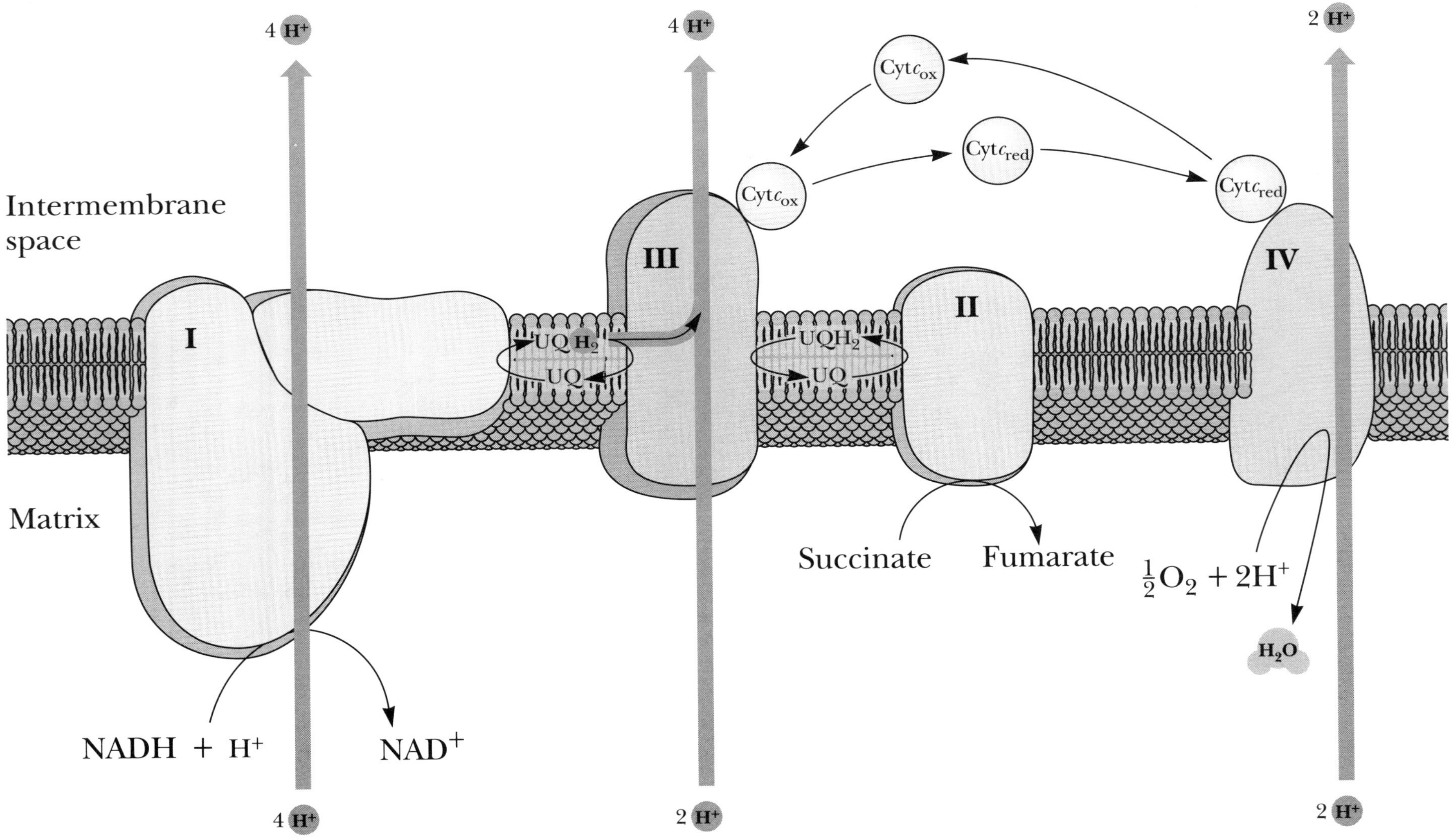

Figure 17.21   A model of the F$_1$ and F$_0$ components of ATP synthase.

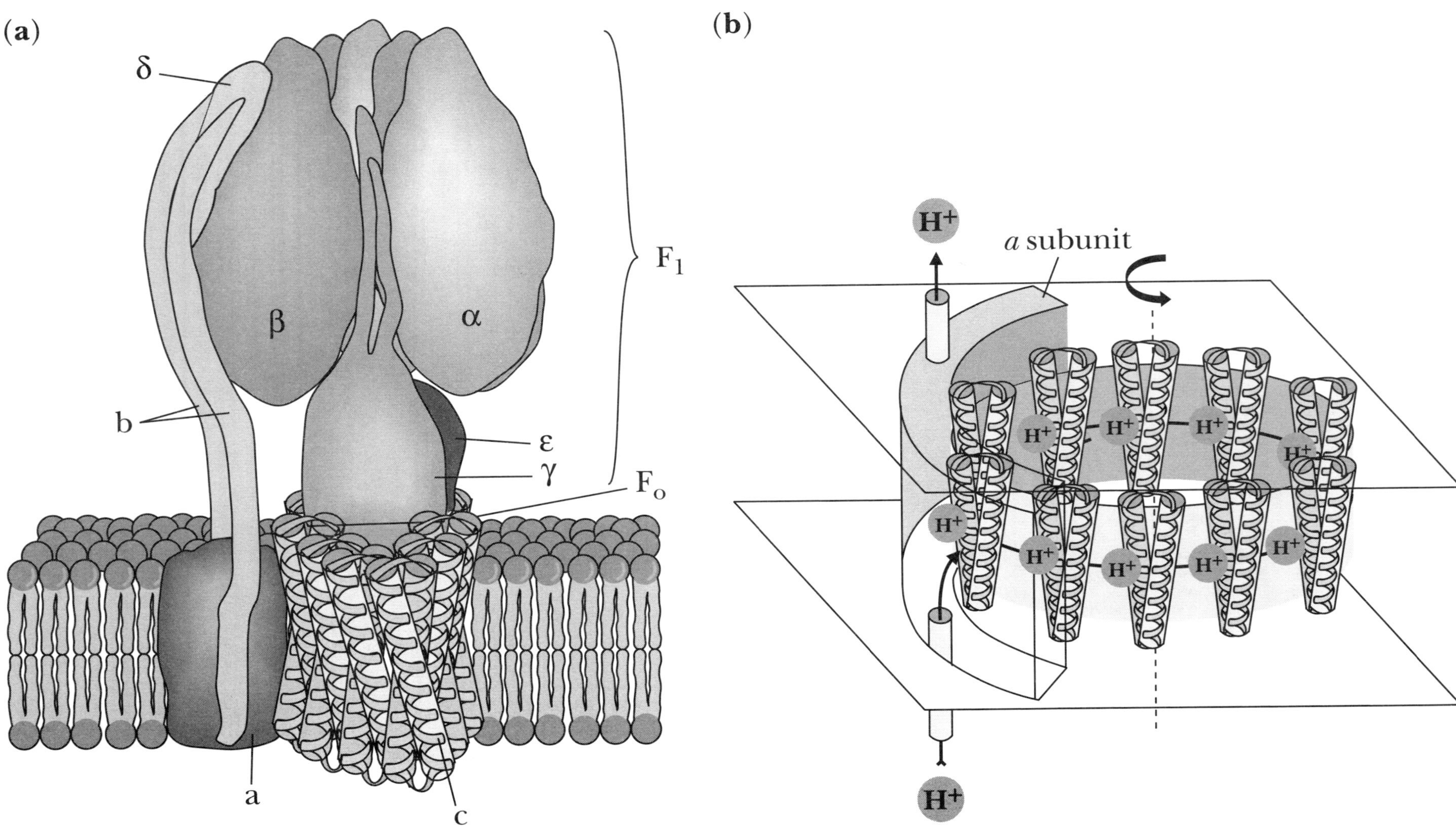

Figure 17.23 Experimental evidence for the Mitchell chemiosmotic hypothesis.

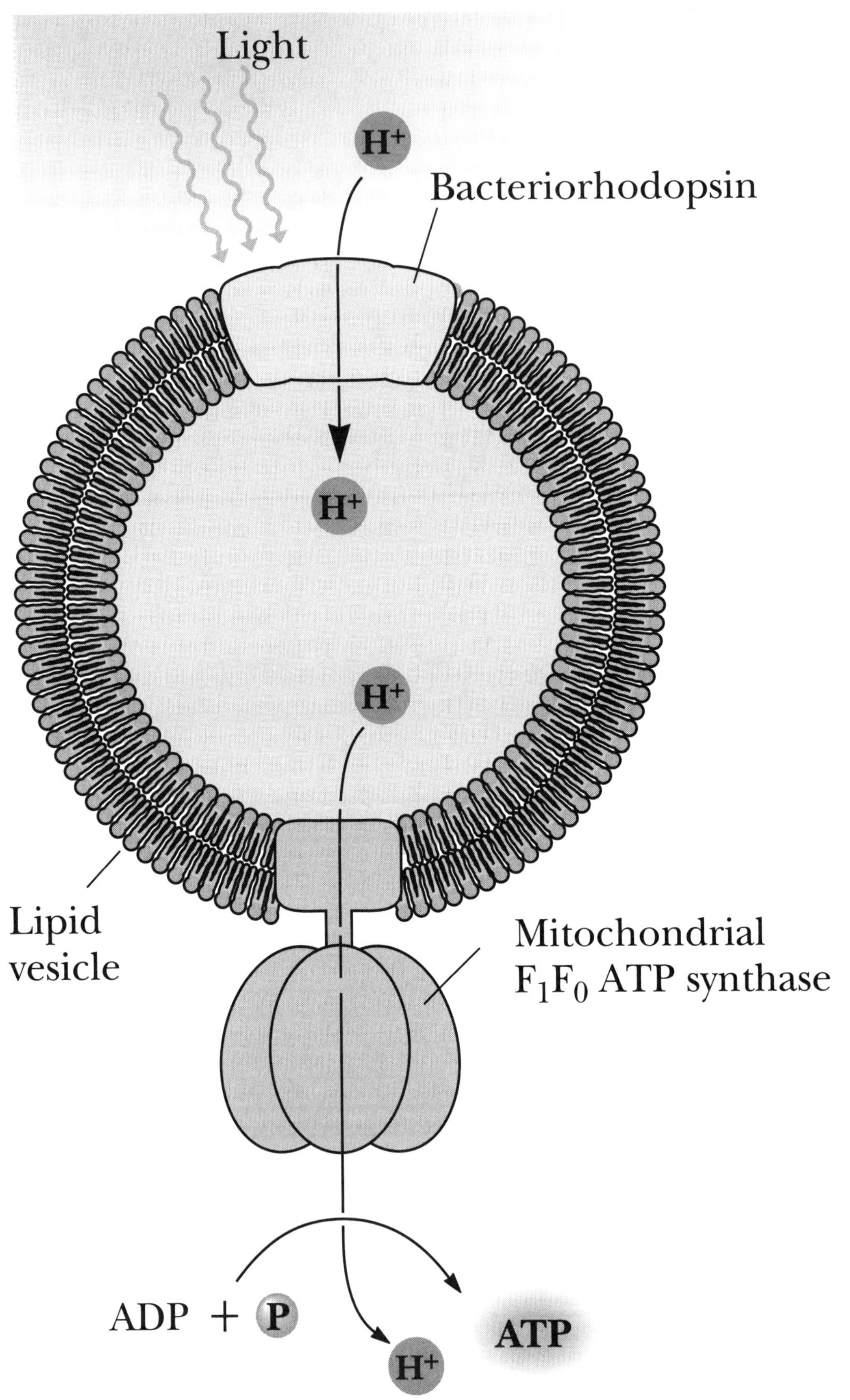

Figure 17.24   Several inhibitors of electron transport.

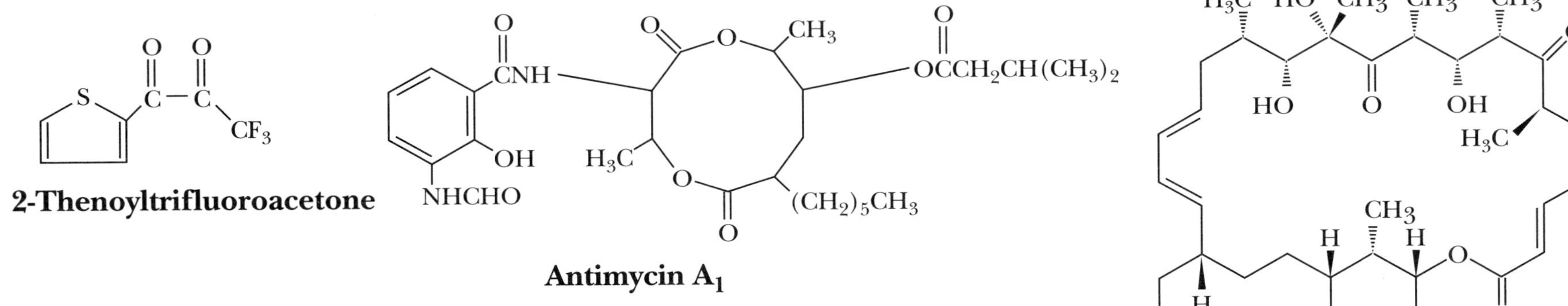
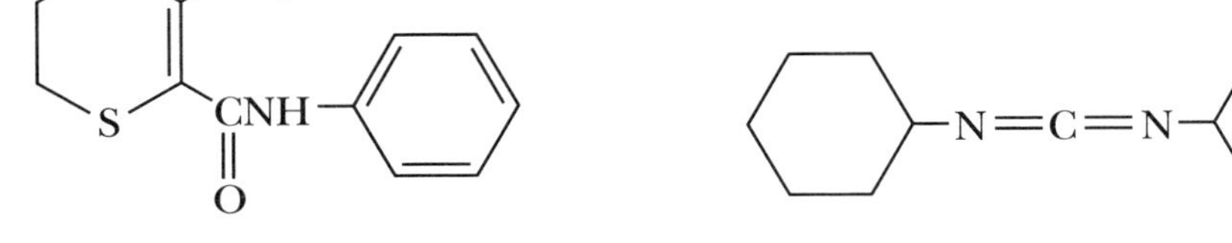

Figure 17.25 Inhibitors of oxidative phosphorylation and their sites of action.

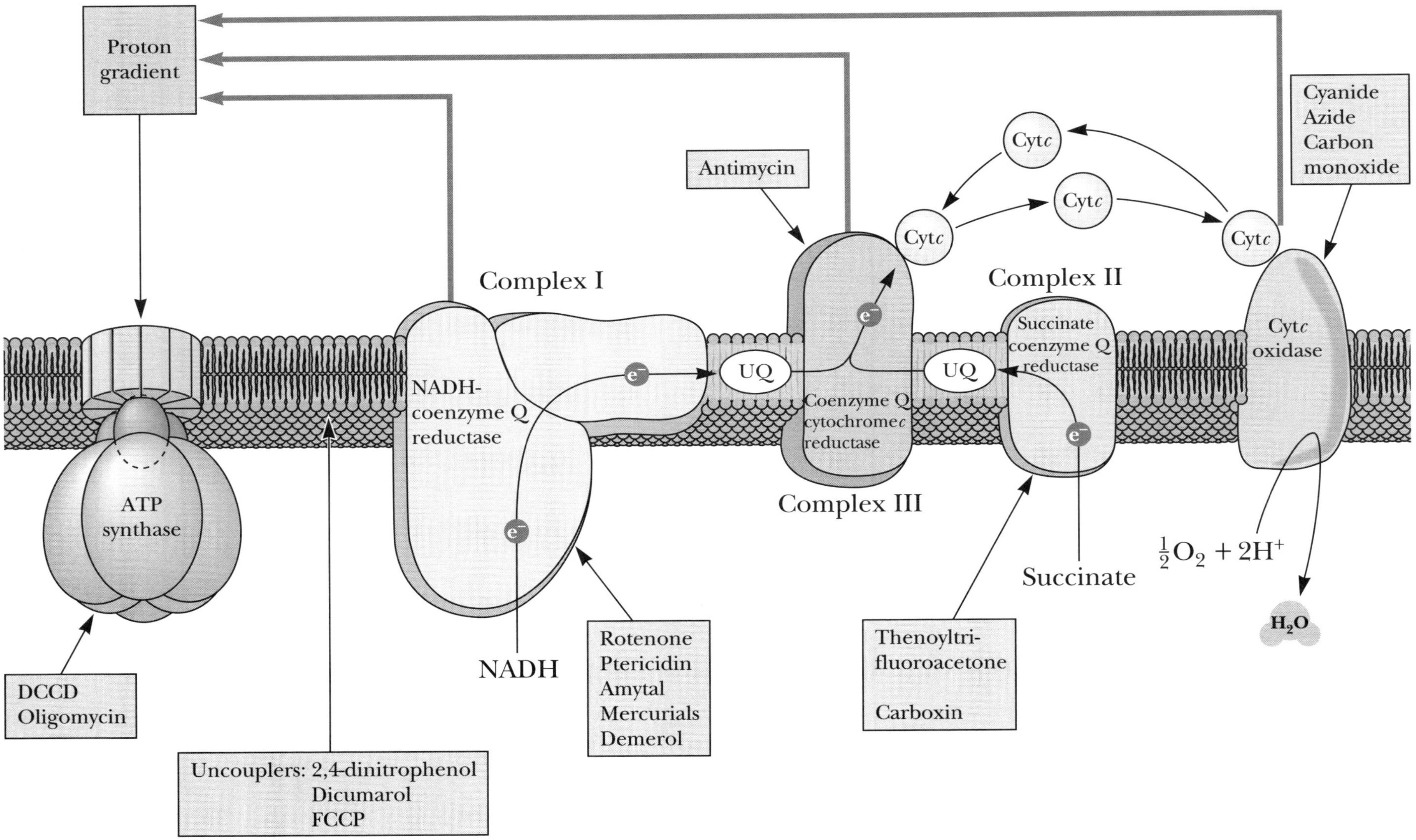

Figure 17.29 The malate-aspartate shuttle.

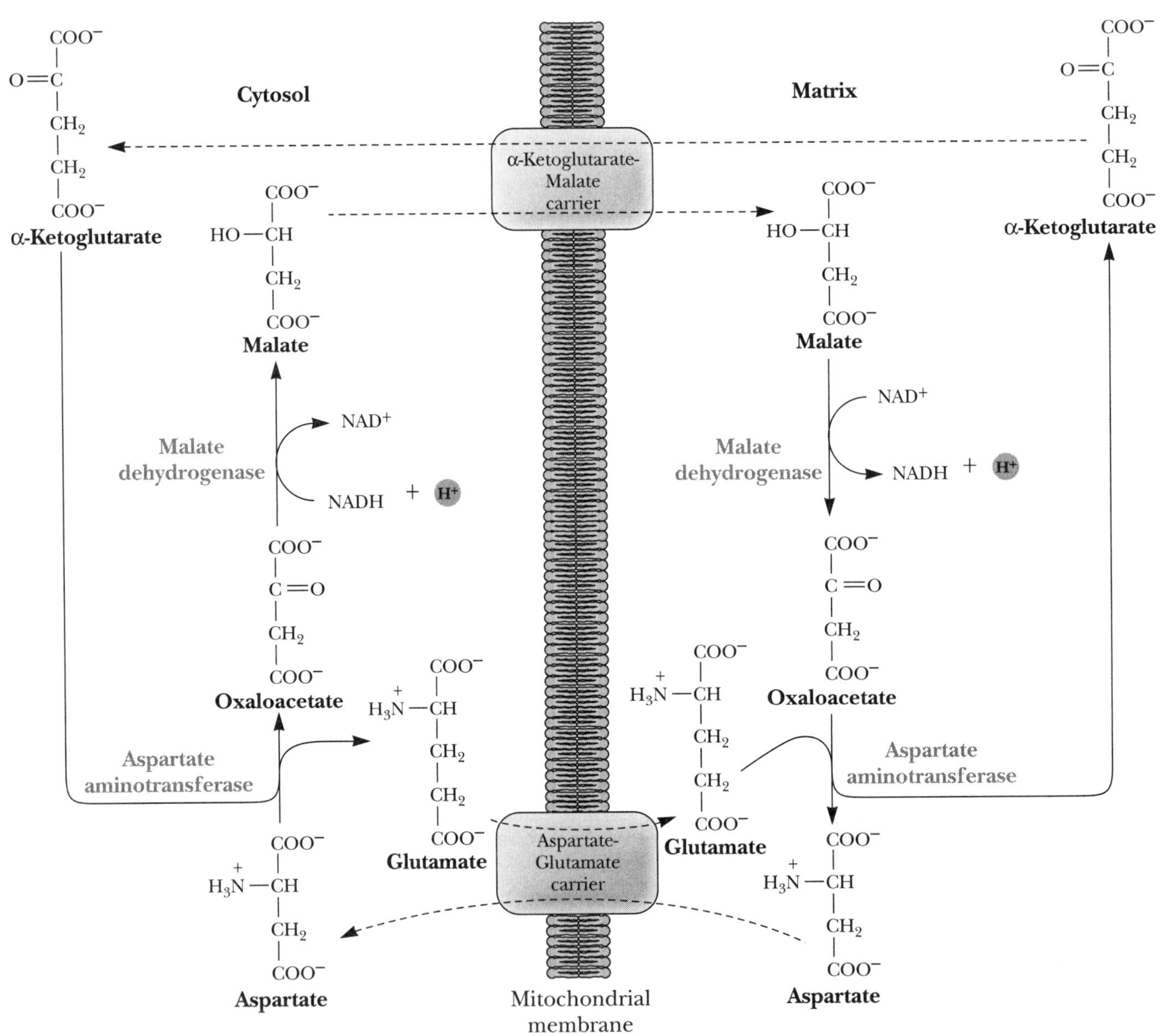

© Harcourt, Inc.

Table 17.4 Yield of ATP from glucose oxidation.

<table>
<tr><th rowspan="3">Pathway</th><th colspan="2">ATP Yield
per Glucose</th></tr>
<tr><th>Glycerol–
Phosphate Shuttle</th><th>Malate–
Aspartate Shuttle</th></tr>
</table>

Table 17.4   Yield of ATP from Glucose Oxidation

Pathway	Glycerol–Phosphate Shuttle	Malate–Aspartate Shuttle
Glycolysis: glucose to pyruvate (cytosol)		
Phosphorylation of glucose	-1	-1
Phosphorylation of fructose-6-phosphate	-1	-1
Dephosphorylation of 2 molecules of 1,3-BPG	$+2$	$+2$
Dephosphorylation of 2 molecules of PEP	$+2$	$+2$
Oxidation of 2 molecules of glyceraldehyde-3-phosphate yields 2 NADH		
Pyruvate conversion to acetyl-CoA (mitochondria)		
2 NADH		
Citric acid cycle (mitochondria)		
2 molecules of GTP from 2 molecules of succinyl-CoA	$+2$	$+2$
Oxidation of 2 molecules each of isocitrate, α-ketoglutarate, and malate yields 6 NADH		
Oxidation of 2 molecules of succinate yields 2 $[FADH_2]$		
Oxidative phosphorylation (mitochondria)		
2 NADH from glycolysis yield 1.5 ATP each if NADH is oxidized by glycerol-phosphate shuttle; 2.5 ATP by malate-aspartate shuttle	$+3$	$+5$
Oxidative decarboxylation of 2 pyruvate to 2 acetyl-CoA: 2 NADH produce 2.5 ATP each	$+5$	$+5$
2 $[FADH_2]$ from citric acid cycle produce 1.5 ATP each	$+3$	$+3$
6 NADH from citric acid cycle produce 2.5 ATP each	$+15$	$+15$
Net Yield	30	32

Note: These P/O ratios of 2.5 and 1.5 for mitochondrial oxidation of NADH and $[FADH_2]$ are "consensus values." Because they may not reflect actual values and because these ratios may change depending on metabolic conditions, these estimates of ATP yield from glucose oxidation are approximate.

Figure 18.12a   The Z-scheme of photosynthesis.

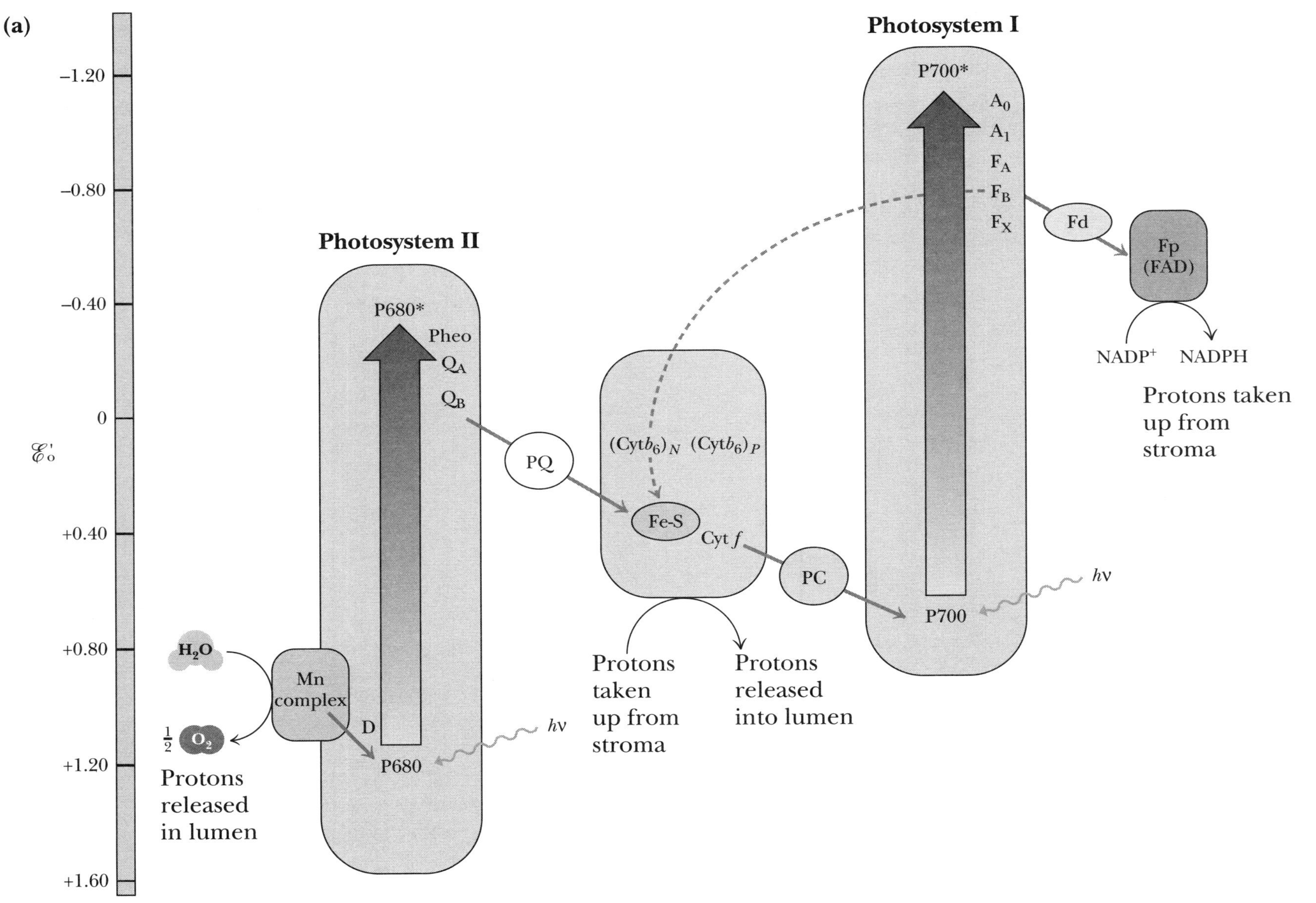

© Harcourt, Inc.

264

Figure 18.12b   The Z-scheme of photosynthesis.

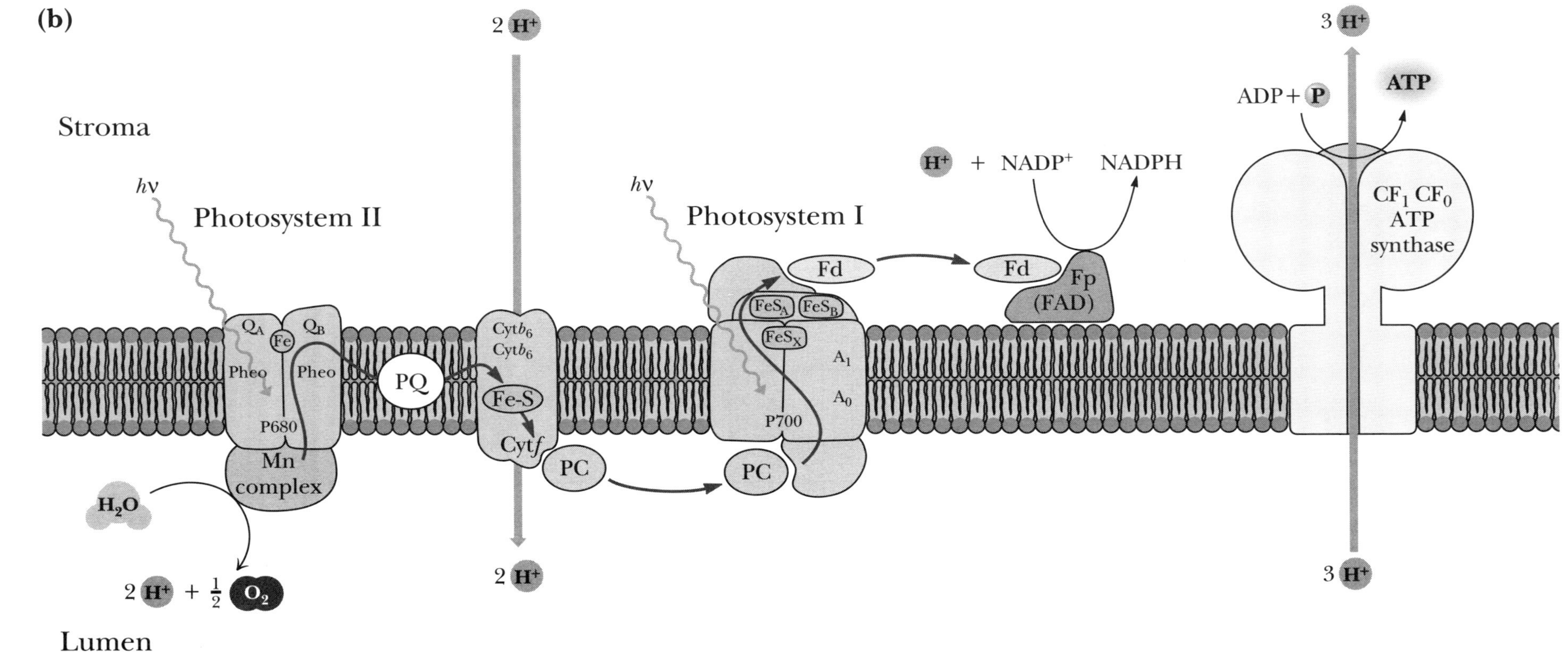

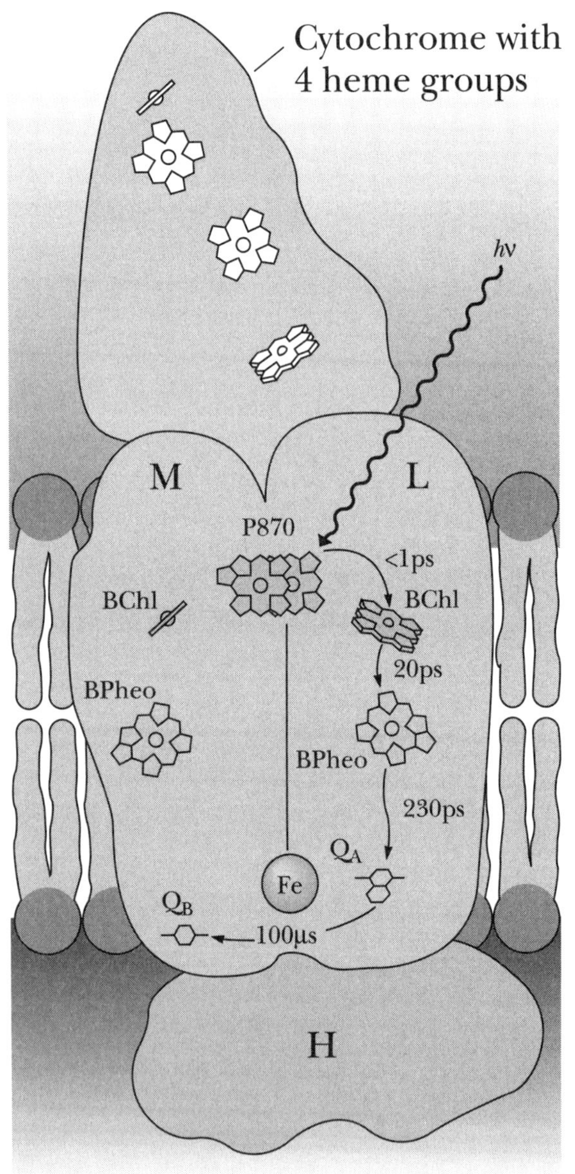

Note: The cytochrome subunit is membrane-associated via a diacylglycerol moiety on its N-terminal Cys residue:

Figure 18.15 The *R. viridis* reaction center is coupled to the cytochrome b/c_1 complex through the quinone pool.

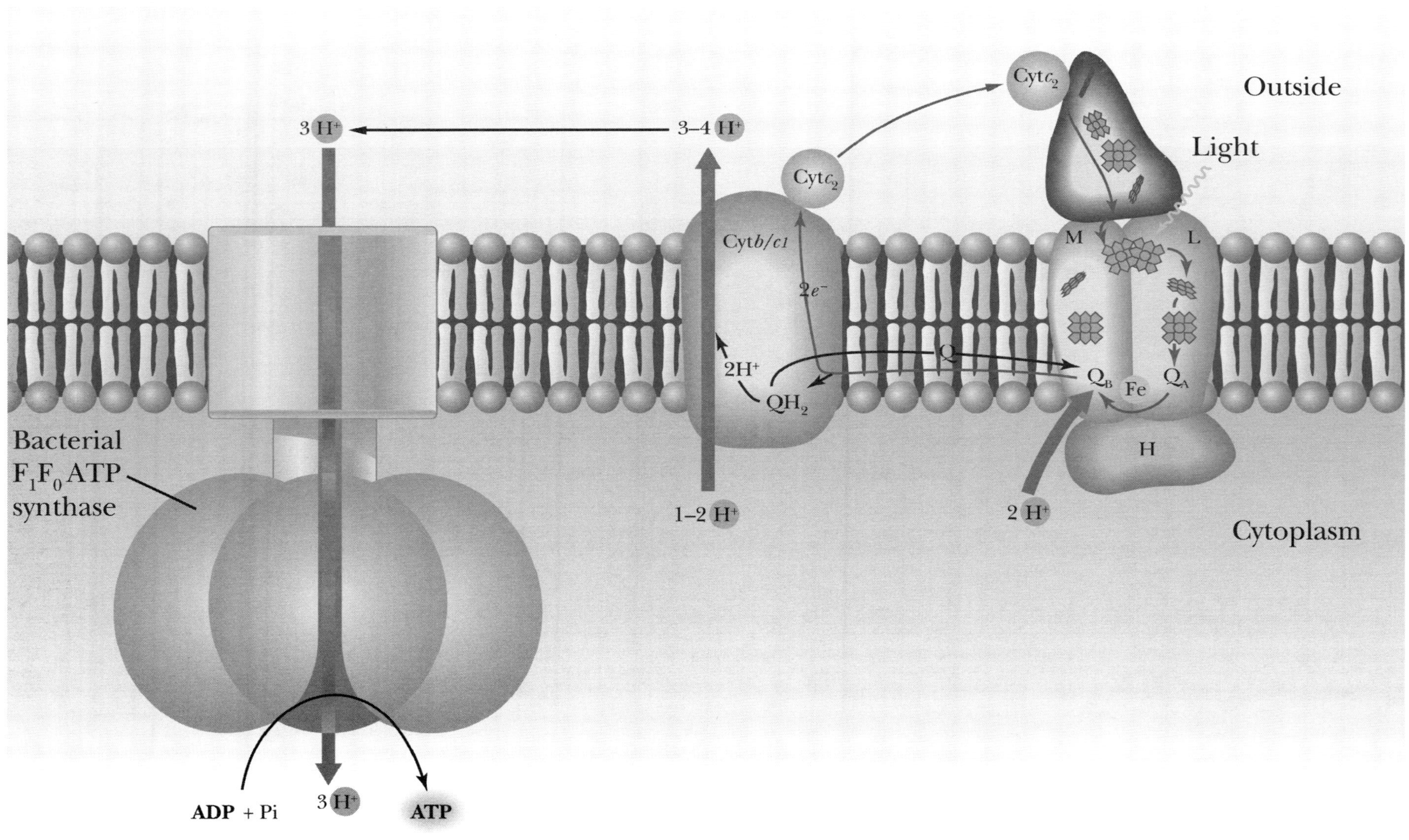

© Harcourt, Inc.

Figure 18.17   The mechanism of photophosphorylation.

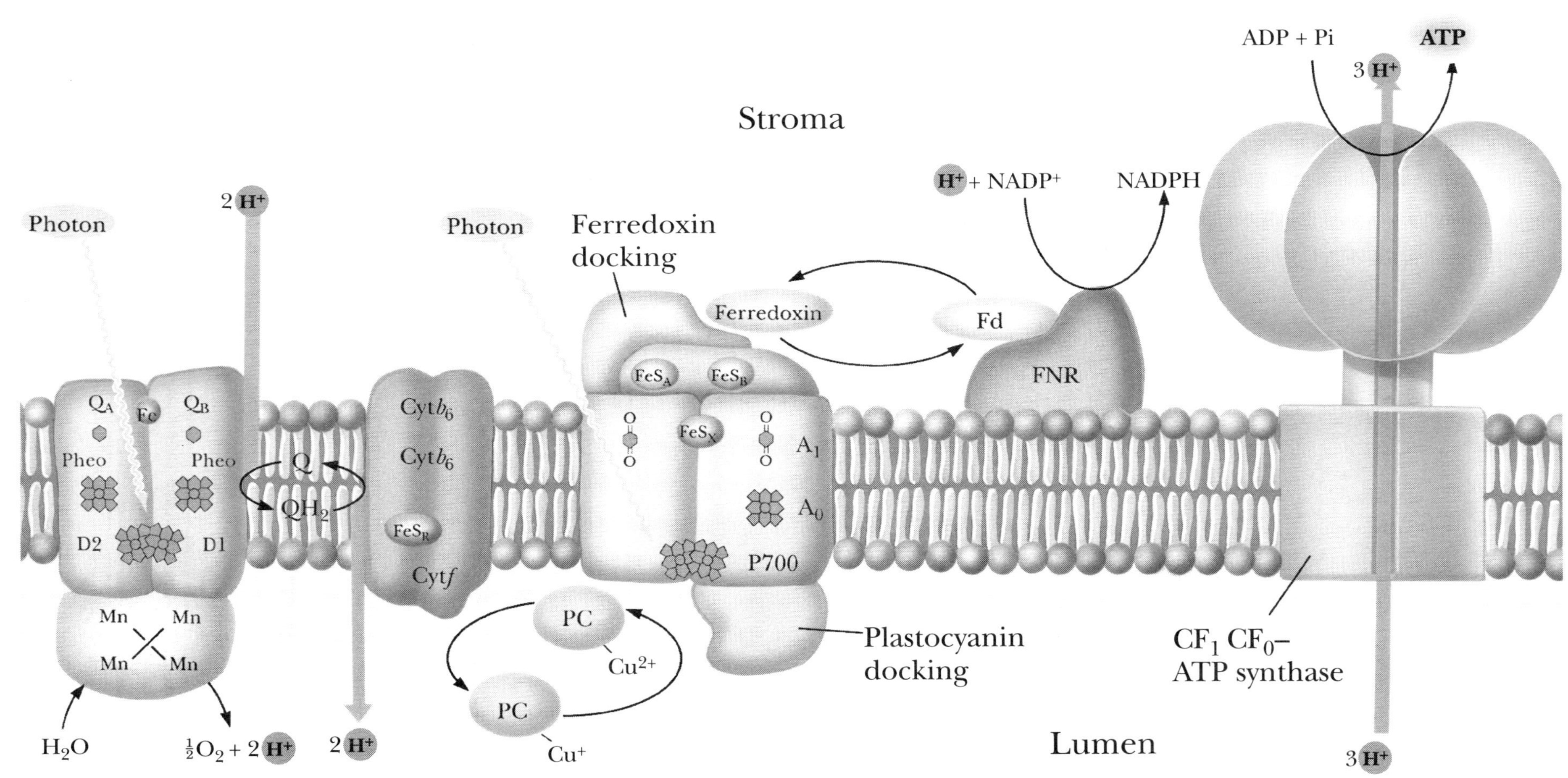

Figure 18.18   The pathway of cyclic photophosphorylation by PSI.

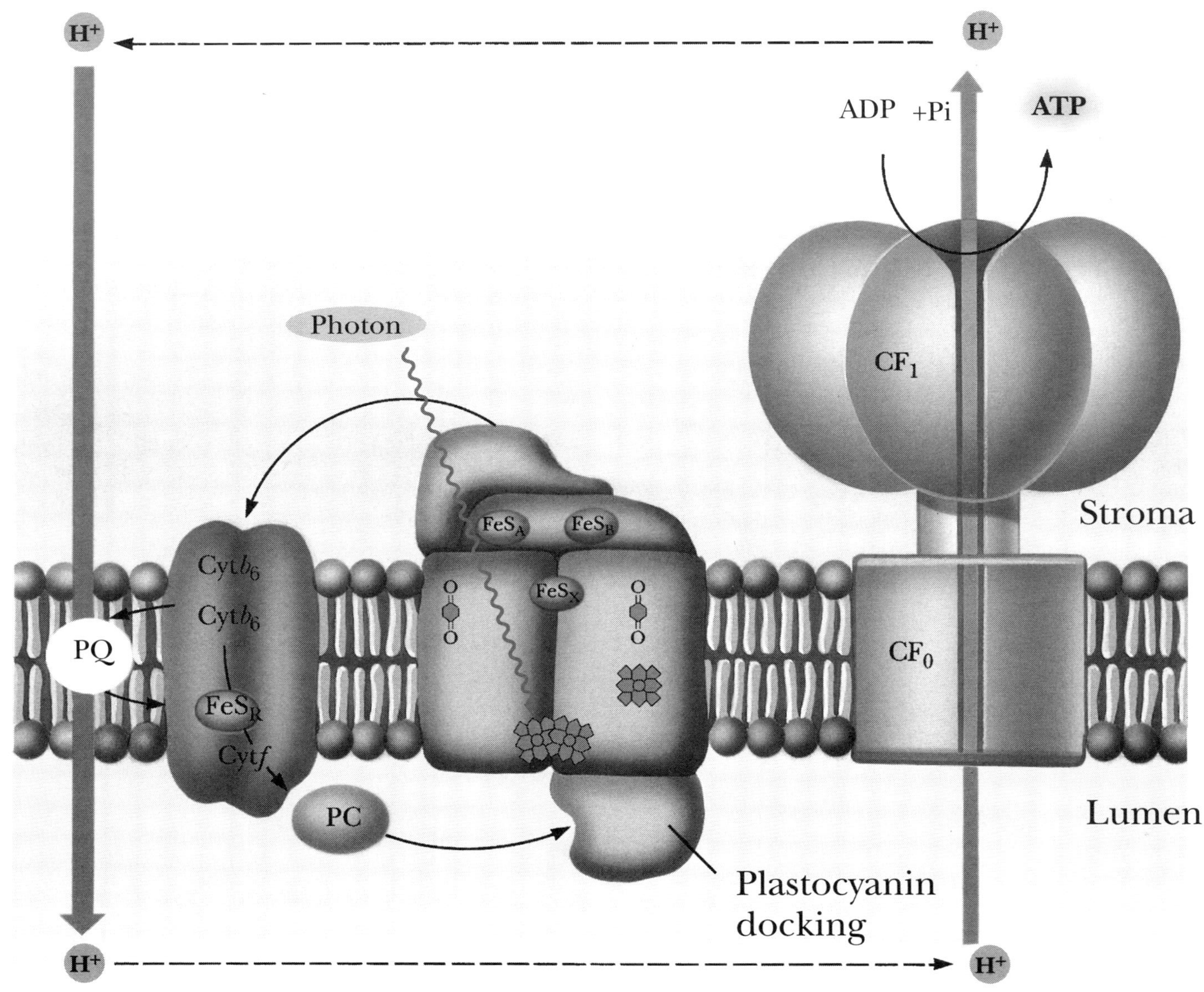

Figure 18.21 The Calvin-Benson cycle of reactions.

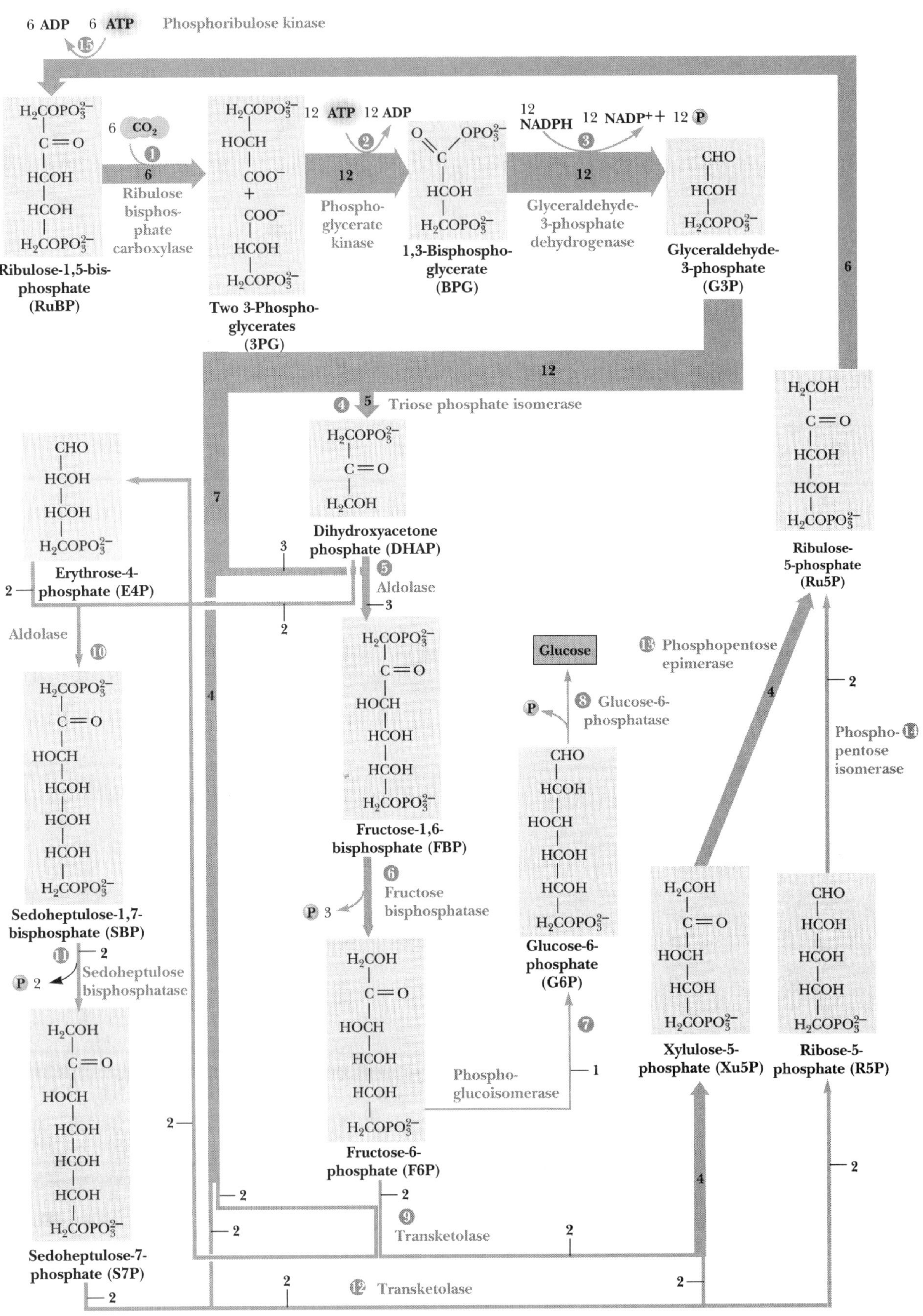

© Harcourt, Inc.

Table 18.2 The Calvin cycle series of reactions.

Table 18.2 The Calvin Cycle Series of Reactions

Reactions 1 through 15 constitute the cycle that leads to the formation of one equivalent of glucose. The enzyme catalyzing each step, a concise reaction, and the overall carbon balance is given. Numbers in parentheses show the numbers of carbon atoms in the substrate and product molecules. Prefix numbers indicate in a stoichiometric fashion how many times each step is carried out in order to provide a balanced net reaction.

1. Ribulose bisphosphate carboxylase: $6\ CO_2 + 6\ H_2O + 6\ RuBP \longrightarrow 12\ 3\text{-PG}$ $6(1) + 6(5) \longrightarrow 12(3)$
2. 3-Phosphoglycerate kinase: $12\ 3\text{-PG} + 12\ ATP \longrightarrow 12\ 1,3\text{-BPG} + 12\ ADP$ $12(3) \longrightarrow 12(3)$
3. $NADP^+$-glyceraldehyde-3-P dehydrogenase:
     $12\ 1,3\text{-BPG} + 12\ NADPH \longrightarrow 12\ NADP^+ + 12\ G3P + 12\ P_i$ $12(3) \longrightarrow 12(3)$
4. Triose-P isomerase: $5\ G3P \longrightarrow 5\ DHAP$ $5(3) \longrightarrow 5(3)$
5. Aldolase: $3\ G3P + 3\ DHAP \longrightarrow 3\ FBP$ $3(3) + 3(3) \longrightarrow 3(6)$
6. Fructose bisphosphatase: $3\ FBP + 3\ H_2O \longrightarrow 3\ F6P + 3\ P_1$ $3(6) \longrightarrow 3(6)$
7. Phosphoglucoisomerase: $1\ F6P \longrightarrow 1\ G6P$ $1(6) \longrightarrow 1(6)$
8. Glucose phosphatase: $1\ G6P + 1\ H_2O \longrightarrow 1\ GLUCOSE + 1\ P_i$ $1(6) \longrightarrow 1(6)$

    The remainder of the pathway involves regenerating six RuBP acceptors ($= 30$ C) from the leftover two F6P (12 C), four G3P (12 C), and two DHAP (6 C).

9. Transketolase: $2\ F6P + 2\ G3P \longrightarrow 2\ Xu5P + 2\ E4P$ $2(6) + 2(3) \longrightarrow 2(5) + 2(4)$
10. Aldolase: $2\ E4P + 2\ DHAP \longrightarrow 2$ sedoheptulose-1,7-bisphosphate (SBP) $2(4) + 2(3) \longrightarrow 2(7)$
11. Sedoheptulose bisphosphatase: $2\ SBP + 2\ H_2O \longrightarrow 2\ S7P + 2\ P_i$ $2(7) \longrightarrow 2(7)$
12. Transketolase: $2\ S7P + 2\ G3P \longrightarrow 2\ Xu5P + 2\ R5P$ $2(7) + 2(3) \longrightarrow 4(5)$
13. Phosphopentose epimerase: $4\ Xu5P \longrightarrow 4\ Ru5P$ $4(5) \longrightarrow 4(5)$
14. Phosphopentose isomerase: $2\ R5P \longrightarrow 2\ Ru5P$ $2(5) \longrightarrow 2(5)$
15. Phosphoribulose kinase: $6\ Ru5P + 6\ ATP \longrightarrow 6\ RuBP + 6\ ADP$ $6(5) \longrightarrow 6(5)$

Net: $6\ CO_2 + 18\ ATP + 12\ NADPH + 12\ H^+ + 12\ H_2O \longrightarrow$ glucose $+ 18\ ADP + 18\ P_i + 12\ NADP^+$ $6(1) \longrightarrow 1(6)$

Figure 18.24 Essential features of the compartmentation and biochemistry of the Hatch-Slack pathway.

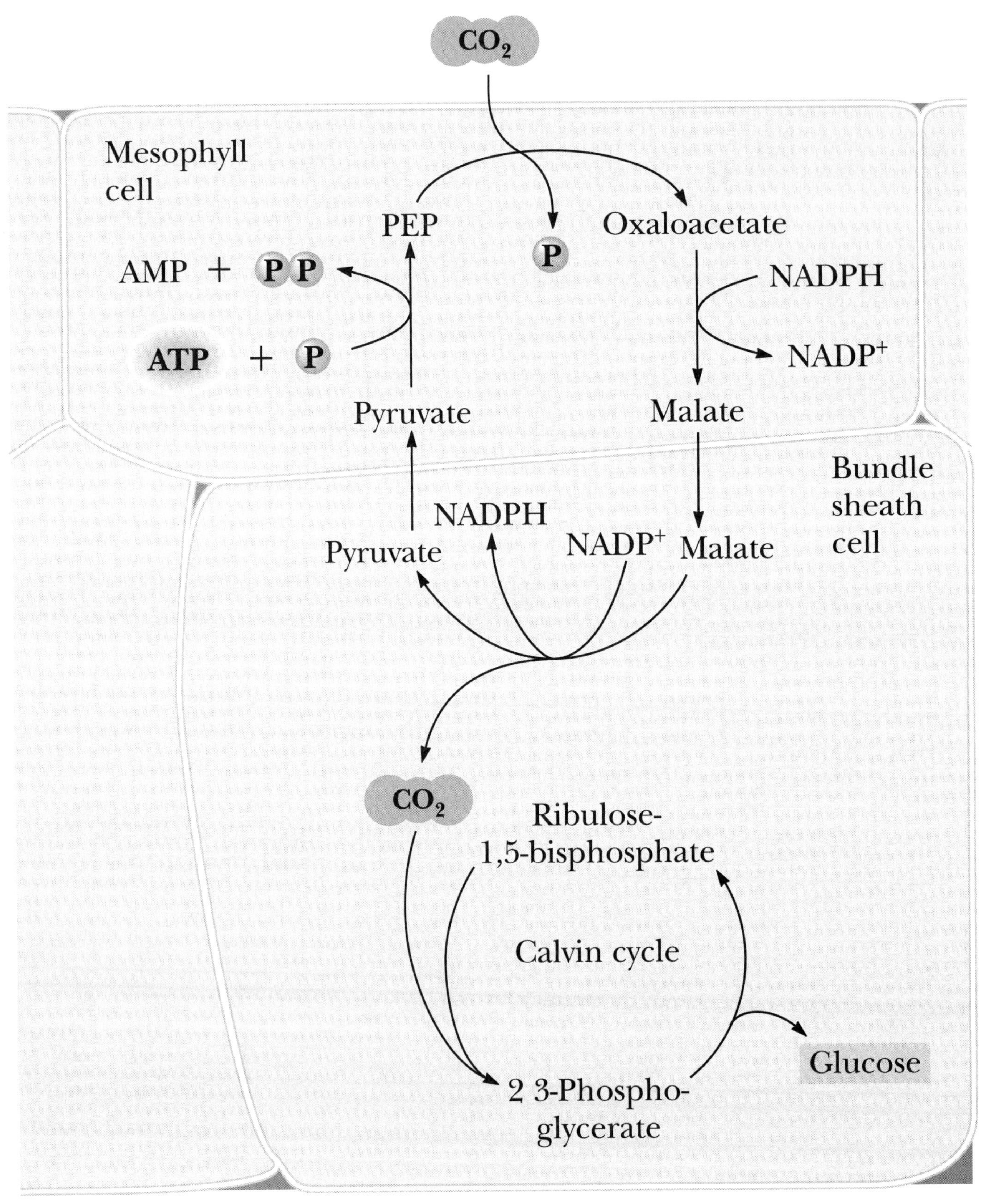

CO₂
Mesophyll cell
PEP
P
Oxaloacetate
AMP + P P
NADPH
ATP + P
NADP⁺
Pyruvate
Malate
NADPH
Bundle sheath cell
Pyruvate
NADP⁺ Malate
CO₂
Ribulose-1,5-bisphosphate
Calvin cycle
Glucose
2 3-Phospho-glycerate

Figure 19.1 The pathways of gluconeogenesis and glycolysis.

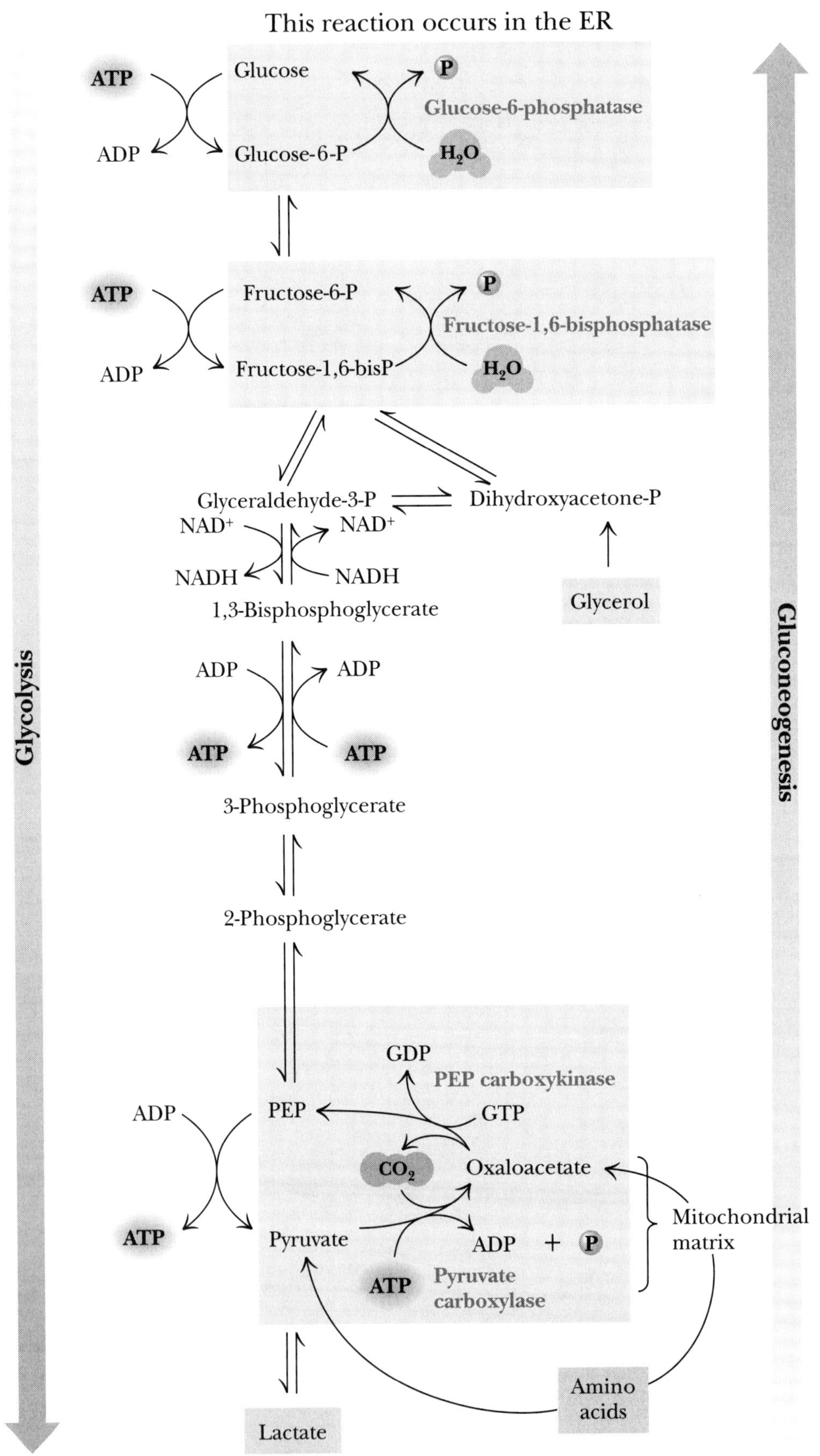

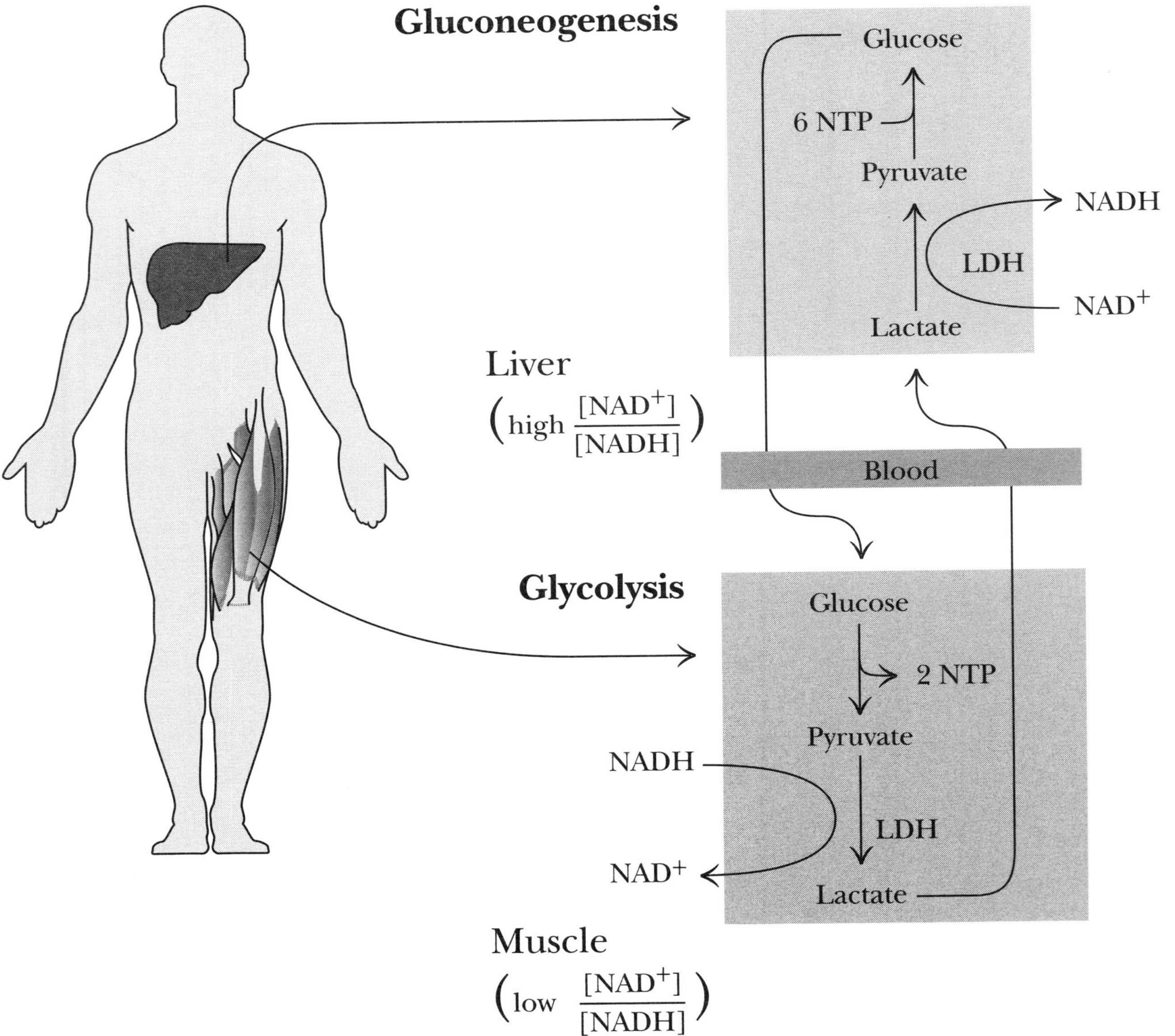

Figure 19.9   The Cori cycle.

Figure 19.10 The principal regulatory mechanisms in glycolysis and gluconeogenesis.

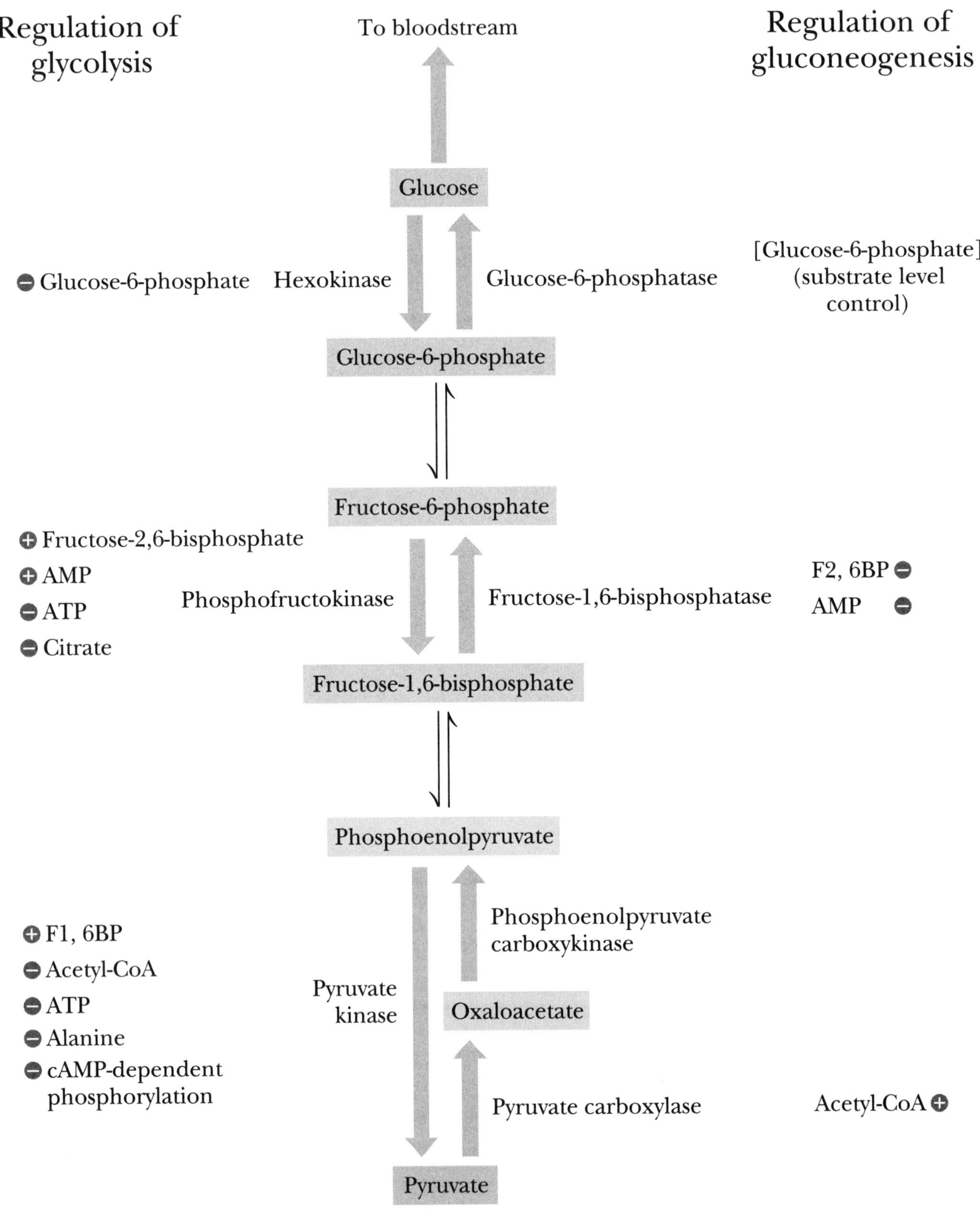

Figure 19.11   The inhibition of fructose-1,6-biphosphate by fructose-2,6-biphosphate.

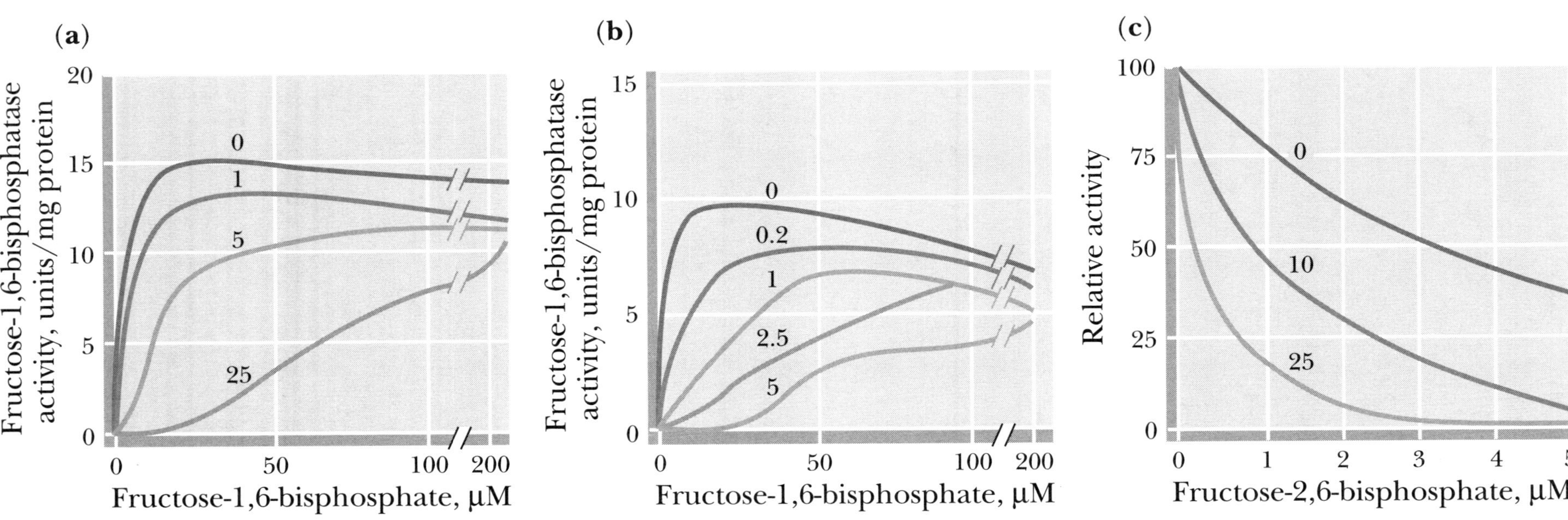

Figure 19.19 The pentose phosphate pathway.

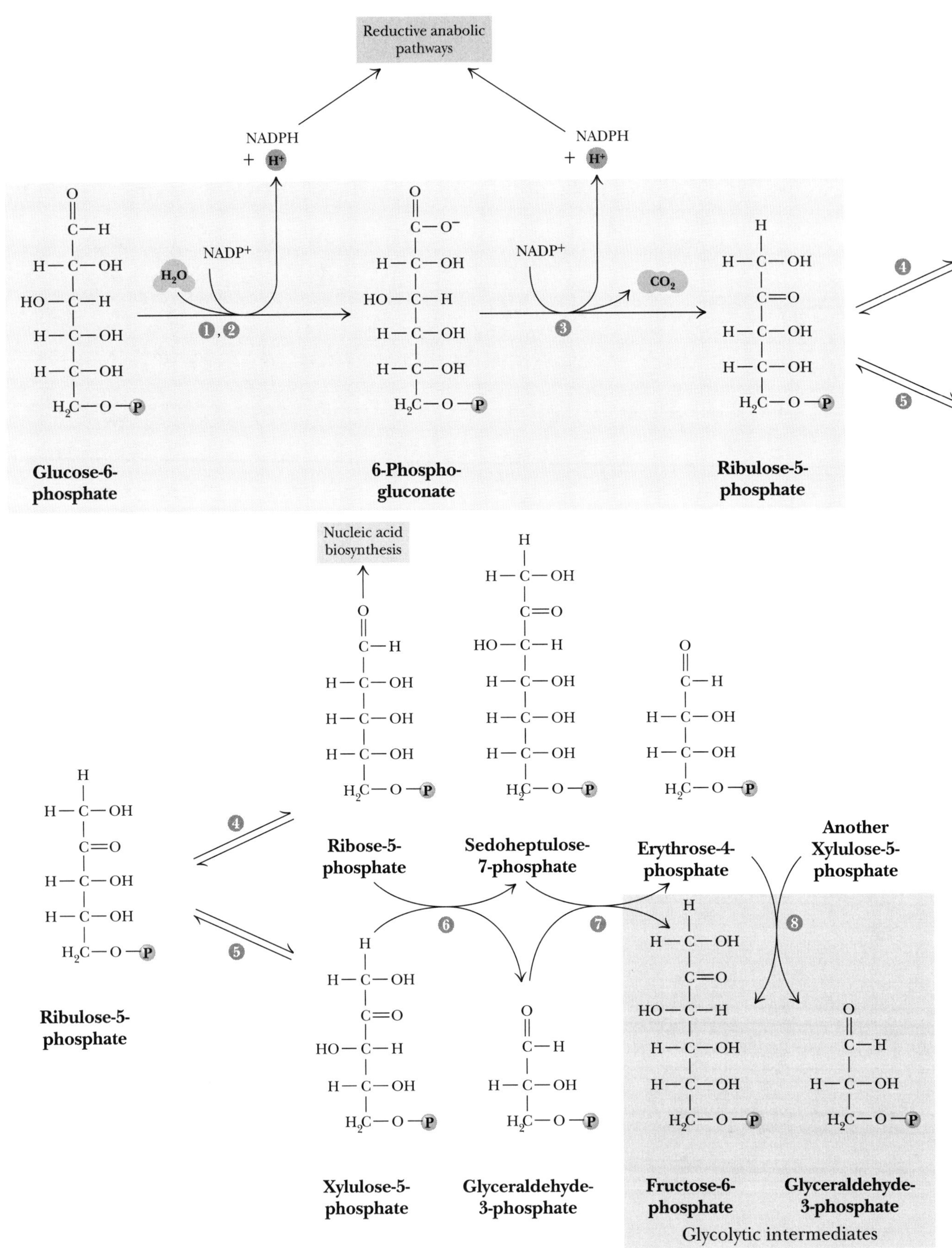

Figure 20.2 The hormone-dependent liberation of fatty acids from triacylglycerols in adipose tissue.

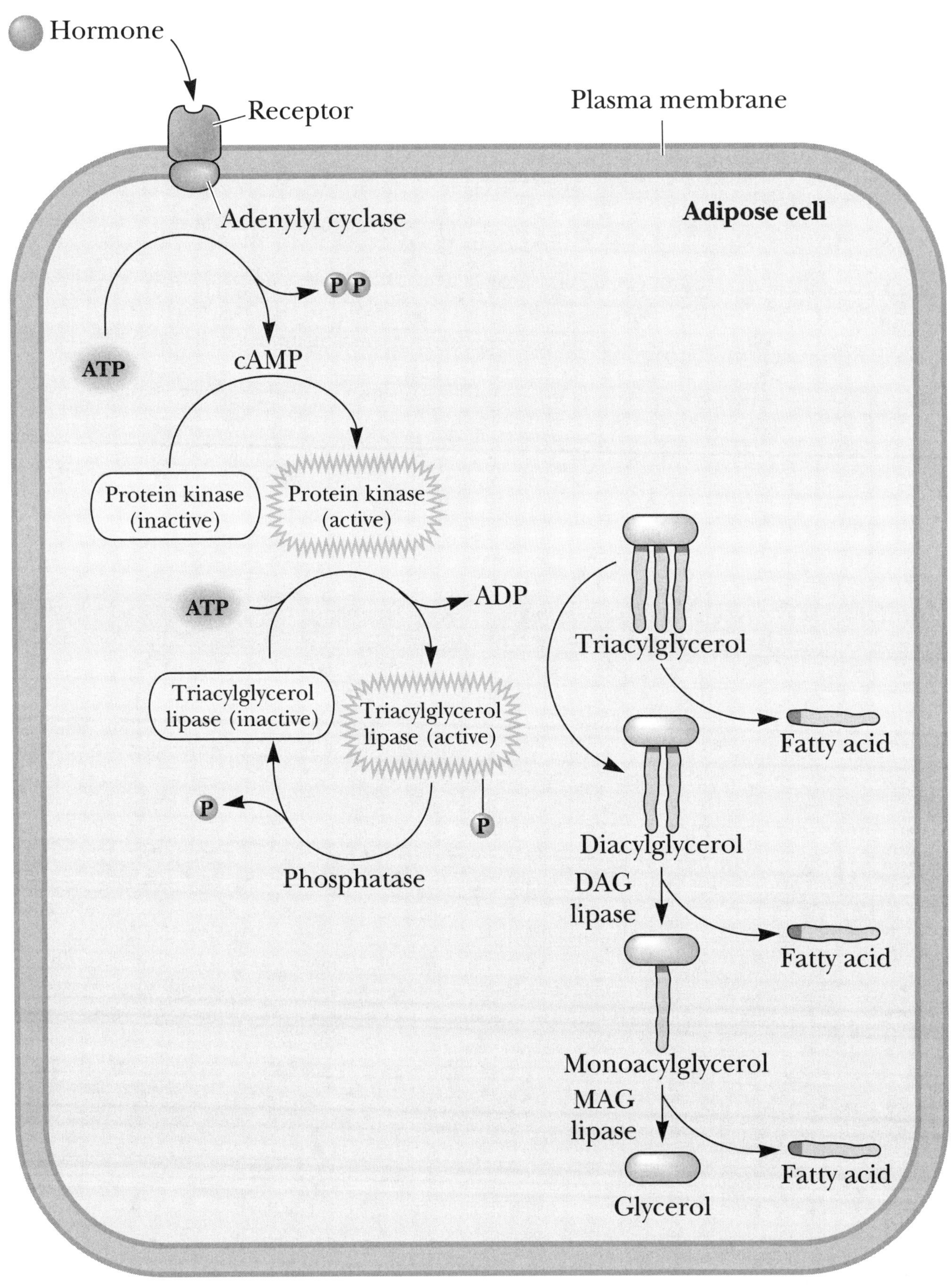

Figure 20.3 Degradation of dietary fatty acids occurs primarily in the duodenum.

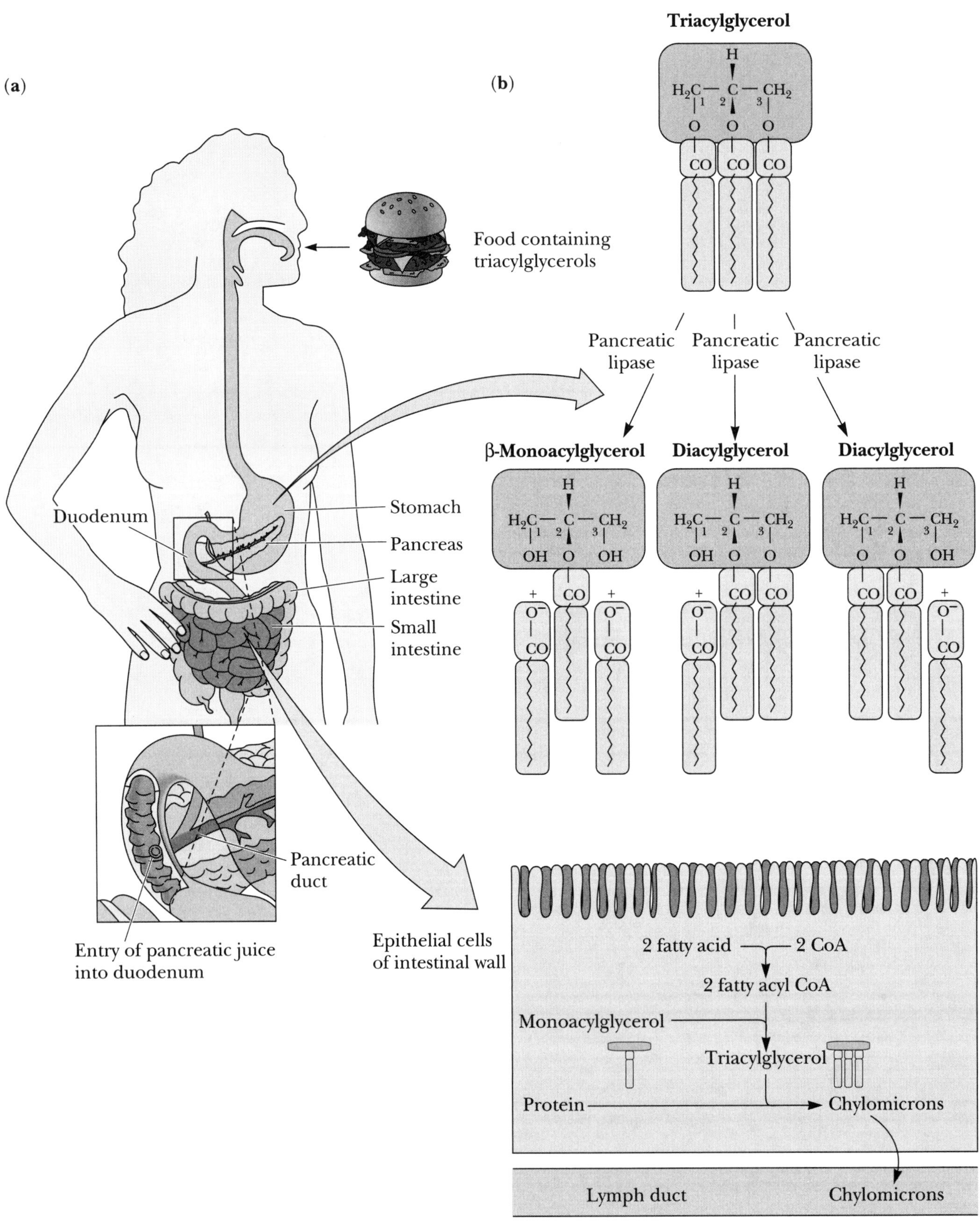

© Harcourt, Inc.

Figure 20.6 The formation of acylcarnitines and their transport across the inner mitochondrial membrane.

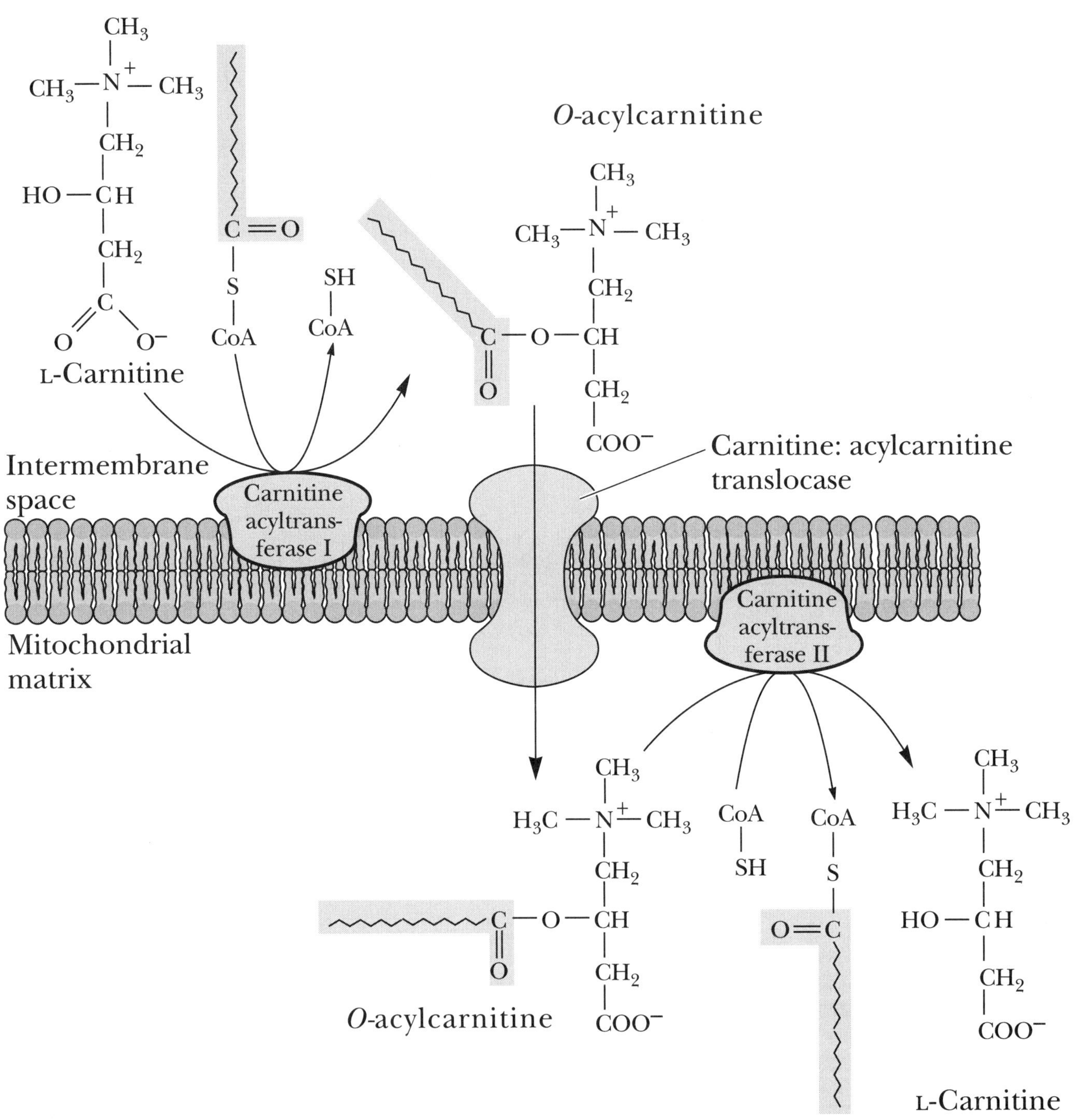

Figure 20.7   The β-oxidation of saturated fatty acids.

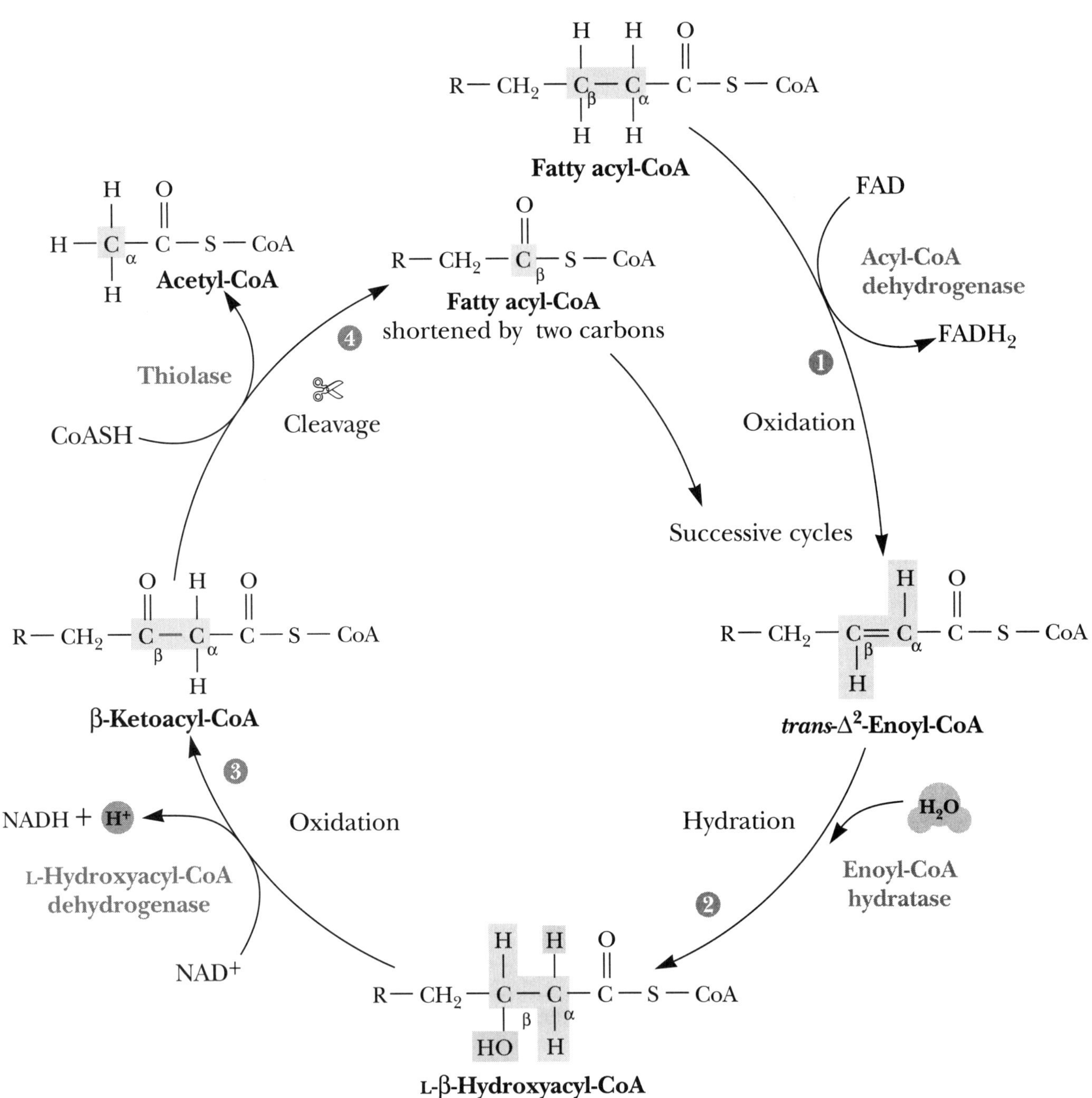

Table 20.2 Equations for the complete oxidation of palmitoyl-CoA to CO_2 and H_2O.

Table 20.2 Equations for the Complete Oxidation of Palmitoyl-CoA to CO_2 and H_2O

Equation	ATP Yield	Free Energy Yield(kJ/mol)
$CH_3(CH_2)_{14}CO\text{-}CoA + 7[FAD] + 7 H_2O + 7 NAD^+ + 7 CoA \longrightarrow 8 CH_3CO\text{-}CoA + 7[FADH_2] + 7 NADH + 7H^+$		
$7[FADH_2] + 10.5 P_i + 10.5 ADP + 3.5 O_2 \longrightarrow 7[FAD] + 17.5 H_2O + 10.5 ATP$	10.5	320
$7 NADH + 7 H^+ + 17.5 P_i + 17.5 ADP + 3.5 O_2 \longrightarrow 7 NAD^+ + 24.5 H_2O + 17.5 ATP$	17.5	534
8-Acetyl-CoA $+ 16 O_2 + 80 ADP + 80 P_i \longrightarrow 8 CoA + 88 H_2O + 16 CO_2 + 80 ATP$	80	2440
$CH_3 - (CH_2)_{14}CO\text{-}CoA + 108 P_i + 108 ADP + 23 O_2 \longrightarrow 108 ATP + 16 CO_2 + 130 H_2O + CoA$	108	3294
Energetic "cost" of forming palmitoyl-CoA from palmitate and CoA	-2	-61
Total	106	3233

Figure 20.15 The oxidation pathway for polyunsaturated fatty acids.

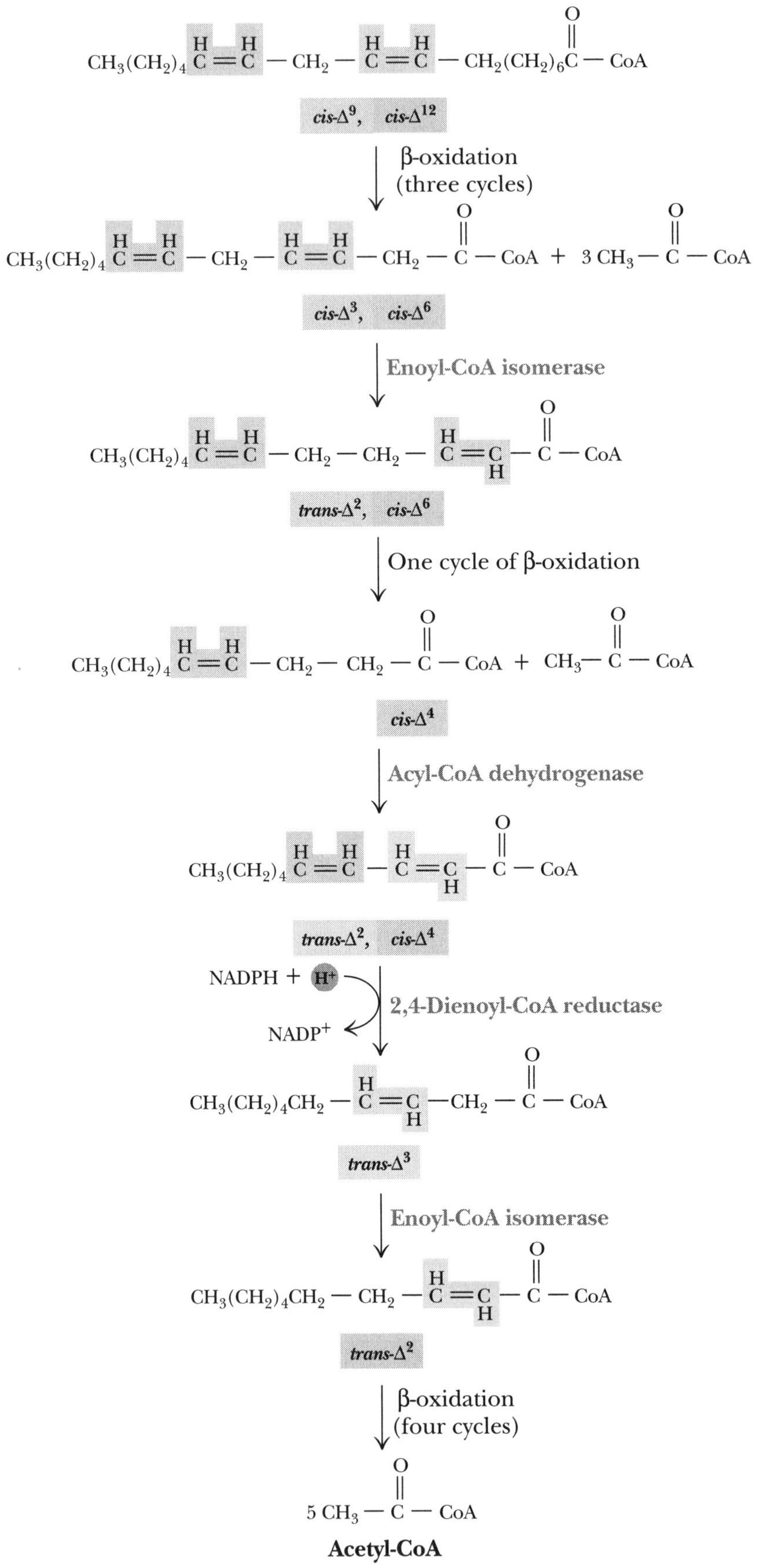

Figure 20.17 Branched-chain fatty acids are oxidized by α-oxidation.

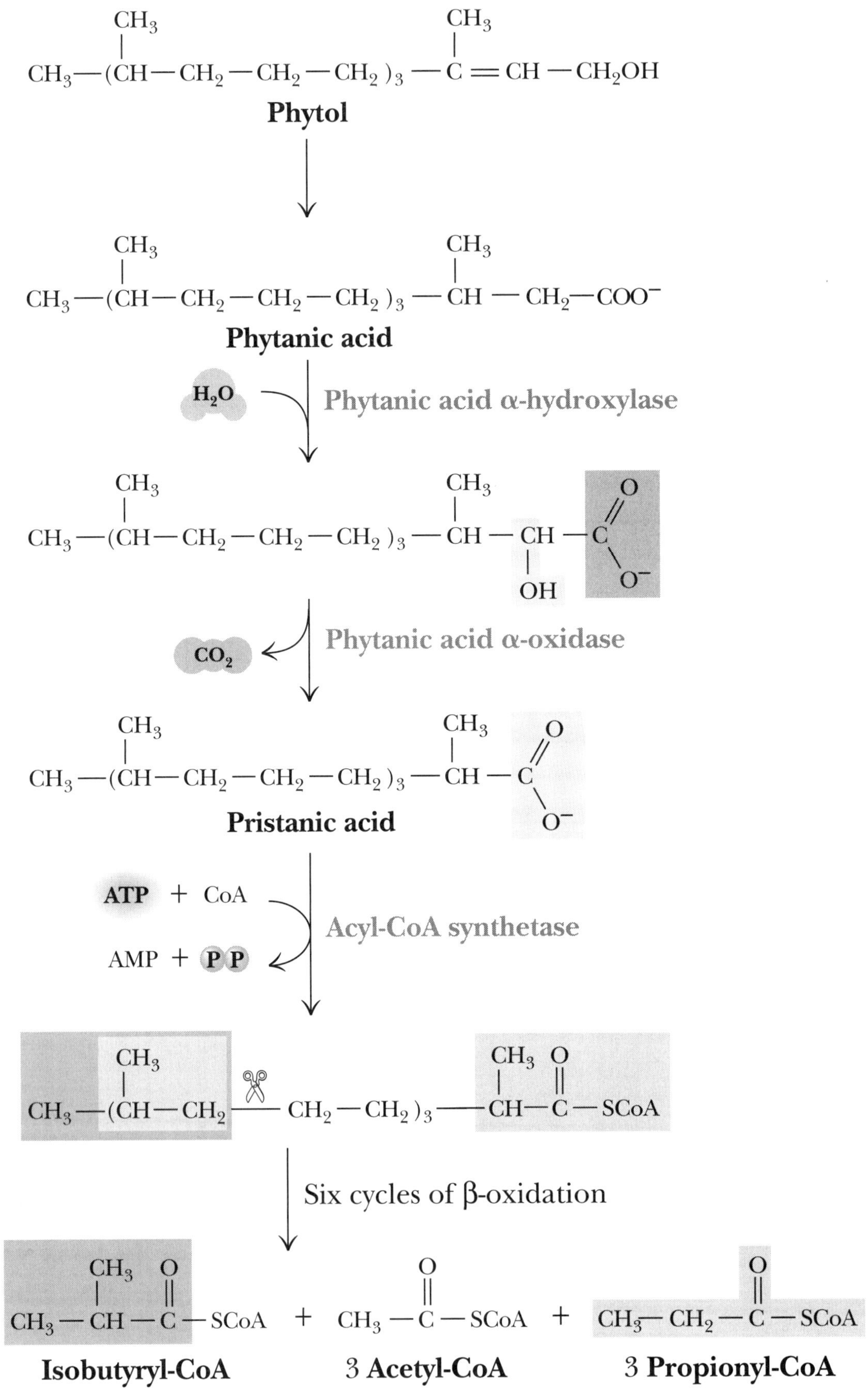

Figure 20.19 The formation of ketone bodies, synthesized primarily in the liver.

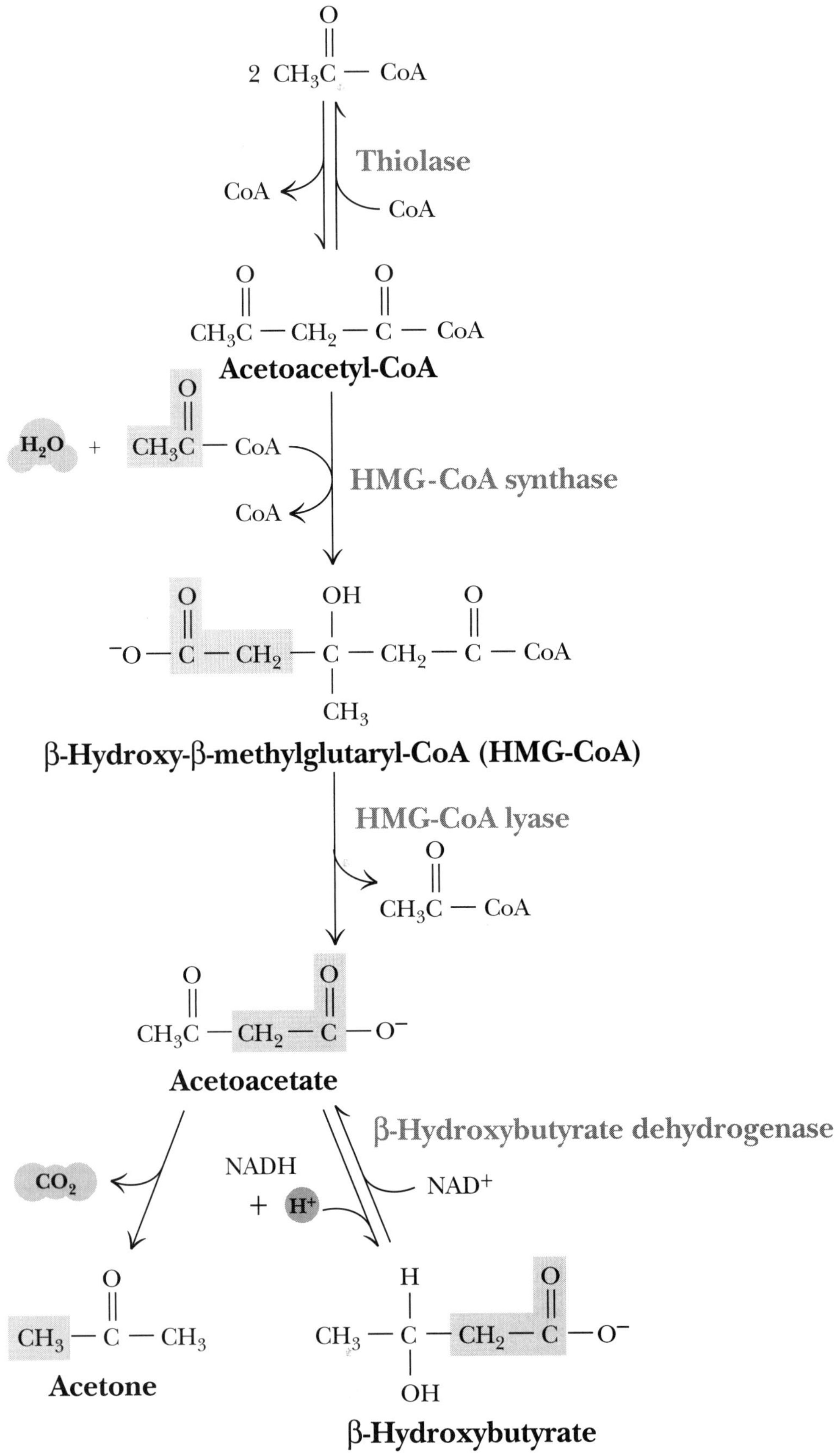

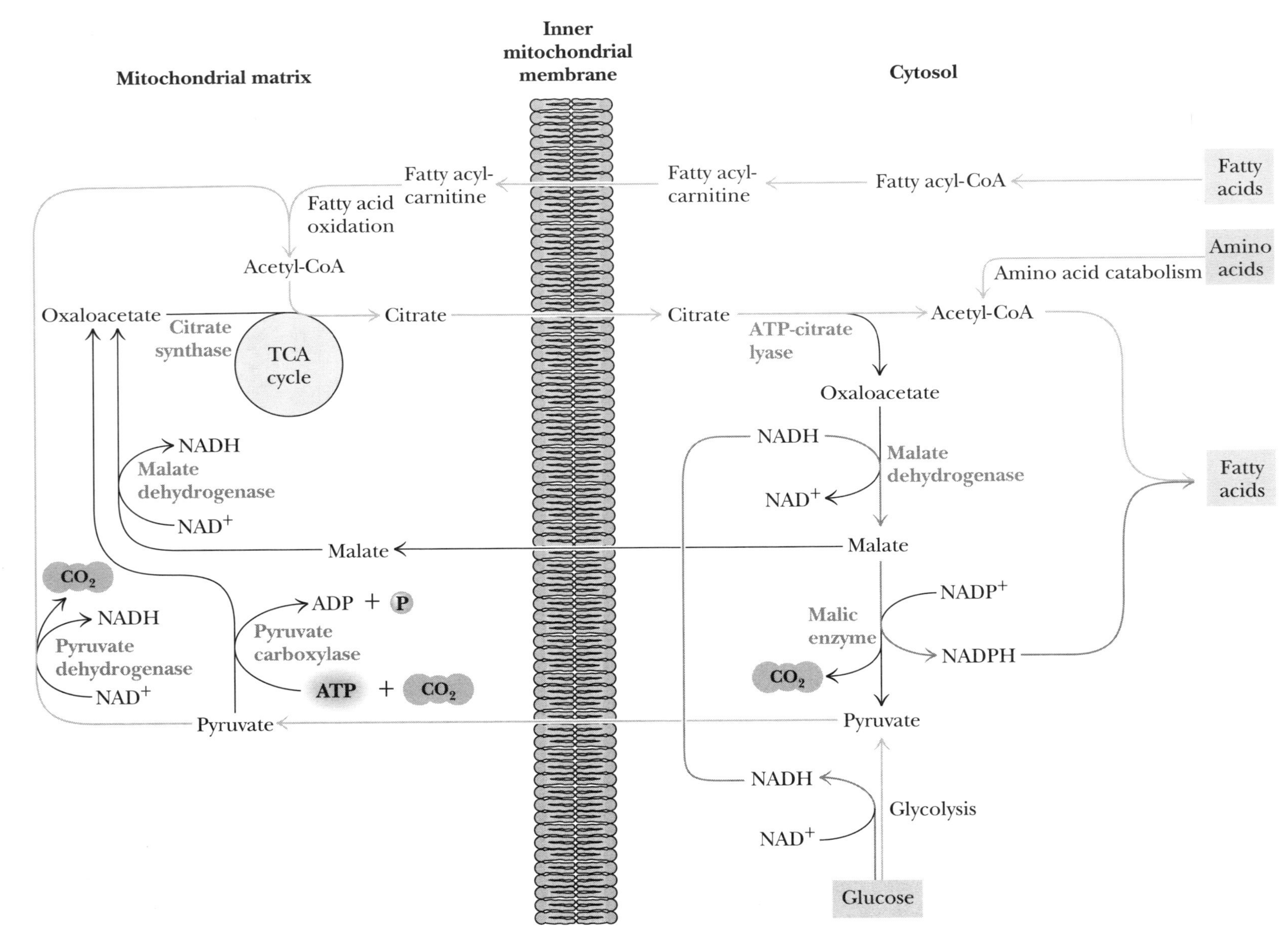

Figure 20.21   The citrate-malate-pyruvate shuttle.

Figure 20.25 The pathway of palmitate synthesis from acetyl-CoA and malonyl-CoA.

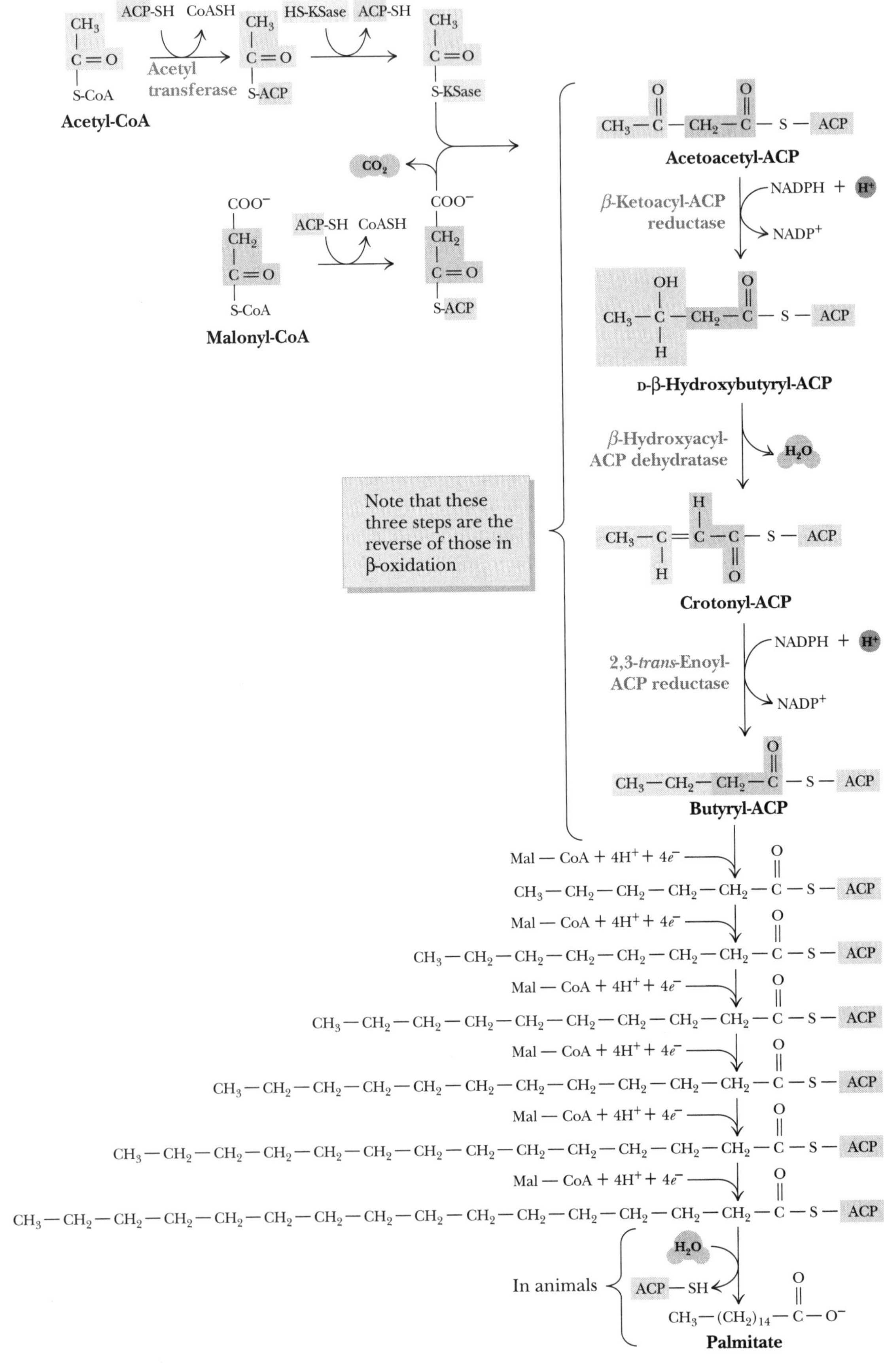

Figure 20.26   Fatty acid synthase in animals.

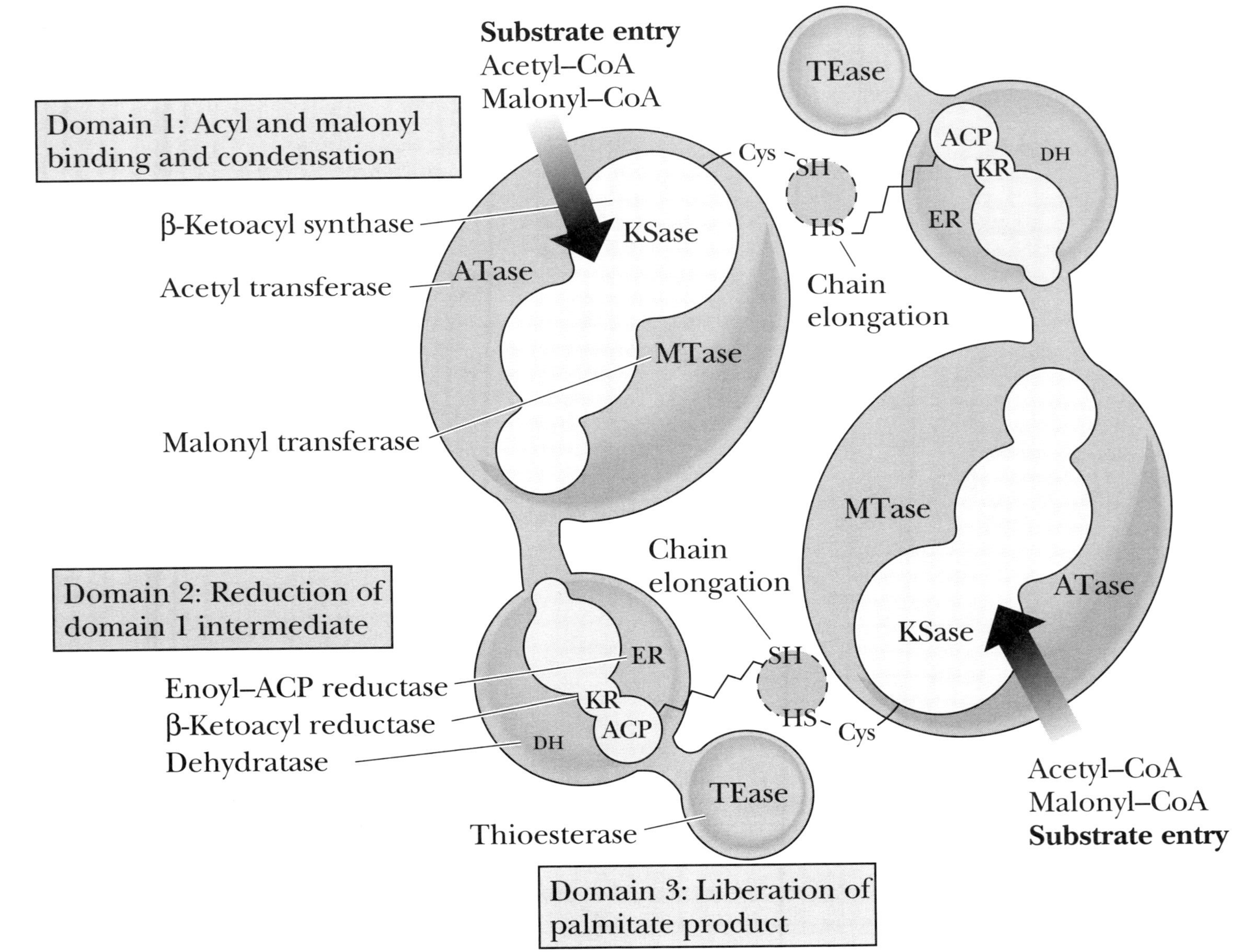

© Harcourt, Inc.

Figure 20.31 Regulation of fatty acid synthesis and fatty acid oxidation are coupled.

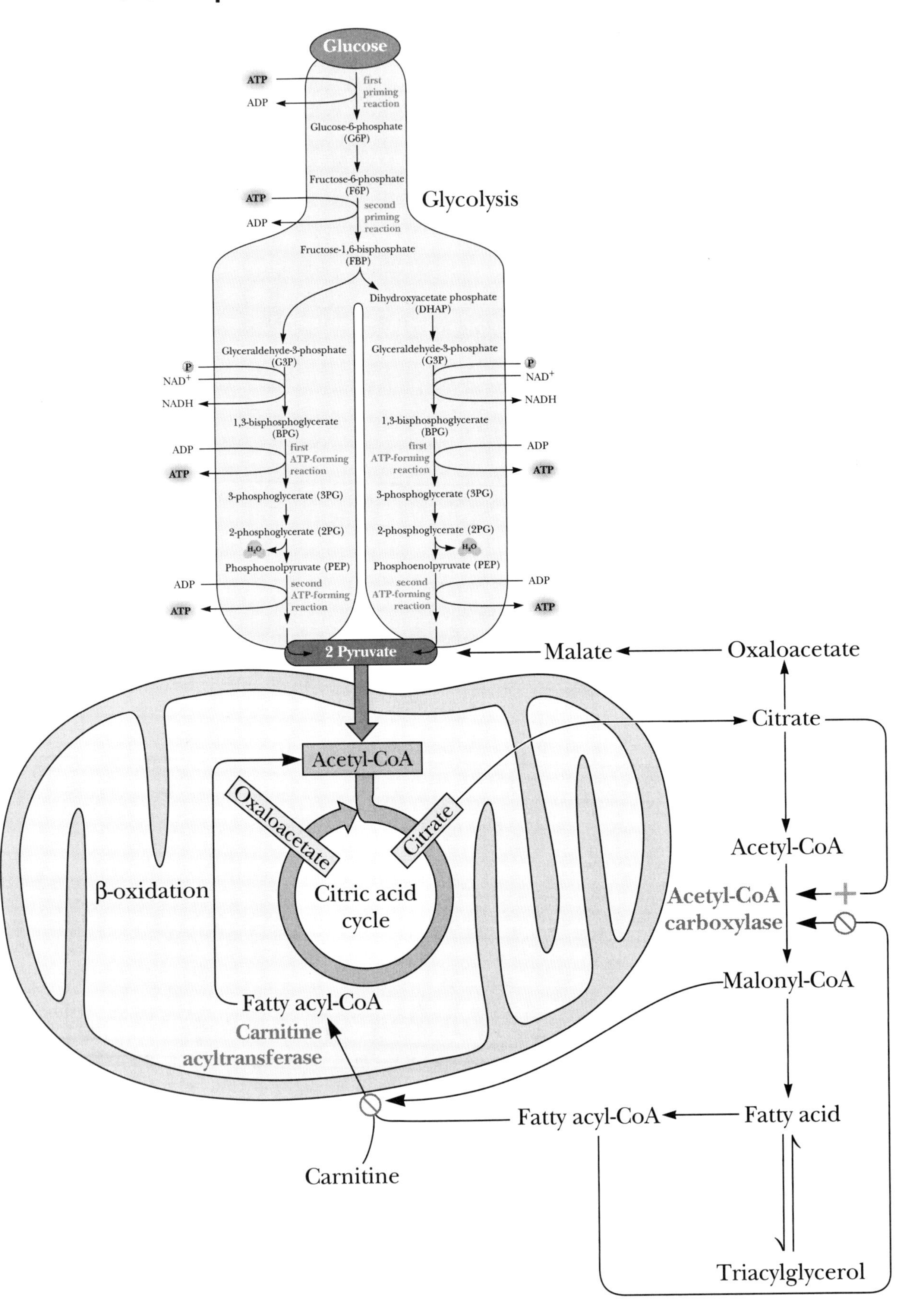

Figure 20.32 Hormonal signals regulating fatty acid synthesis.

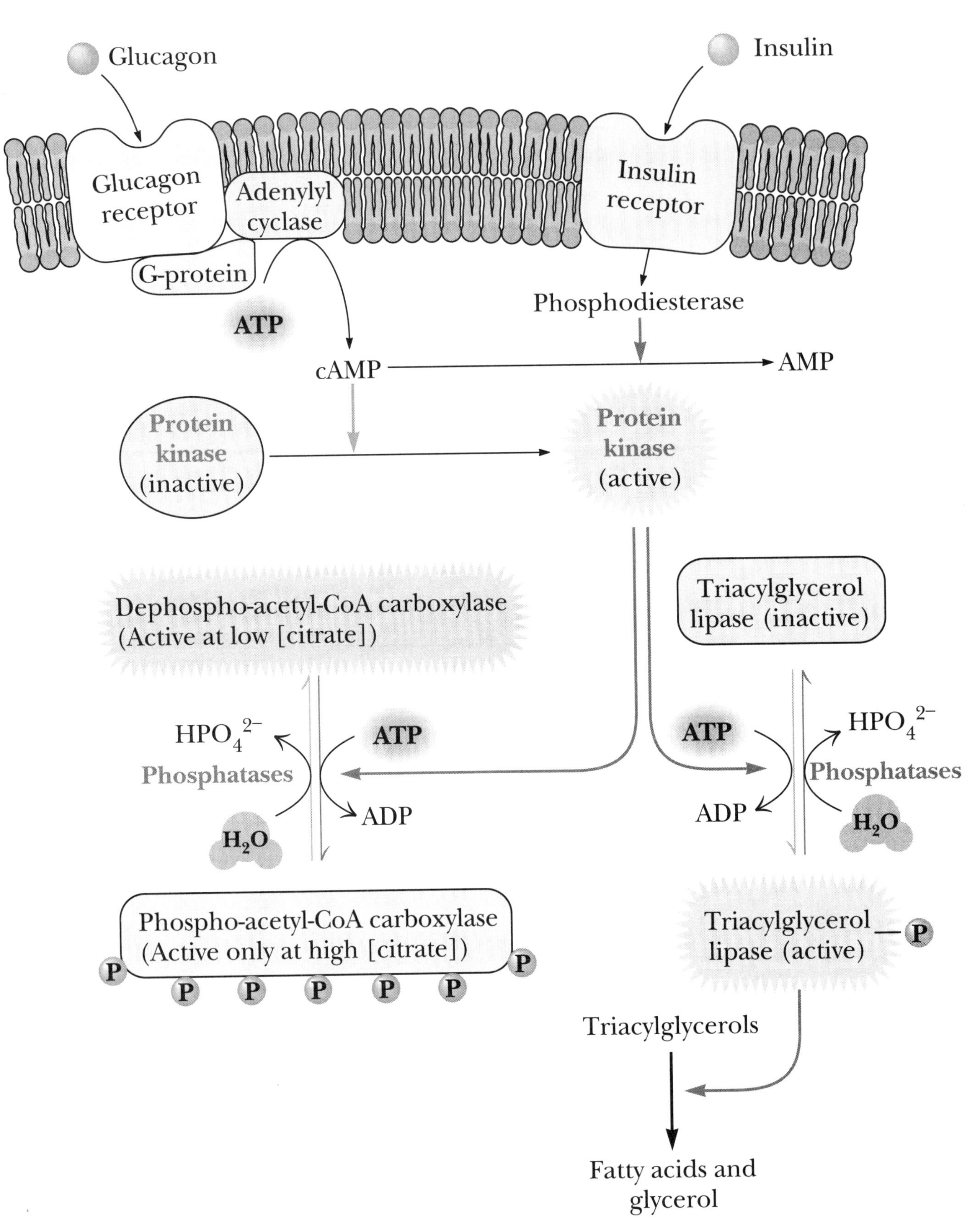

Figure 20.33 Synthesis of glycerolipids in eukaryotes.

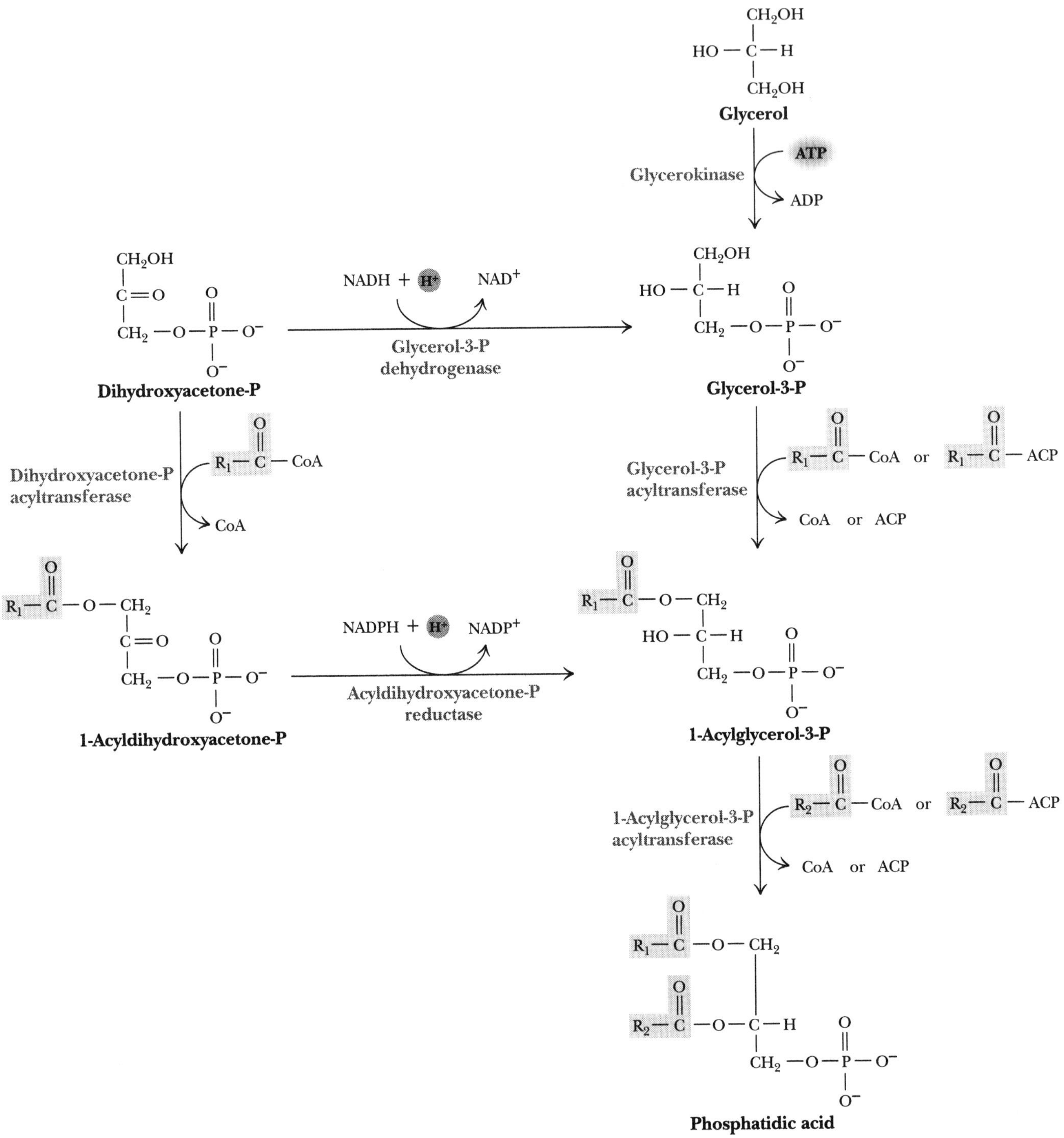

Figure 20.34 Diacylglycerol and CDP-diacylglycerol precursors of glycerolipids in eukaryotes.

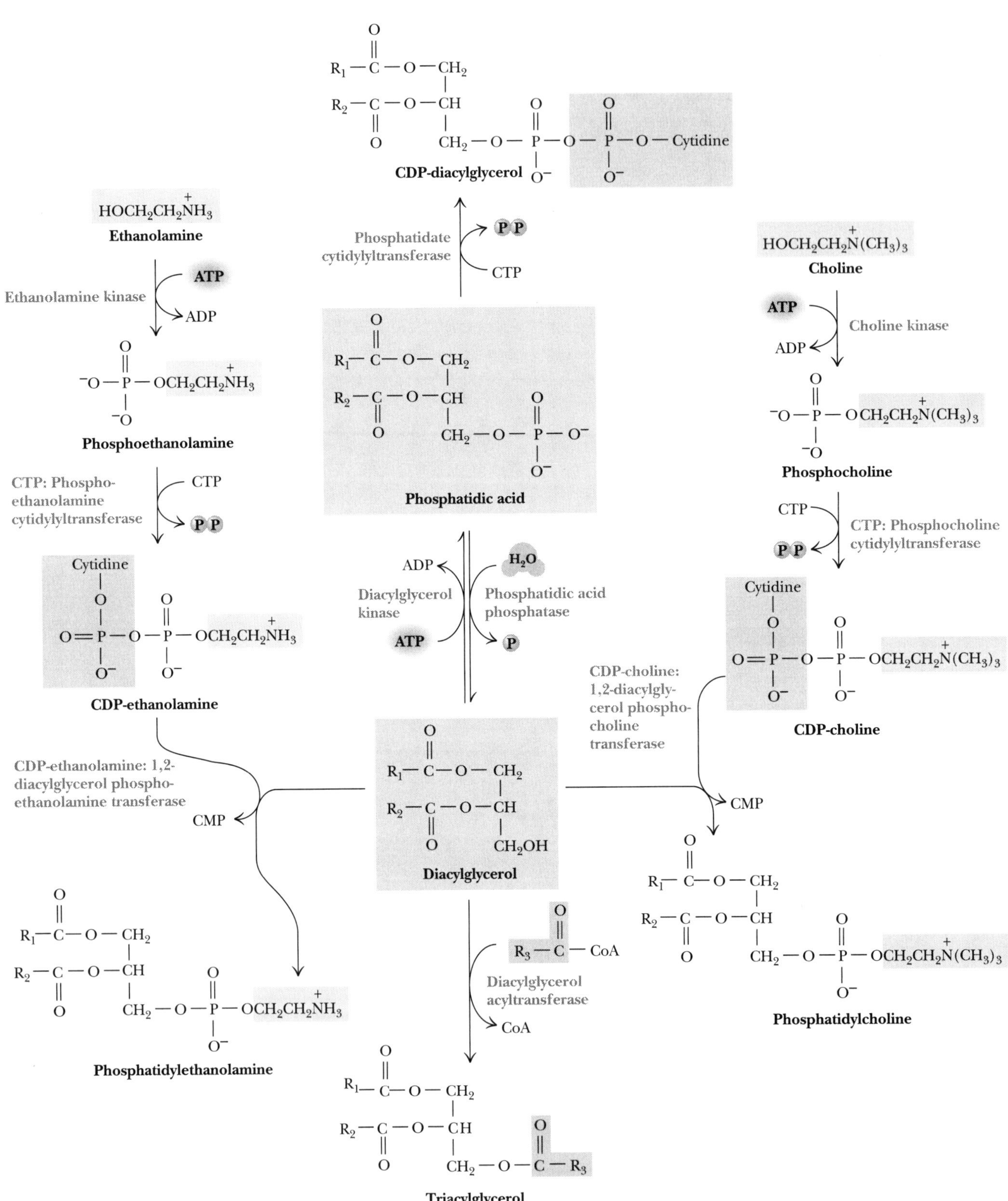

Figure 20.38 Biosynthesis of plasmalogens in animals.

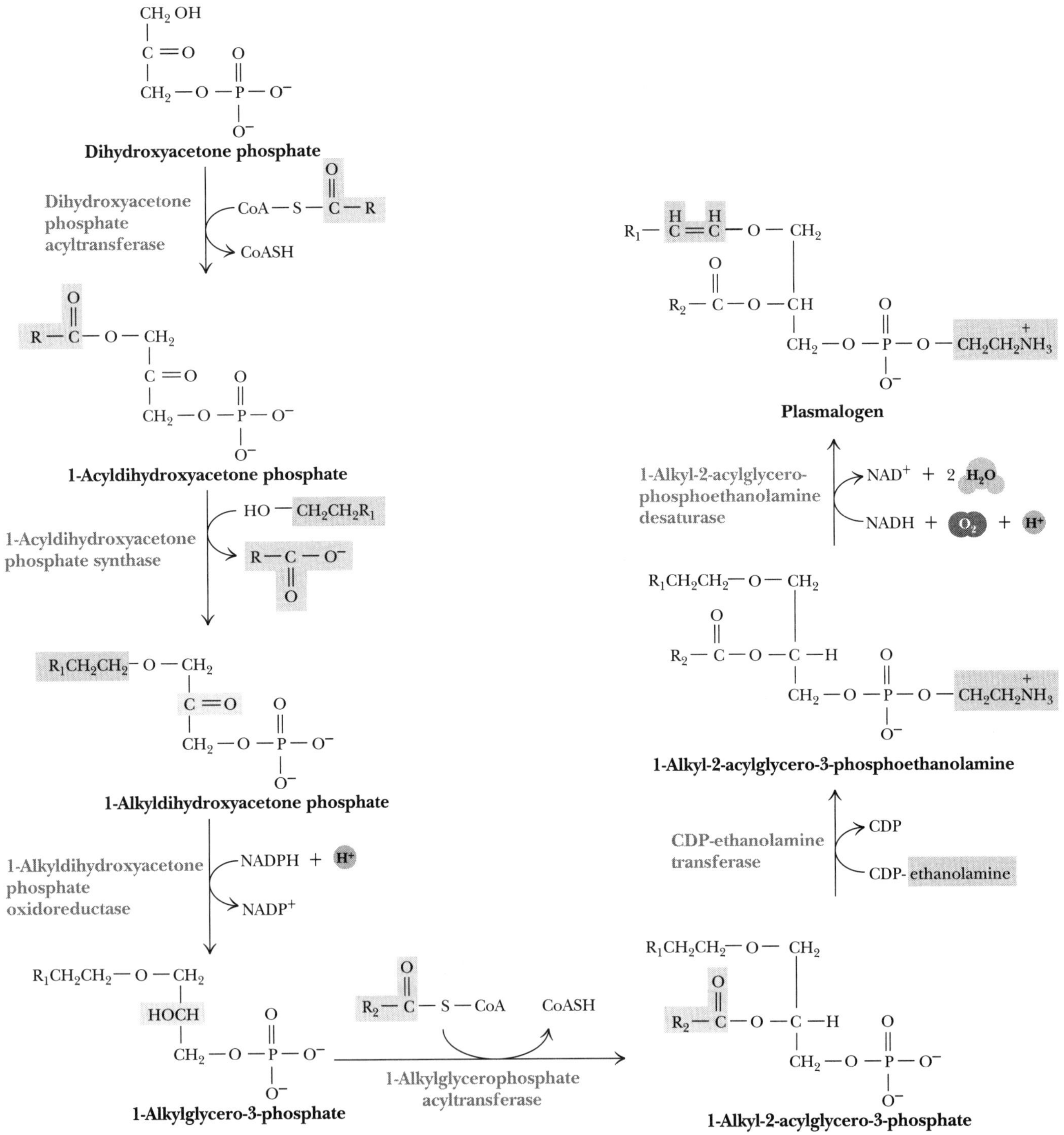

Figure 20.39 Biosynthesis of sphingolipids in animals.

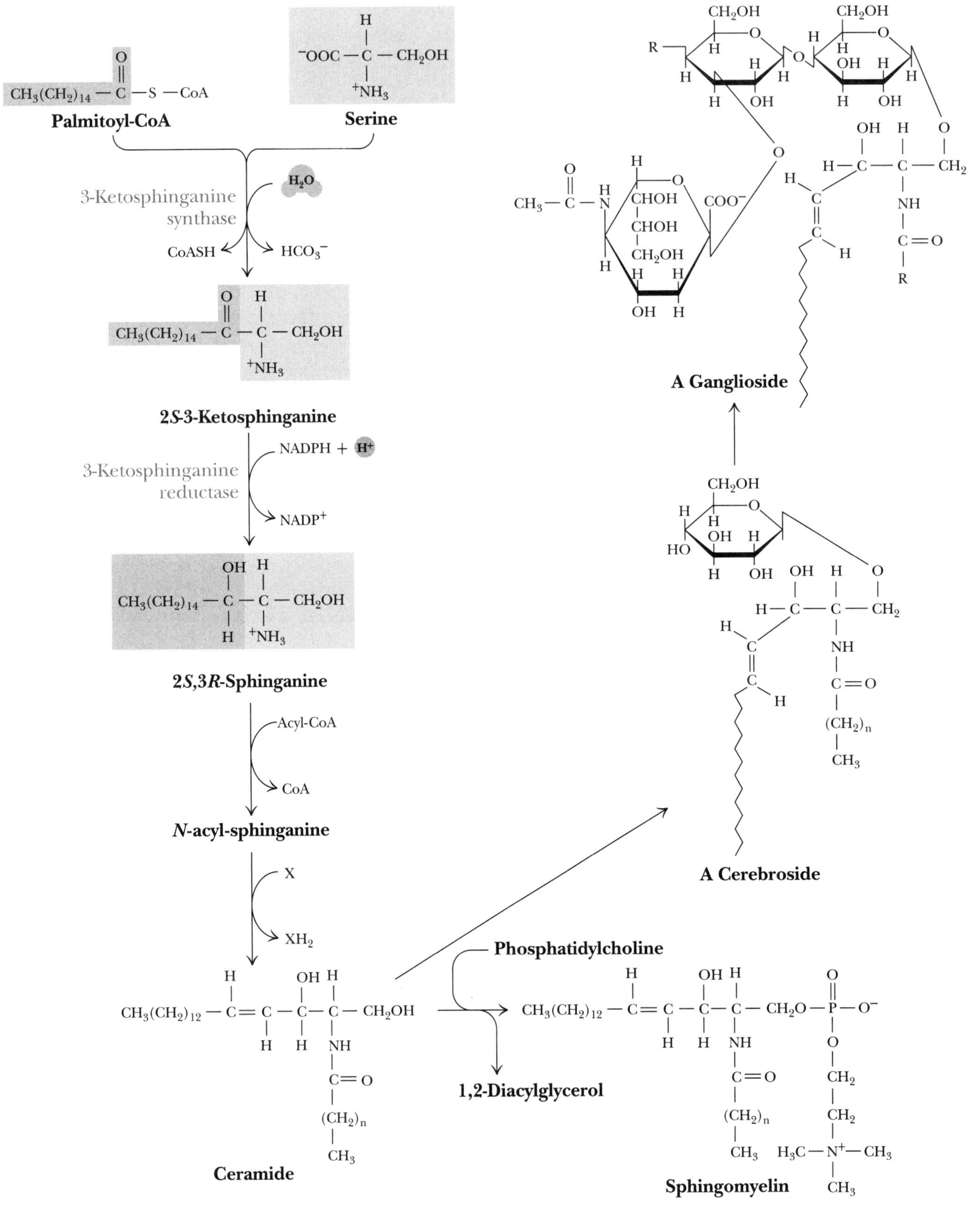

© Harcourt, Inc.

Figure 20.40   Arachidonic acid derived from the breakdown of phospholipids is the precursor of prostaglandins, thromboxanes, and leukotrienes.

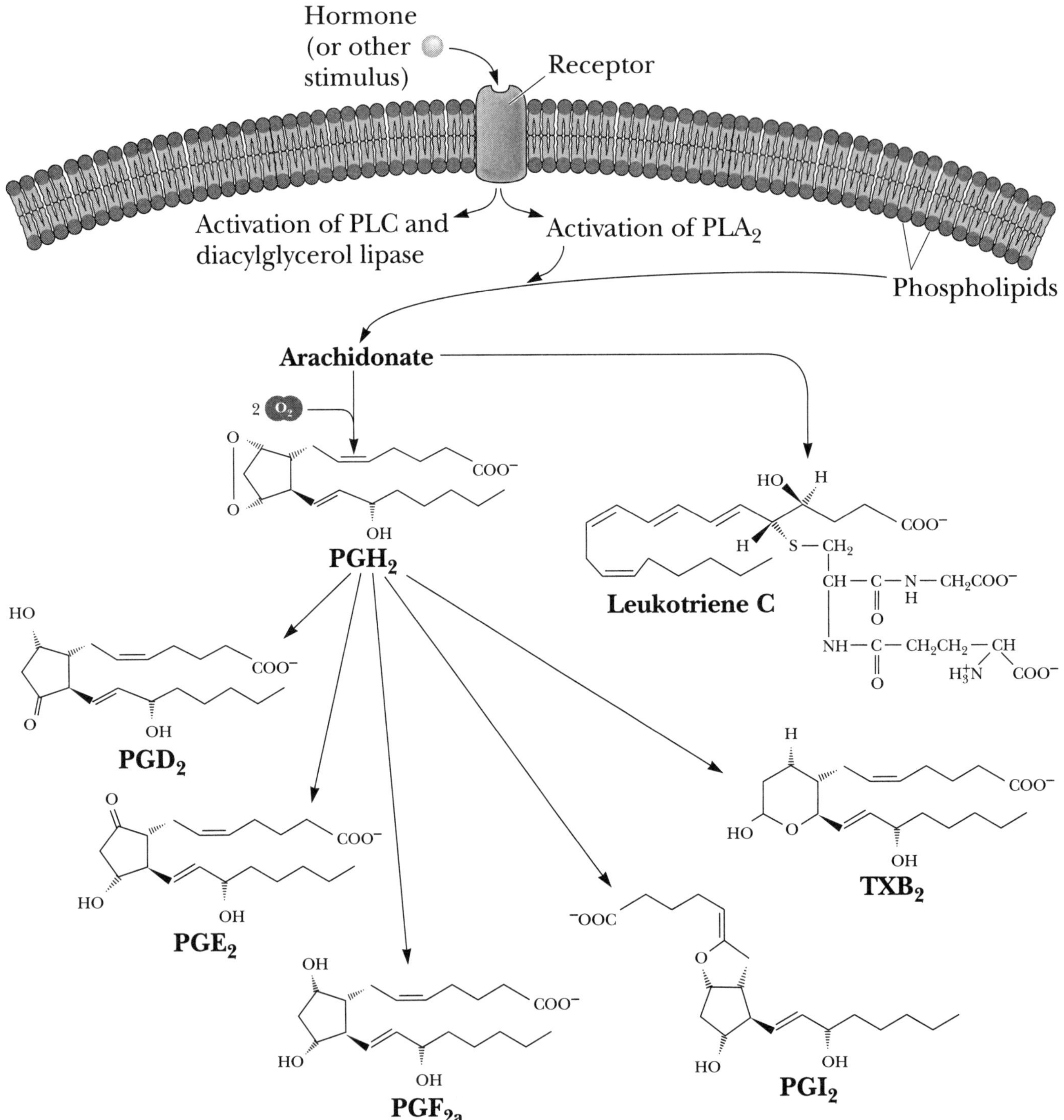

Figure 20.47   Cholesterol is synthesized from squalene via lanosterol.

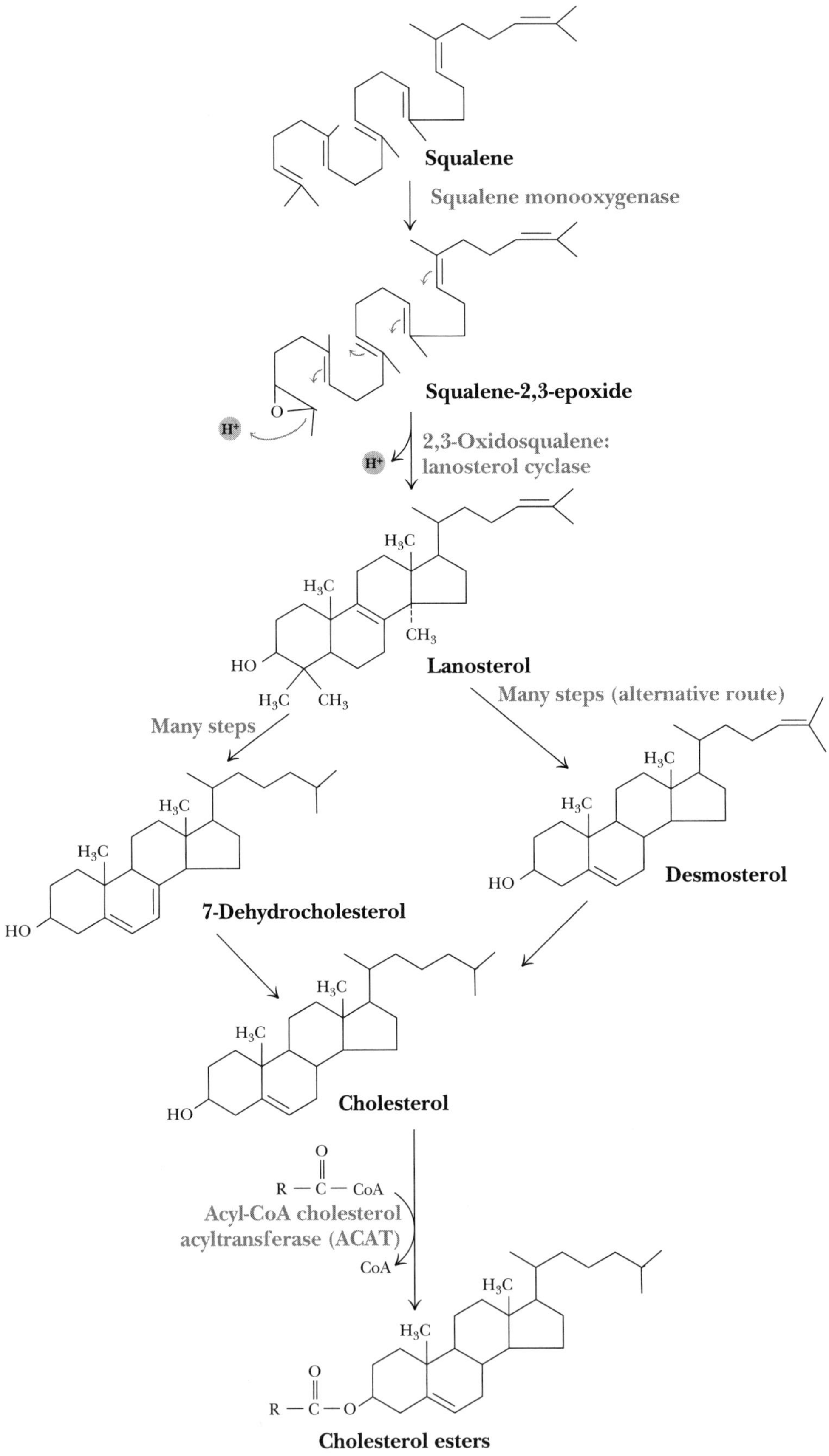

Figure 20.48   Cholic acid is synthesized from cholesterol.

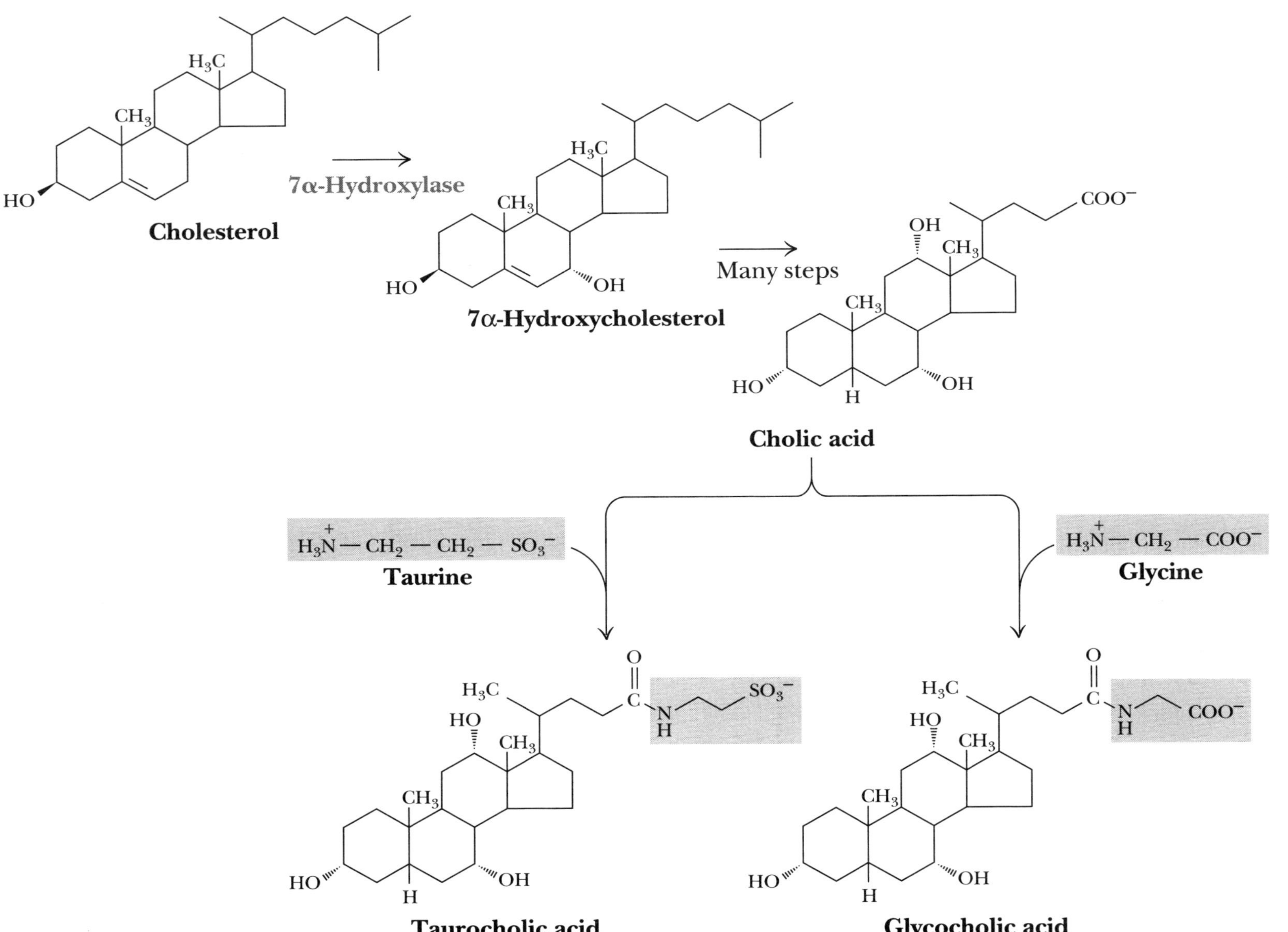

Figure 20.50 Steroid hormones are synthesized from cholesterol.

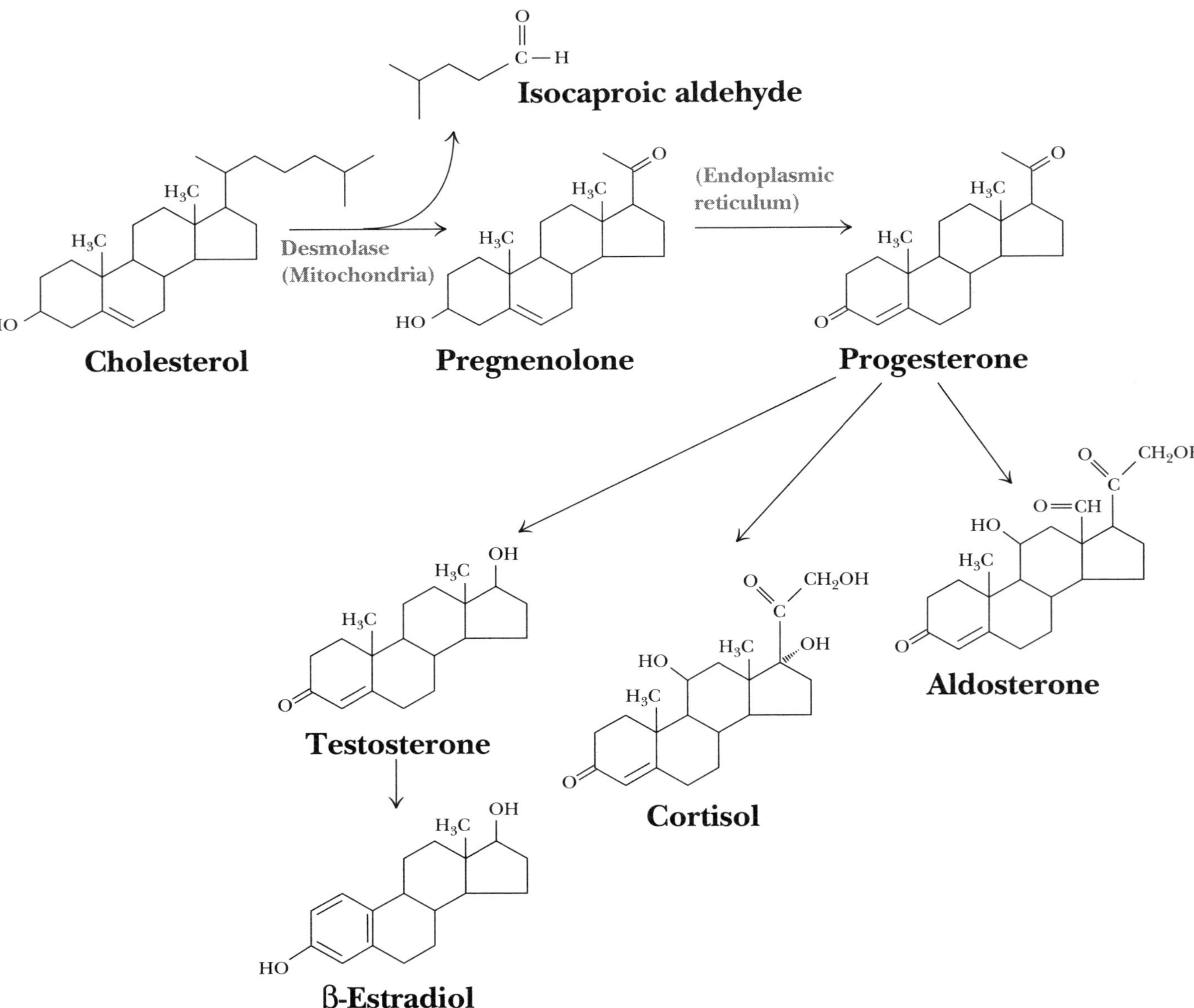

Figure 21.1   The nitrogen cycle.

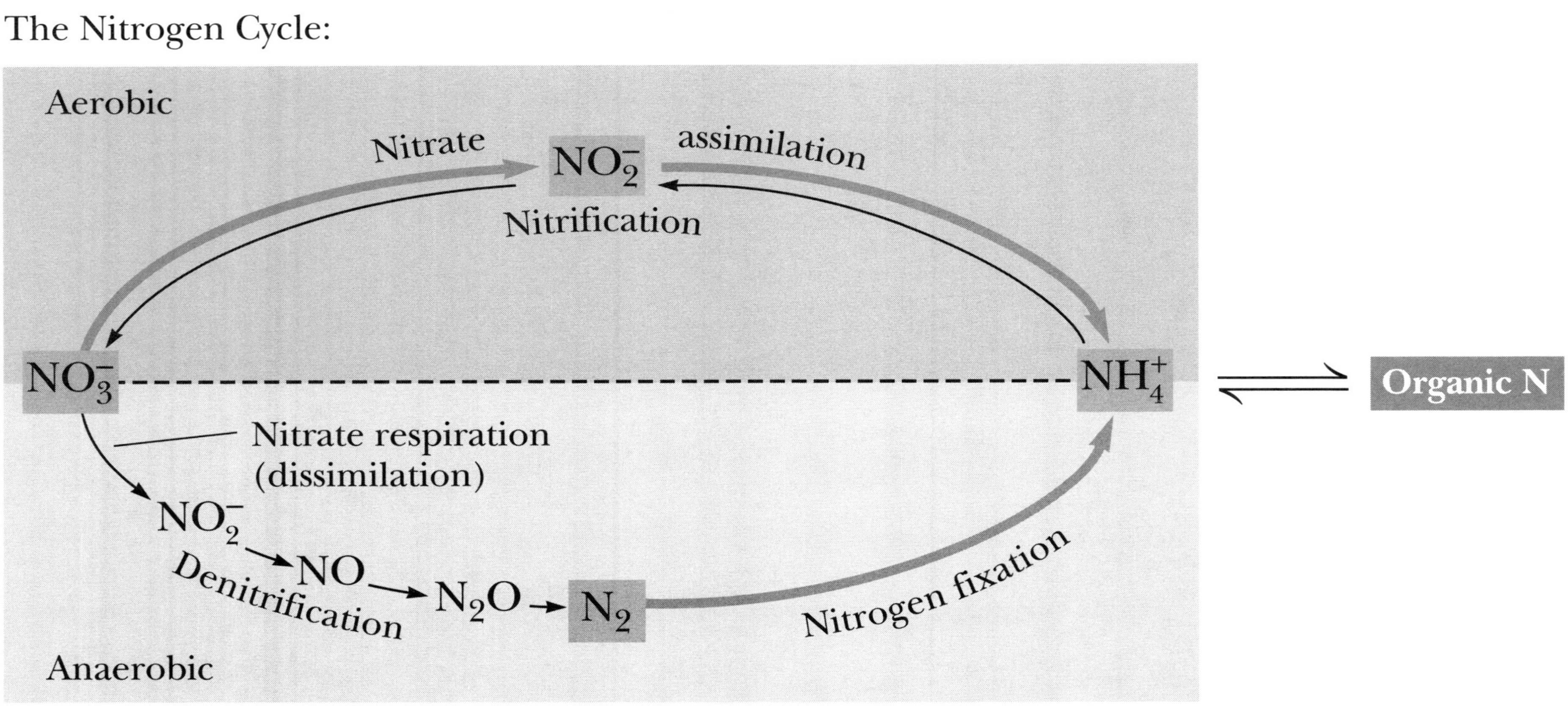

The Nitrogen Cycle:

Figure 21.5 The allosteric regulation of glutamine synthetase activity by cumulative feedback inhibition.

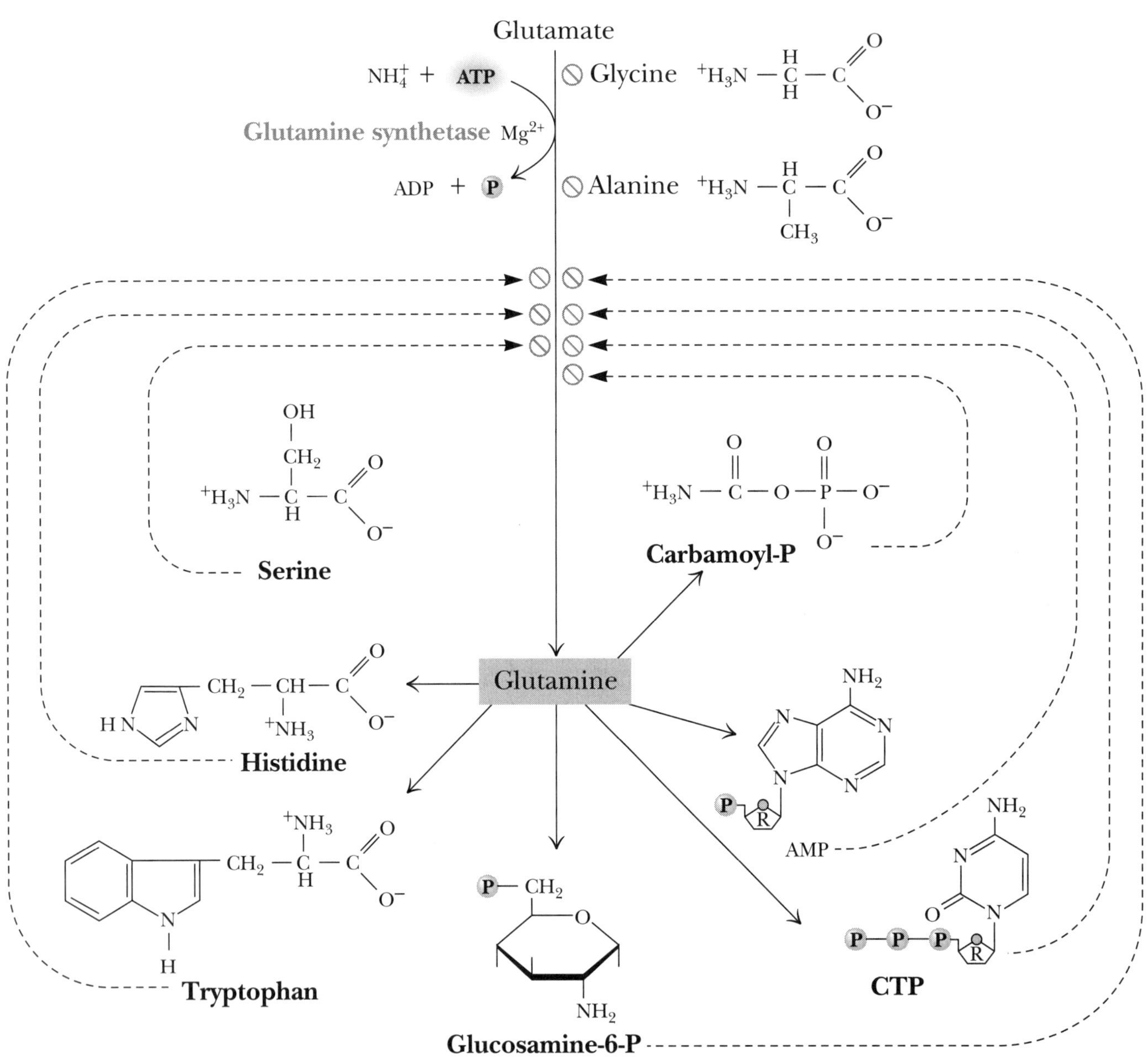

© Harcourt, Inc.

Figure 21.6   Covalent modification of glutamine synthetase.

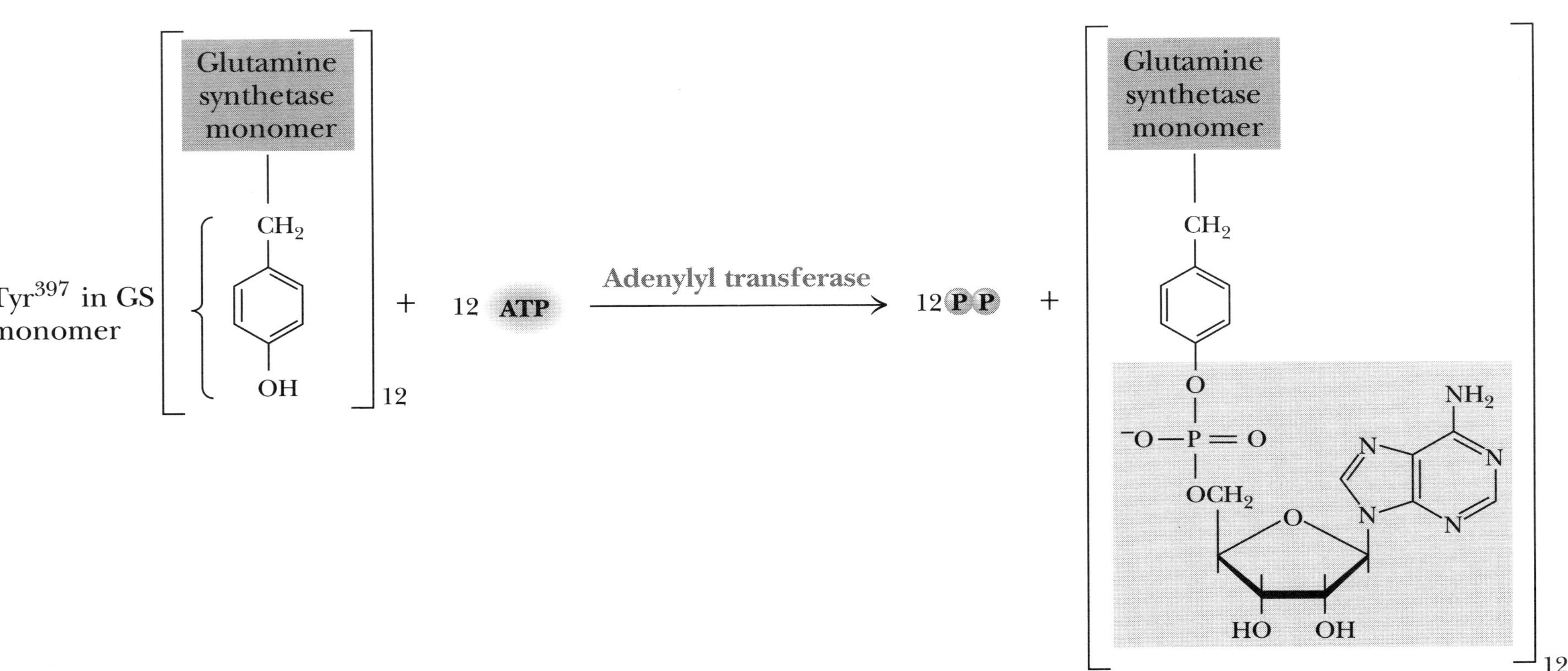

Figure 21.7 The cyclic cascade system regulating the covalent modification of GS.

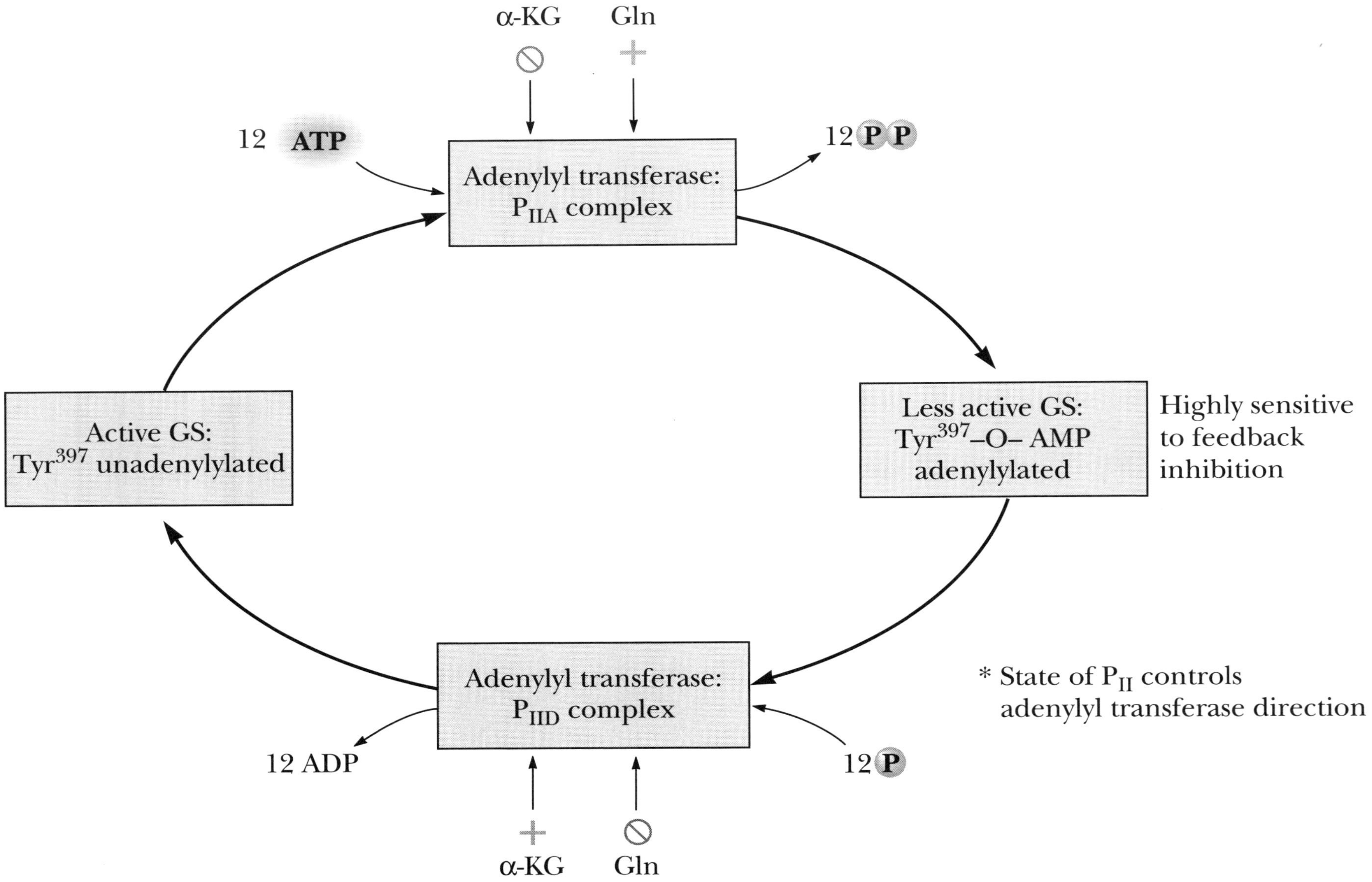

Figure 21.12 The urea cycle series of reactions.

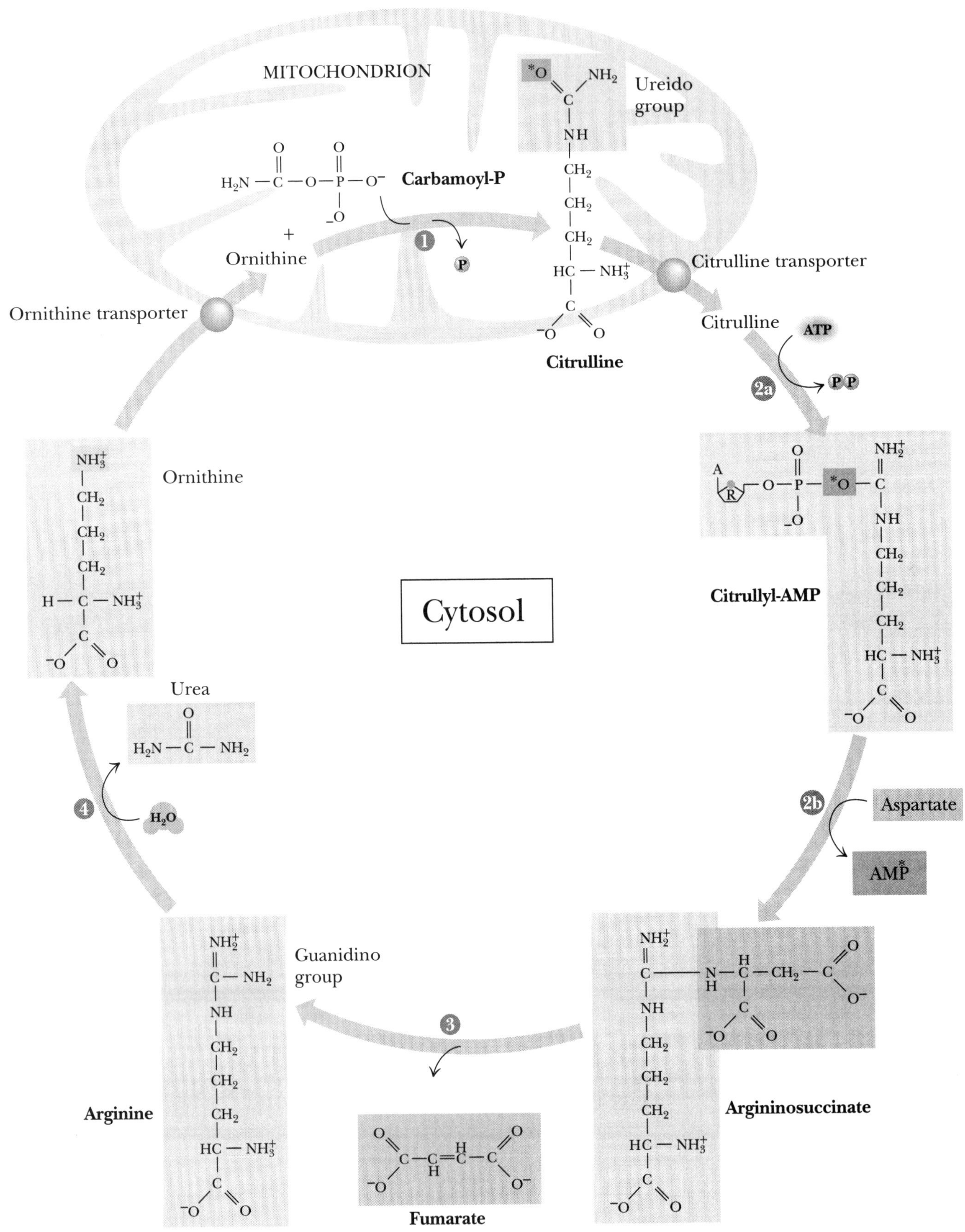

© Harcourt, Inc.

Figure 21.18 The biosynthesis of SAM.

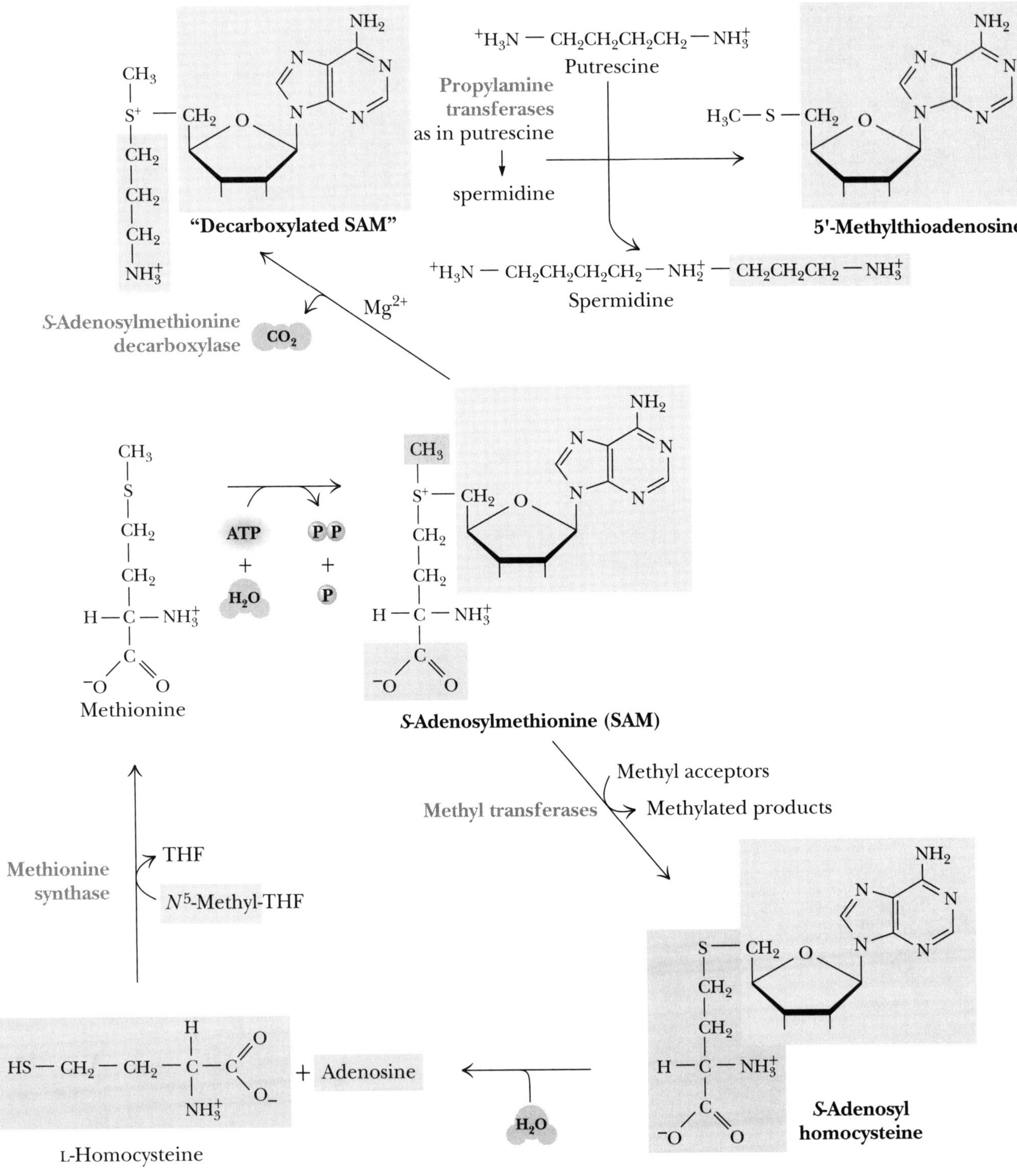

© **Harcourt, Inc.**

Figure 21.21 The biosynthesis of phenylalanine, tyrosine, and tryptophan.

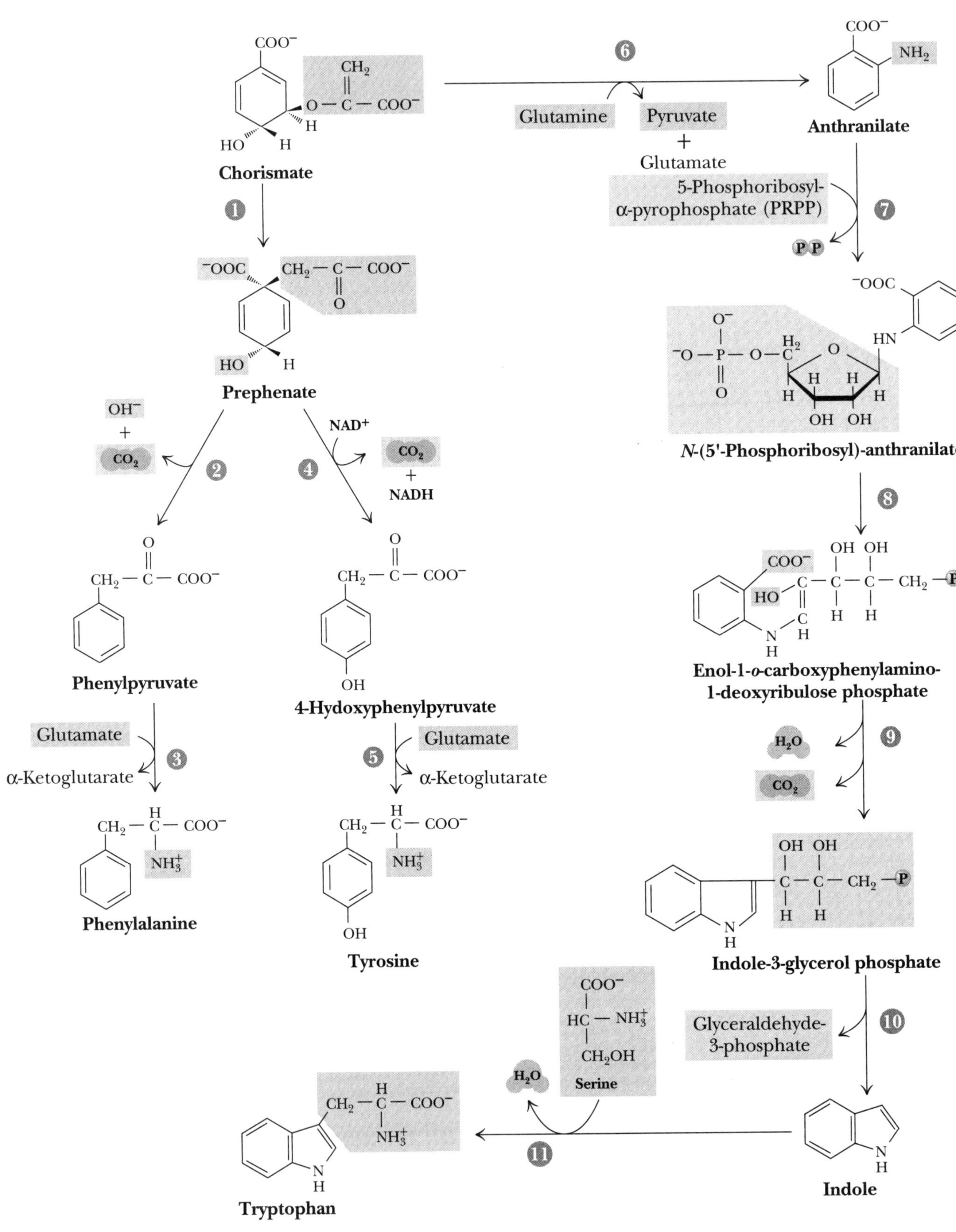

© Harcourt, Inc.

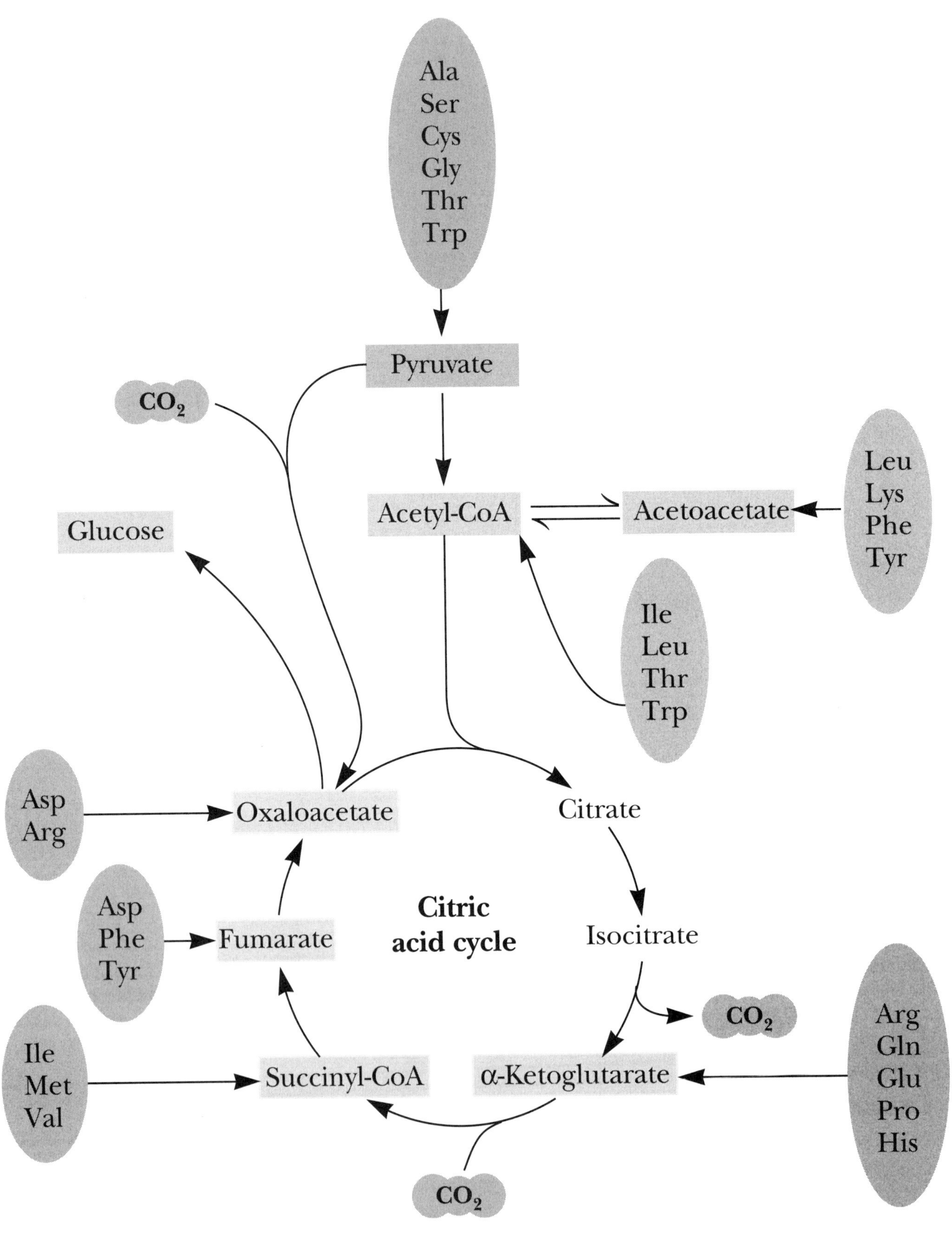
Ala
Ser
Cys
Gly
Thr
Trp
Pyruvate
CO2
Acetyl-CoA
Acetoacetate
Leu
Lys
Phe
Tyr
Glucose
Ile
Leu
Thr
Trp
Asp
Arg
Oxaloacetate
Citrate
Citric
acid cycle
Asp
Phe
Tyr
Fumarate
Isocitrate
CO2
Arg
Gln
Glu
Pro
His
Ile
Met
Val
Succinyl-CoA
α-Ketoglutarate
CO2

Figure 21.26 The *de novo* pathway for purine synthesis.

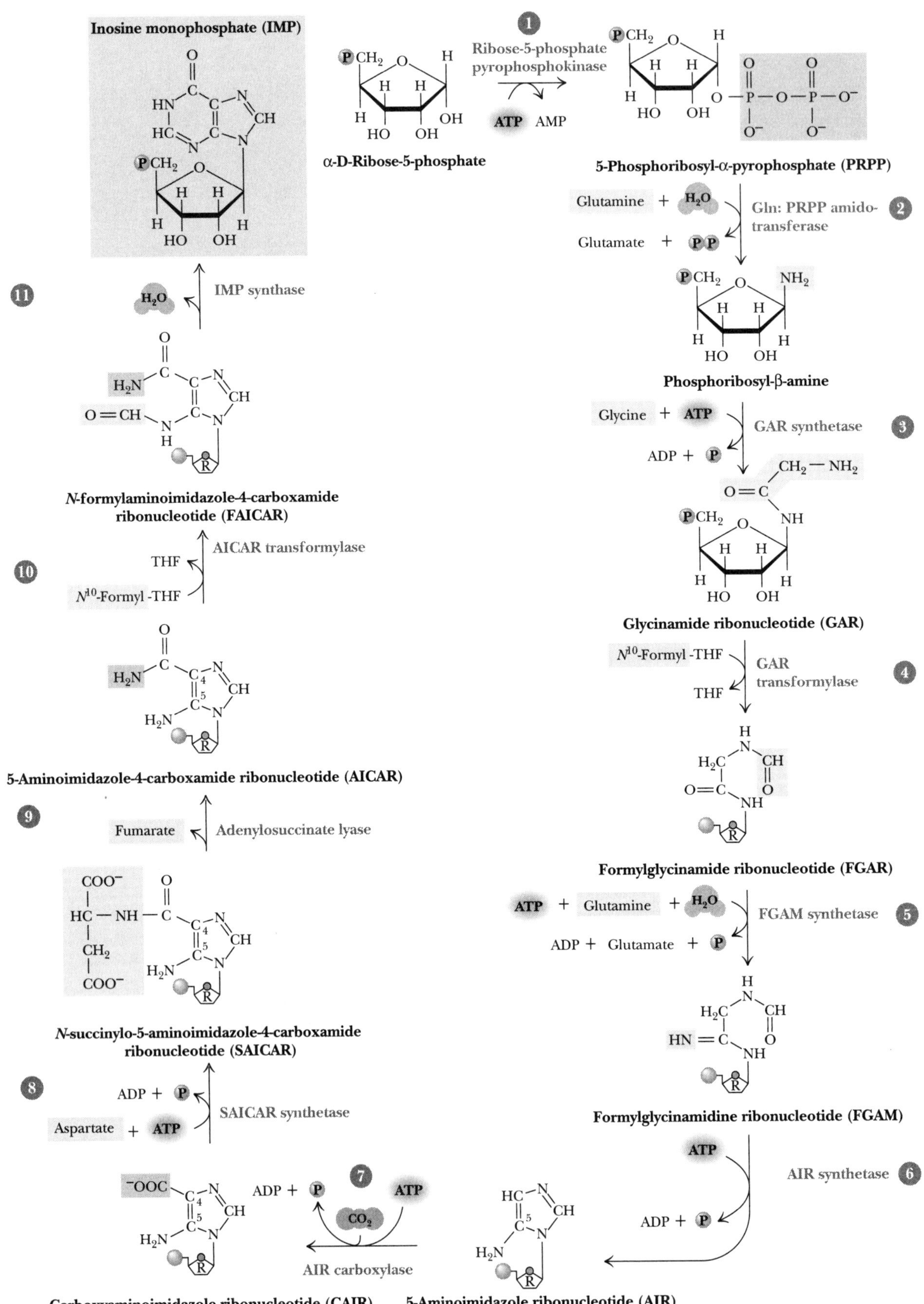

Figure 21.28 The synthesis of AMP and GMP from IMP.

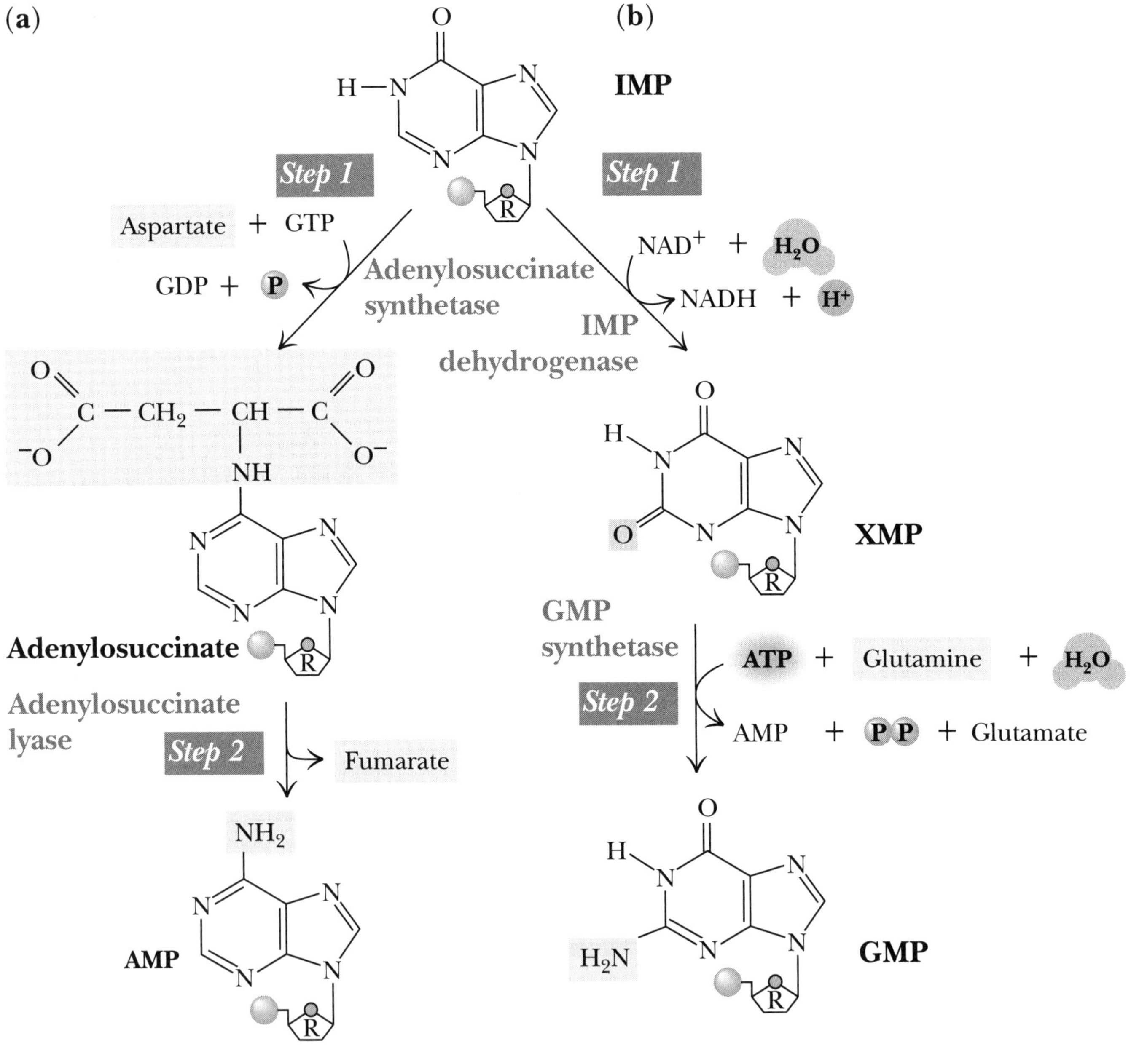

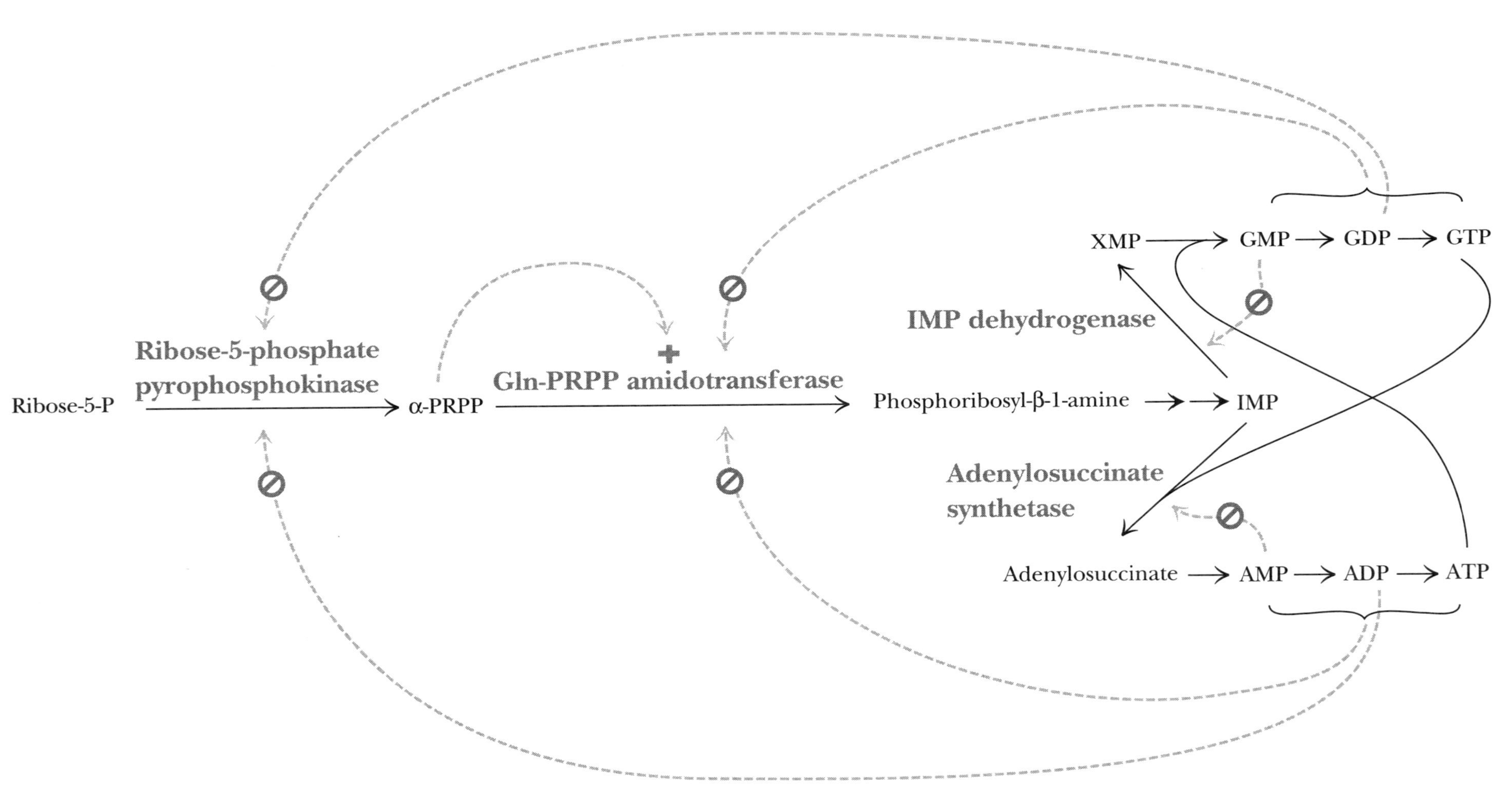

Figure 21.29   The regulatory circuit controlling purine biosynthesis.

Figure 21.31 The major pathways for purine catabolism in animals.

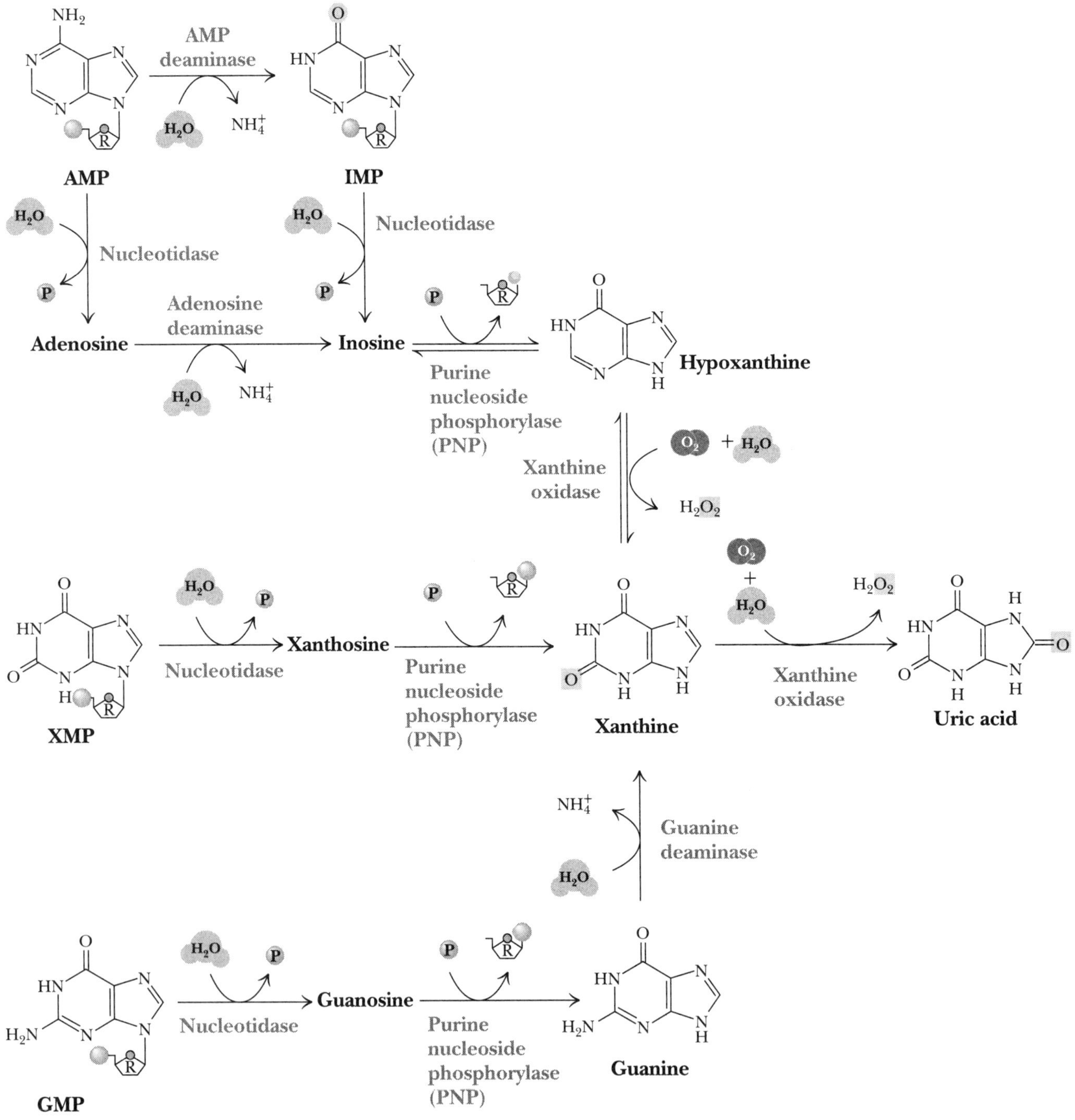

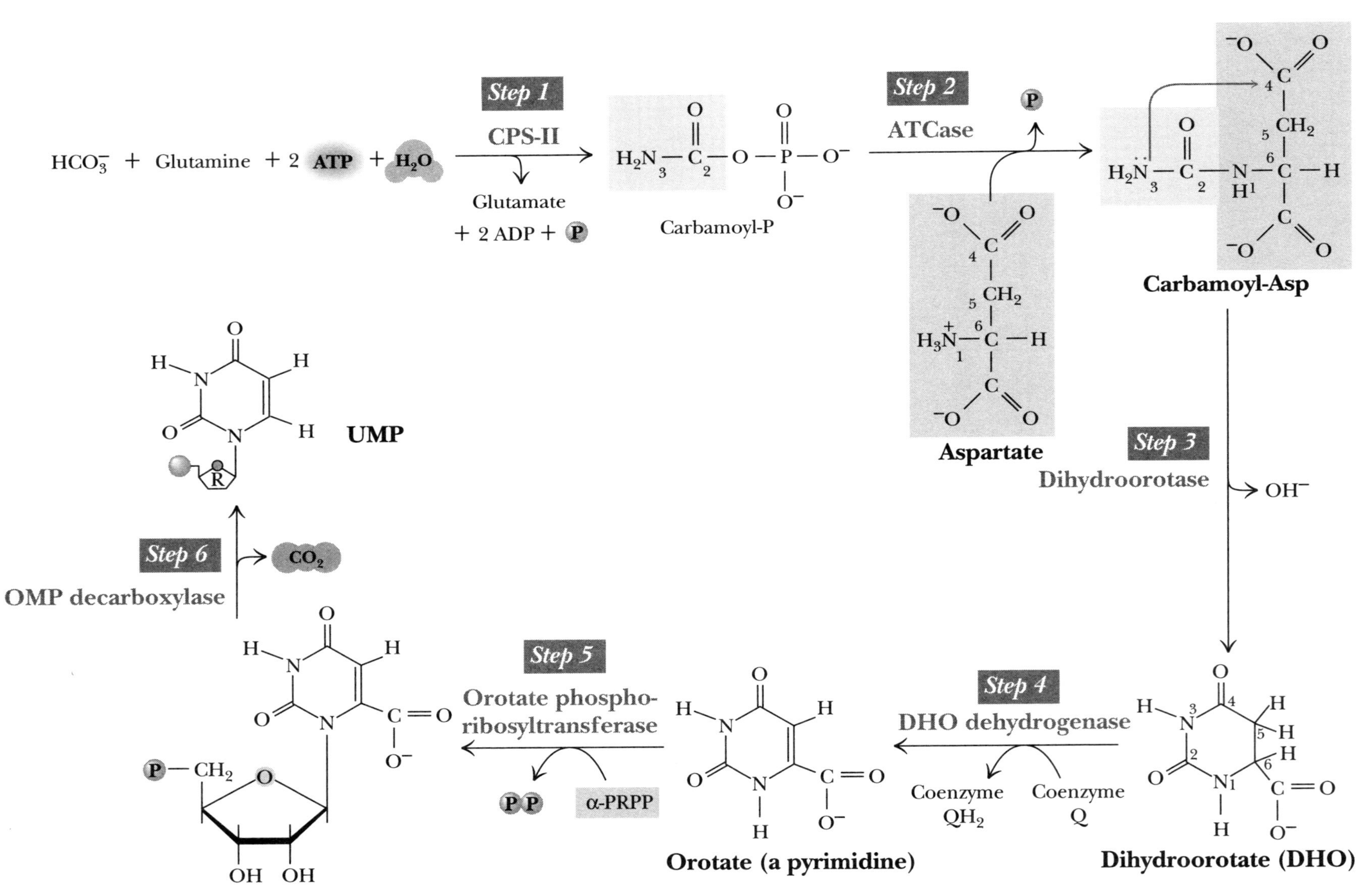

Figure 21.36   The *de novo* pyrimidine biosynthetic pathway.

© Harcourt, Inc.

358

Figure 21.41 *E. Coli* ribonucleotide reductase.

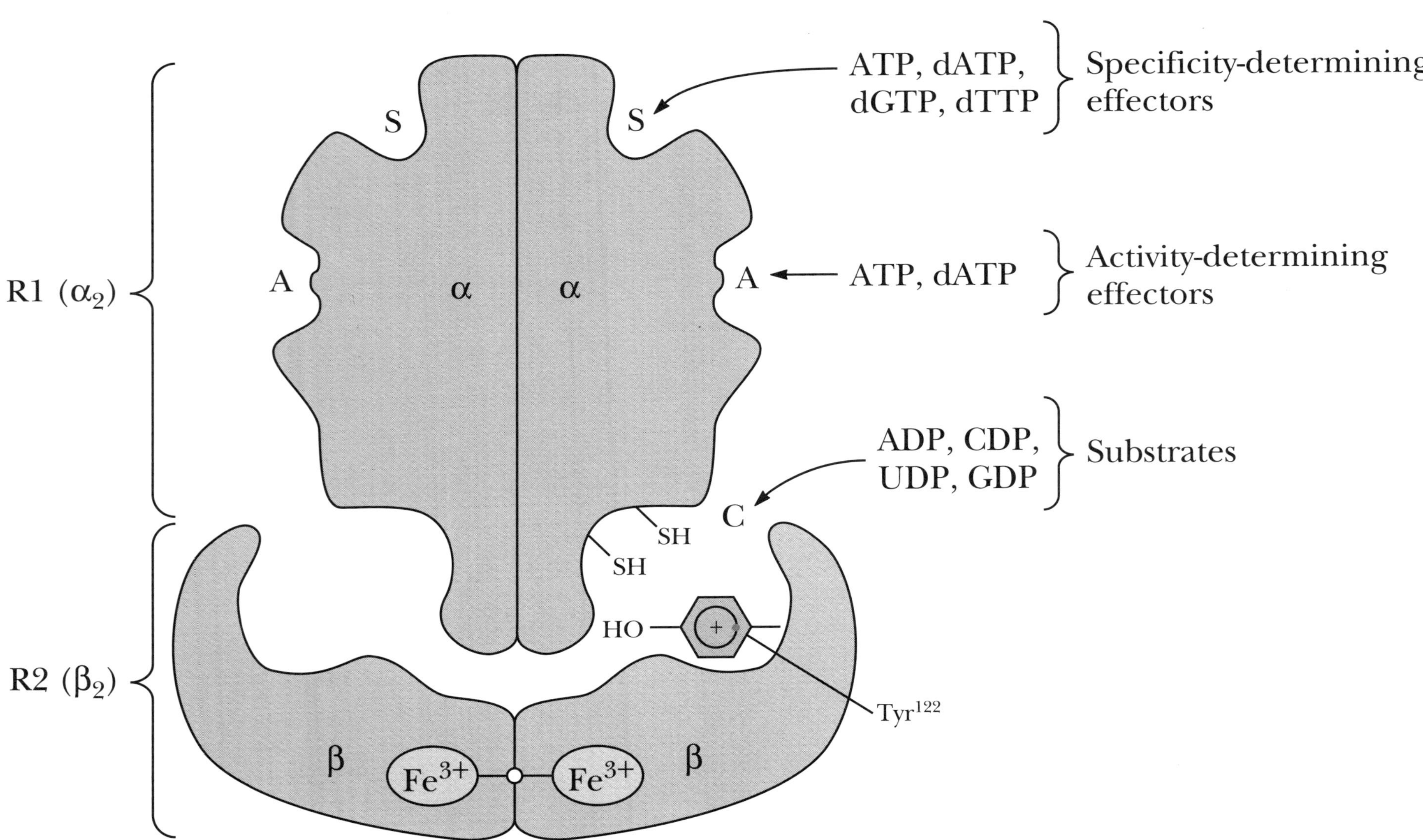

© Harcourt, Inc.

Figure 21.48 The thymidylate synthase reaction.

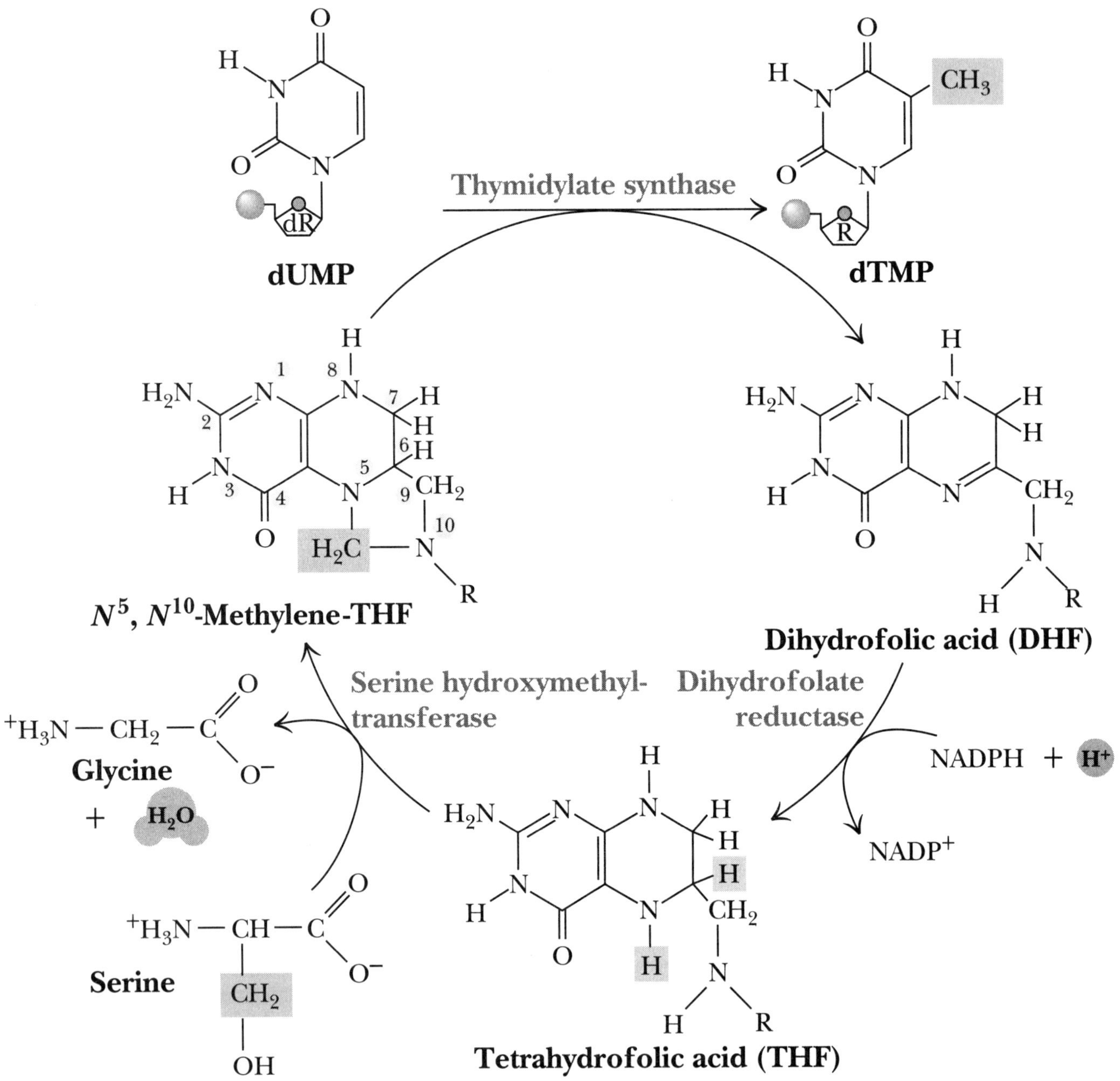

dUMP
Thymidylate synthase
dTMP
CH3

N^5, N^{10}-Methylene-THF

Dihydrofolic acid (DHF)

Serine hydroxymethyl-transferase
Dihydrofolate reductase

Glycine
H2O
Serine

NADPH + H+
NADP+

Tetrahydrofolic acid (THF)

362
© Harcourt, Inc.

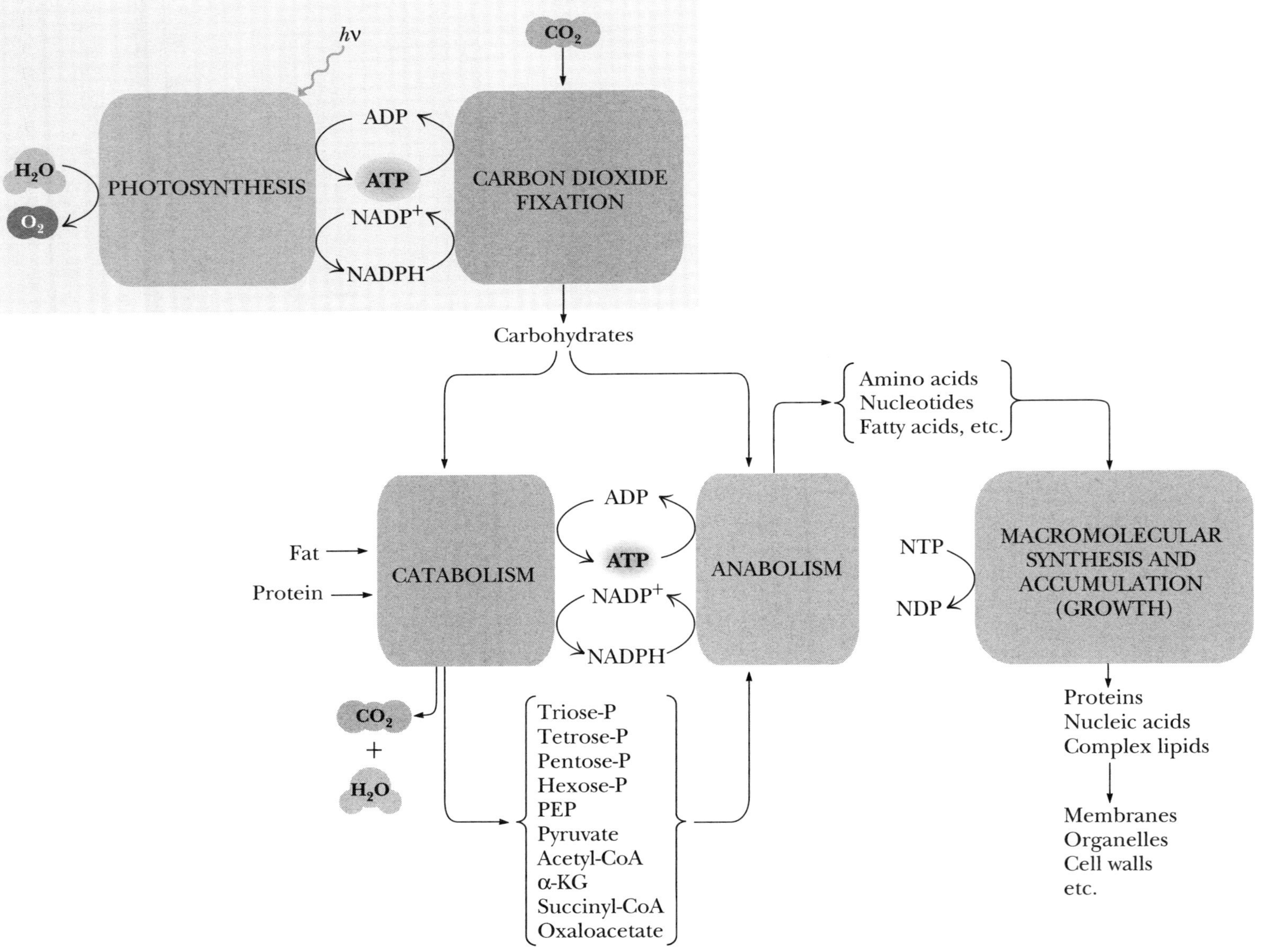

Figure 22.1   A block diagram of intermediary metabolism.

Figure 22.5 Metabolic relationships among the major human organs.

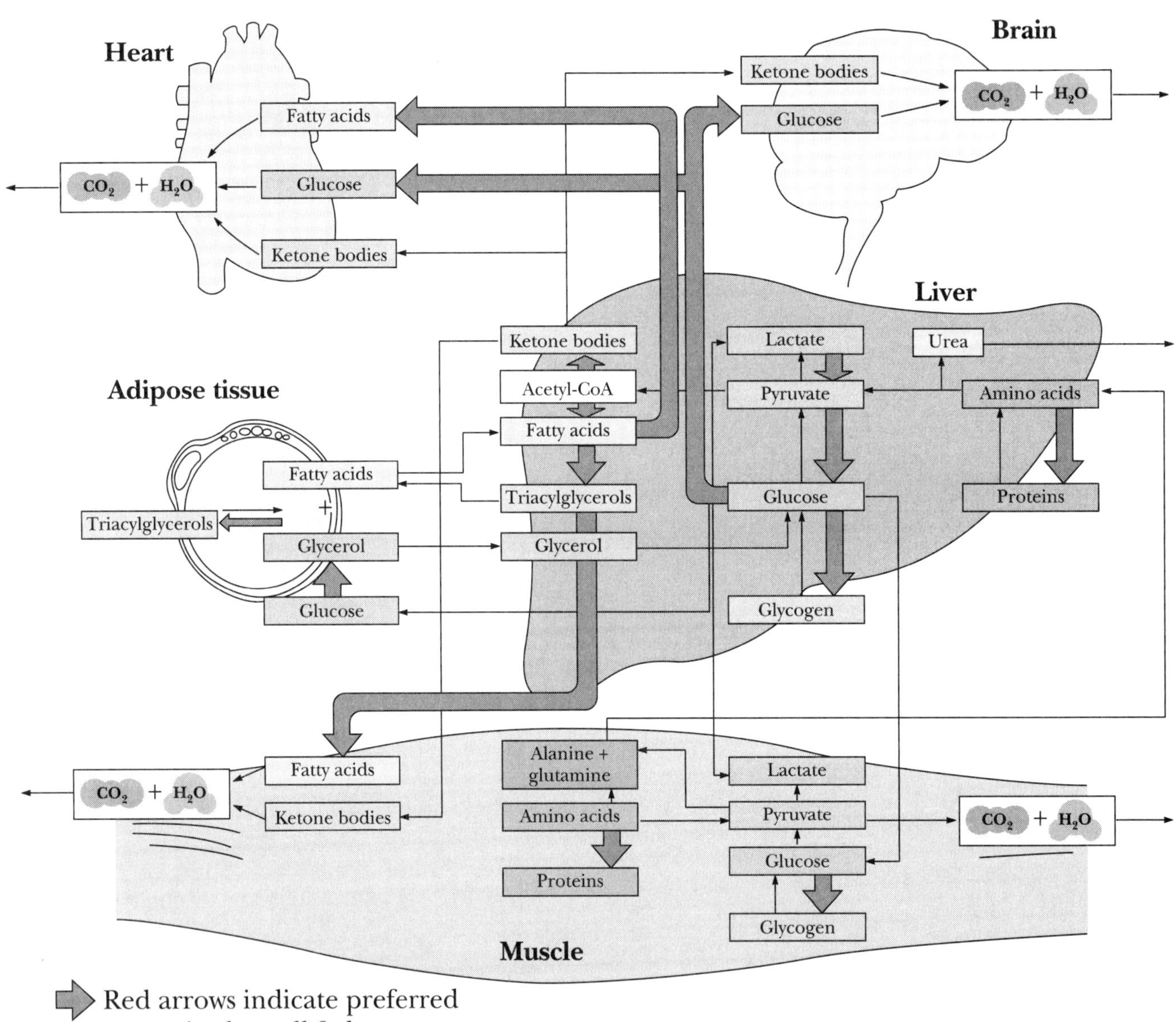

Figure 23.4 The Meselson and Stahl experiment.

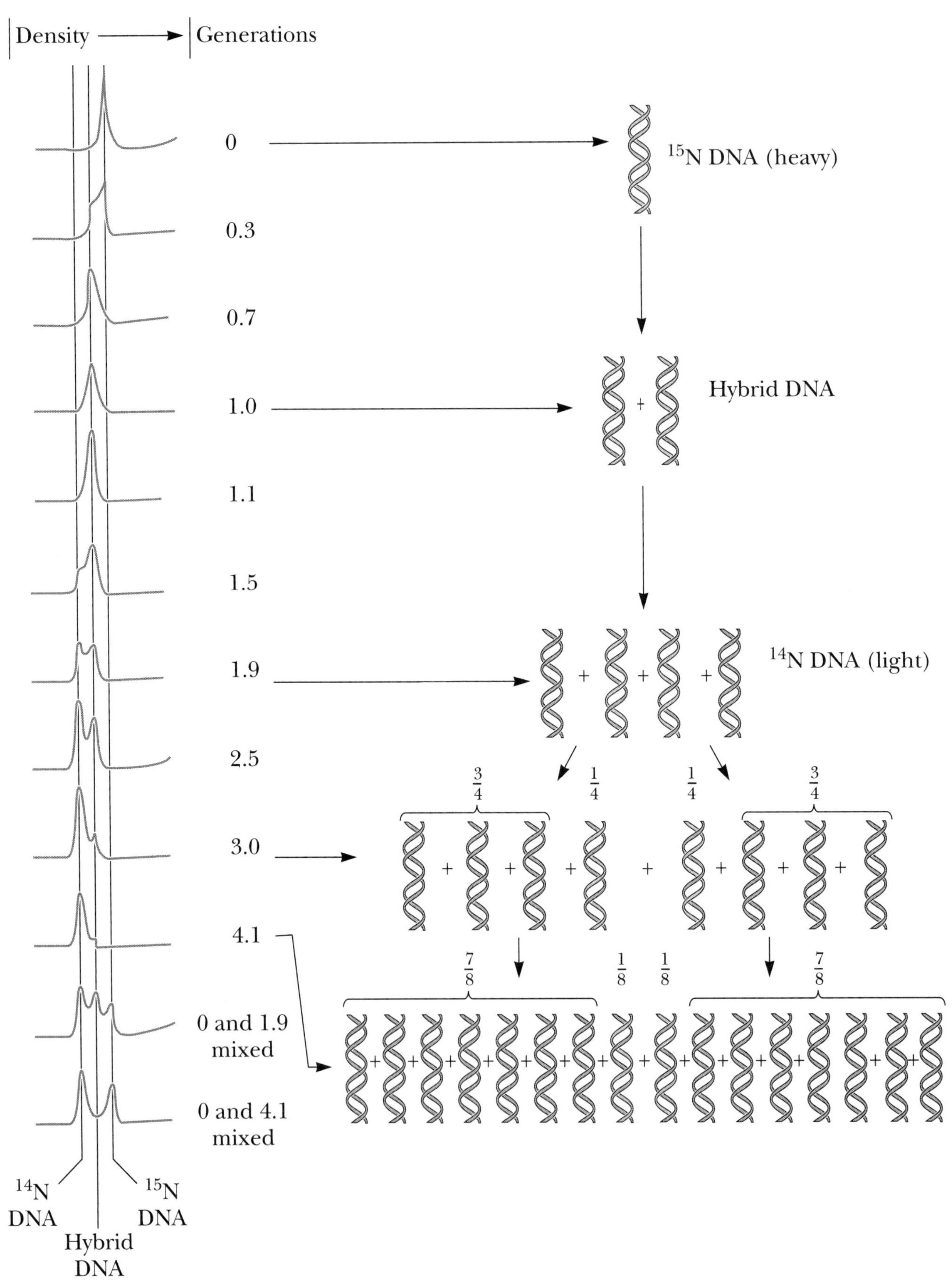

Figure 23.6 The semidiscontinuous model for DNA replication.

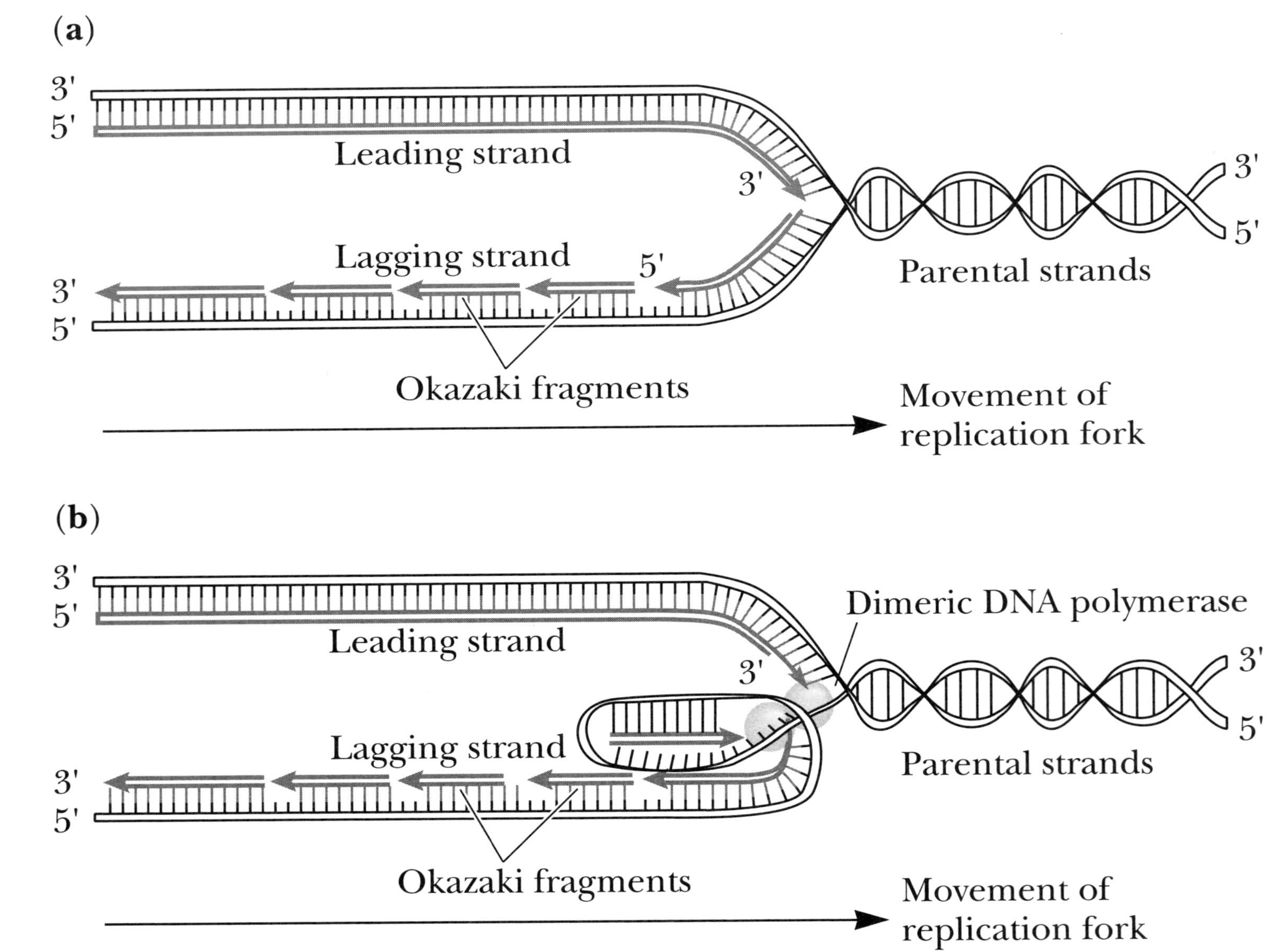

Figure 23.10   General features of the replication fork.

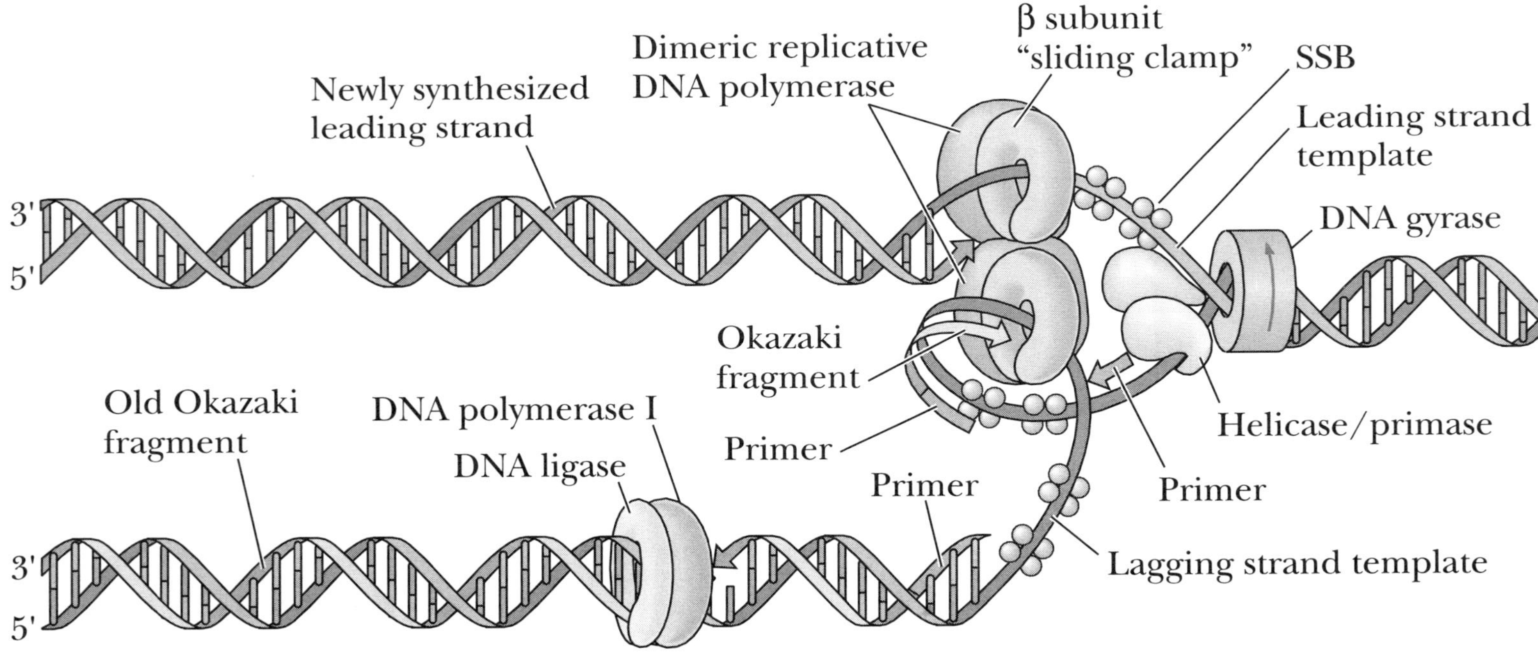

Figure 23.17   Meselson and Weigle's experiment.

Heavy phage ABC
Light phage abc
Infect bacteria with a mixture of heavy and light phage
Recombination occurs
Phage assembly and cell lysis
Phage progeny are centrifuged on CsCl gradient
Light phage abc
Recombinant phage Abc, aBc, abC, etc.
Heavy phage ABC

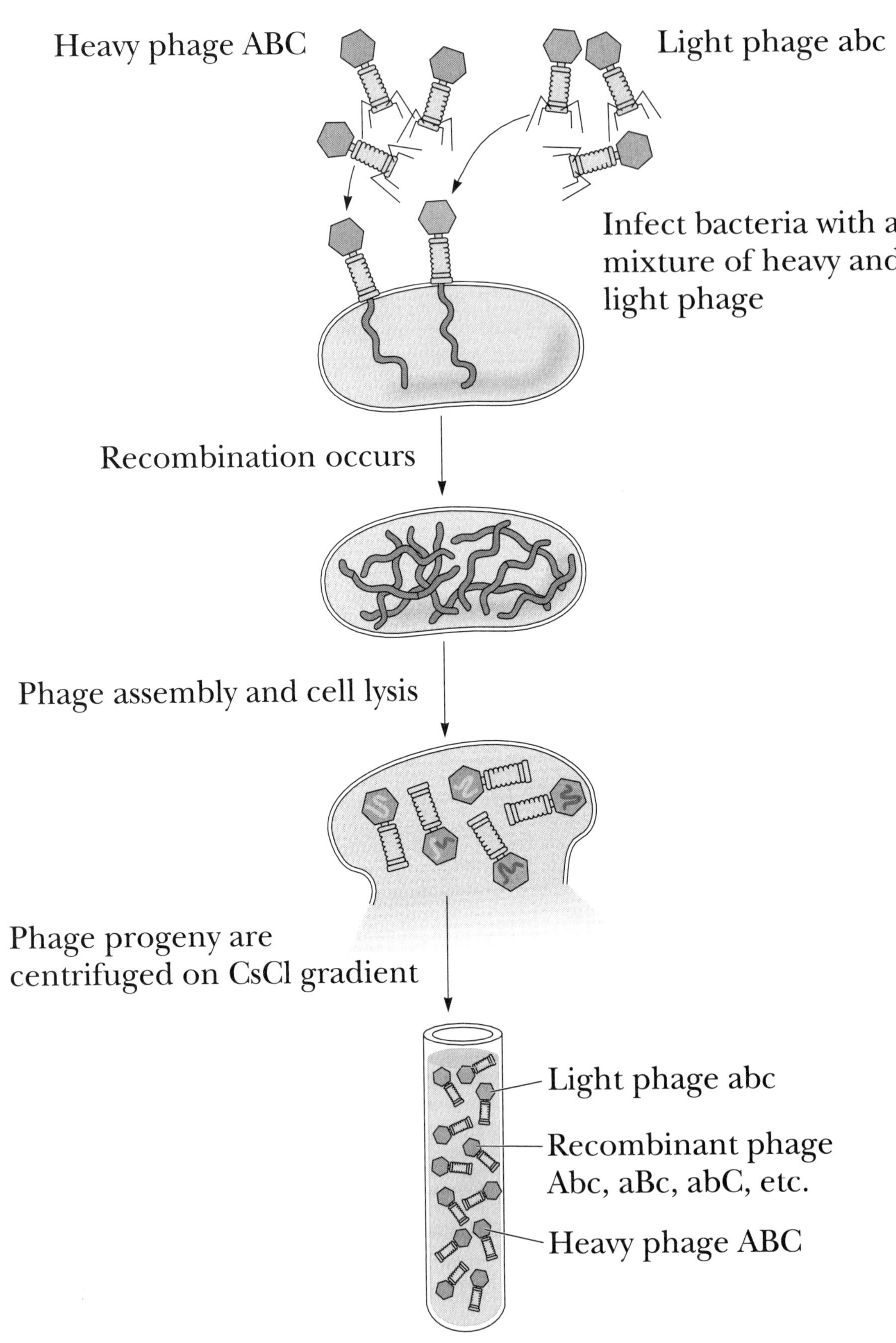

Figure 23.19   The Holliday model for homologous recombination.

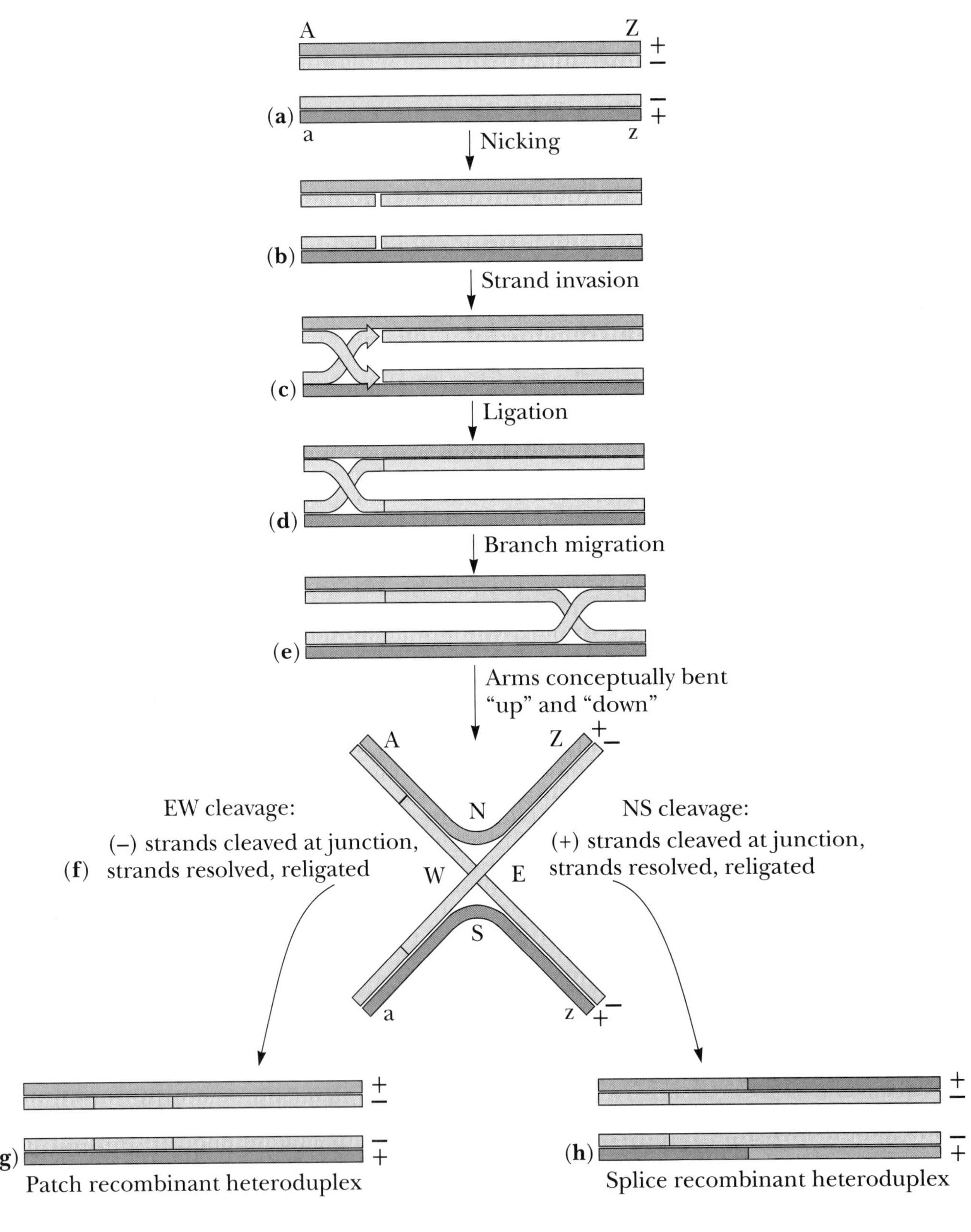

Figure 23.20   Model of RecBCD-dependent initiation of recombination.

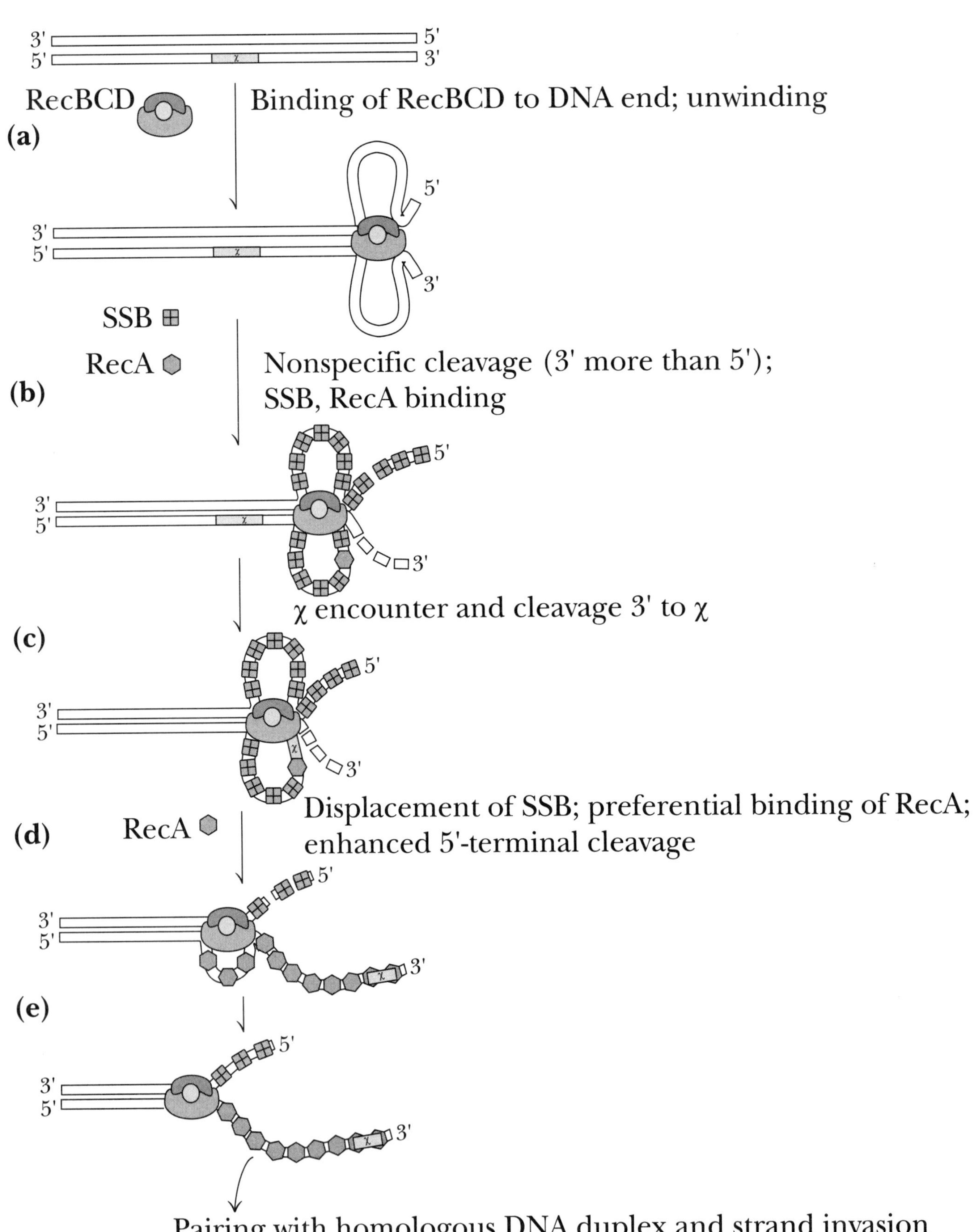

Figure 23.26 Base excision repair.

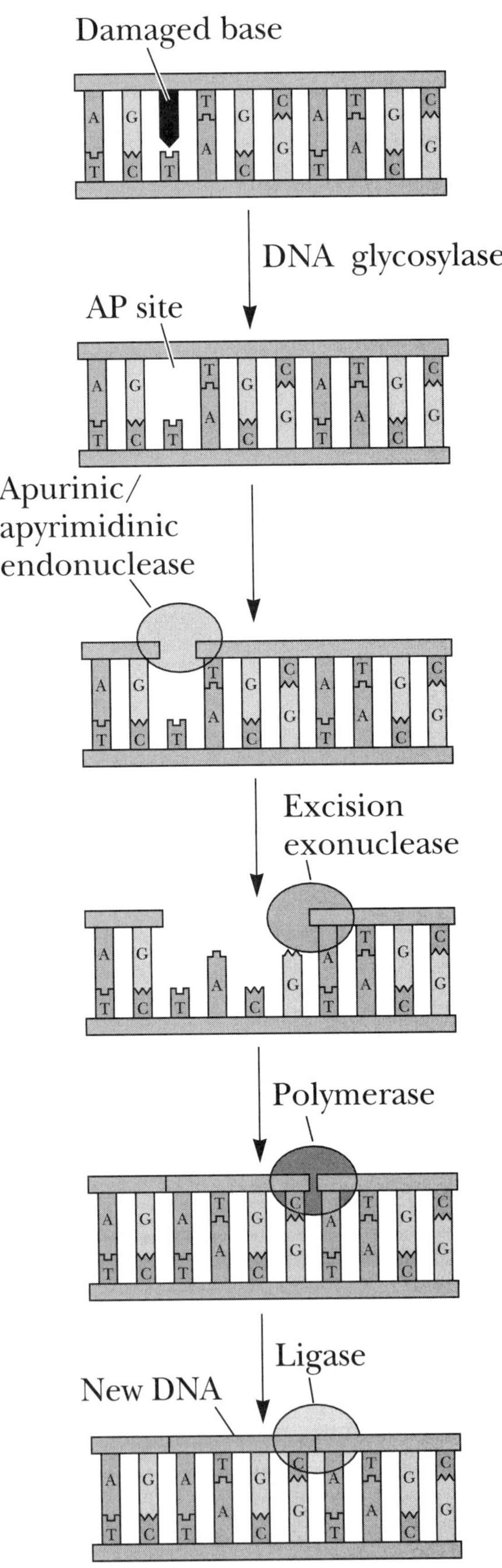

Figure 23.31a, b Chemical mutagens.

(a)

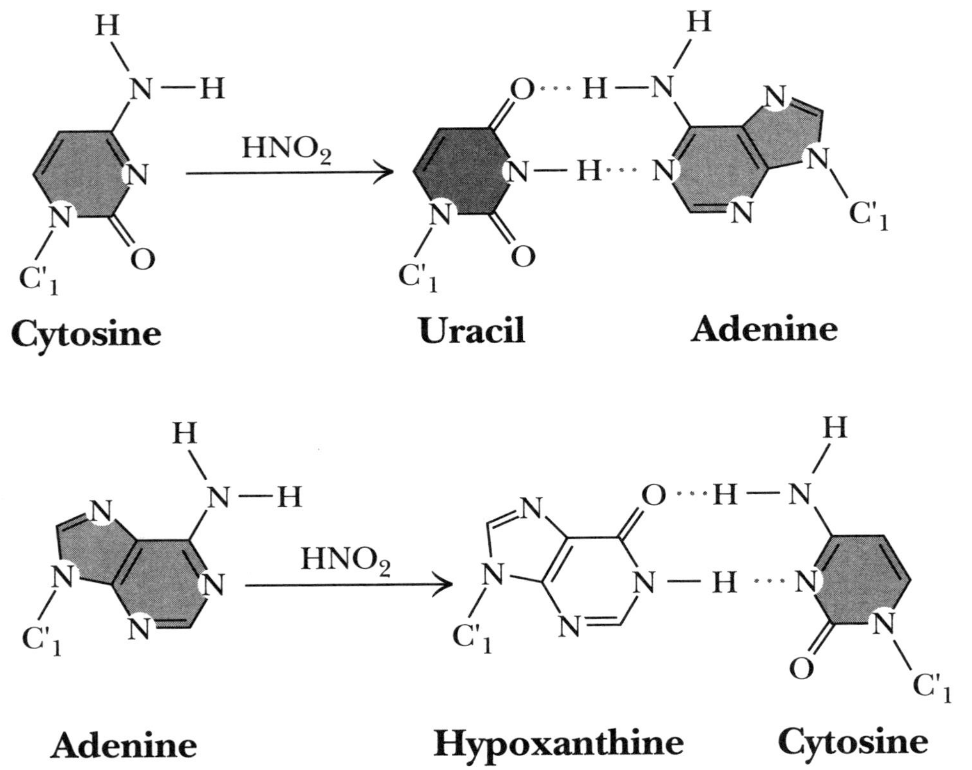

(b) Generic structure of nitrosoamines

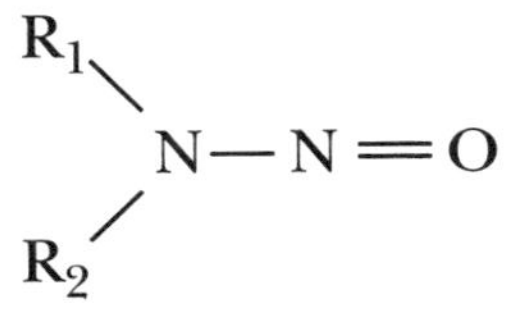

Figure 23.31c, d   Chemical mutagens.

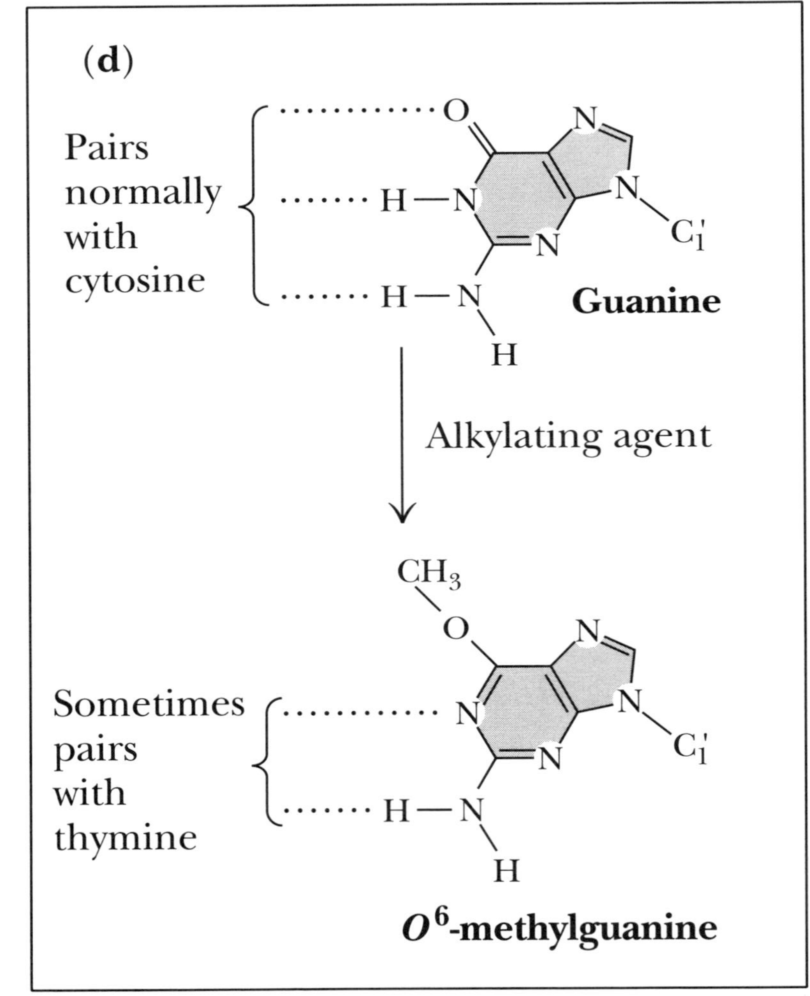

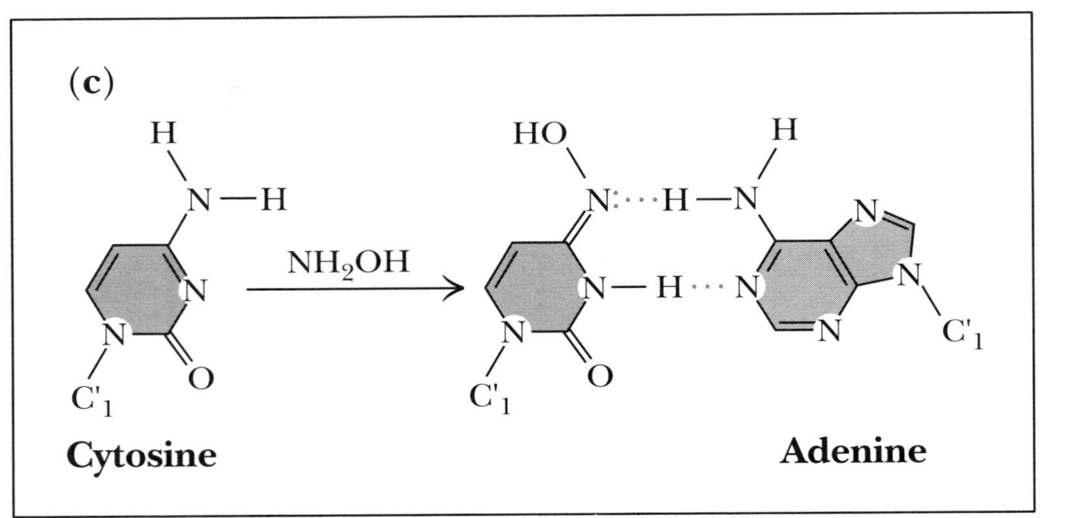

© Harcourt, Inc.

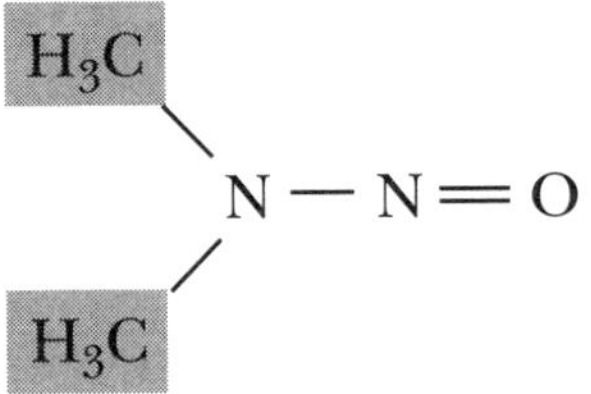

(e)
Nitrosoamines:
Dimethylnitrosoamine
Diethylnitrosoamine
H₃C
N — N = O
H₃C
CH₃CH₂
N — N = O
CH₃CH₂
Nitrosoguanidine:
CH₃
O = N — N
H
C — N
NO₂
NH
N-methyl-N'-nitro-
N-nitrosoguanidine
Nitrosourea:
O
CH₂CH₃
Ethyl nitrosourea
H₂N — C — N
N = O
Alkyl Sulfates:
Ethylmethane
sulfonate
O
H₃C — S — O — CH₂CH₃
O
Dimethyl
sulfate
O
H₃C — O — S — O — CH₃
O
Nitrogen Mustard:
CH₂ — CH₂ — Cl
H₃C — N
NH₂ — CH₂ — Cl

Figure 24.2 Sequence of events in the initiation and elongation phases of transcription in prokaryotes.

RNA pol
σ
Step 1
Recognition of promoter by σ. Binding of polymerase holoenzyme to DNA; migration to promoter
5'
3'
Promoter
3'
5'
DNA template
Step 2
Formation of an RNA polymerase: closed promoter complex
5'
3'
3'
5'
Step 3
Unwinding of DNA at promoter and formation of open promoter complex
5'
3'
3'
5'
Purine NTP
Step 4
RNA polymerase initiates mRNA synthesis, almost always with a purine
5'
3'
3'
5'
P P P Pu
NTPs
Step 5
RNA polymerase holoenzyme-catalyzed elongation of mRNA by about 4 more nucleotides
5'
3'
3'
5'
P P P Pu N N N N
σ
Step 6
Release of σ subunit as core RNA polymerase proceeds down the template, elongating RNA transcript
5'
3'
Promoter
3'
5'
P P P Pu N N N N
5'

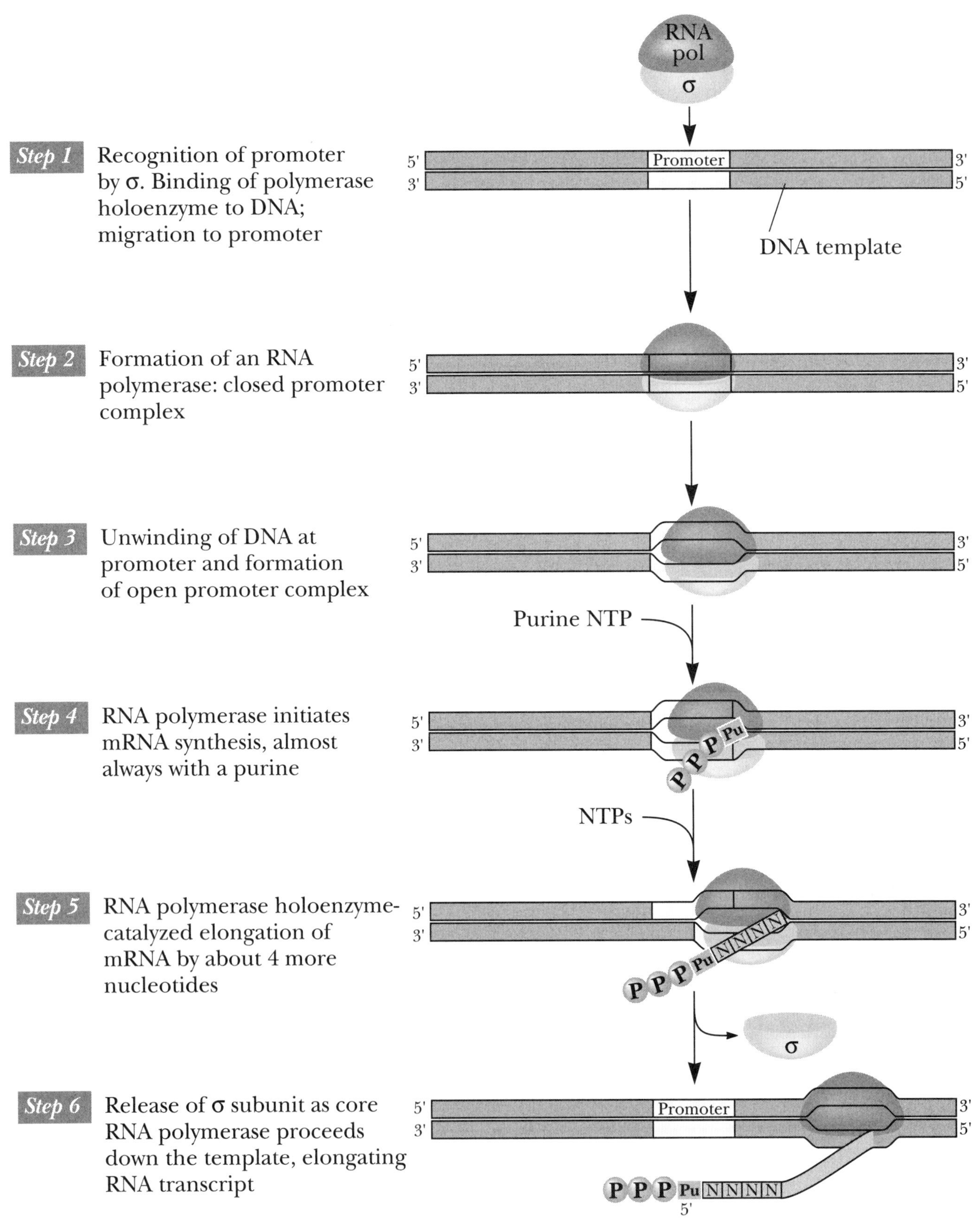

Figure 24.4   Supercoiling versus transcription.

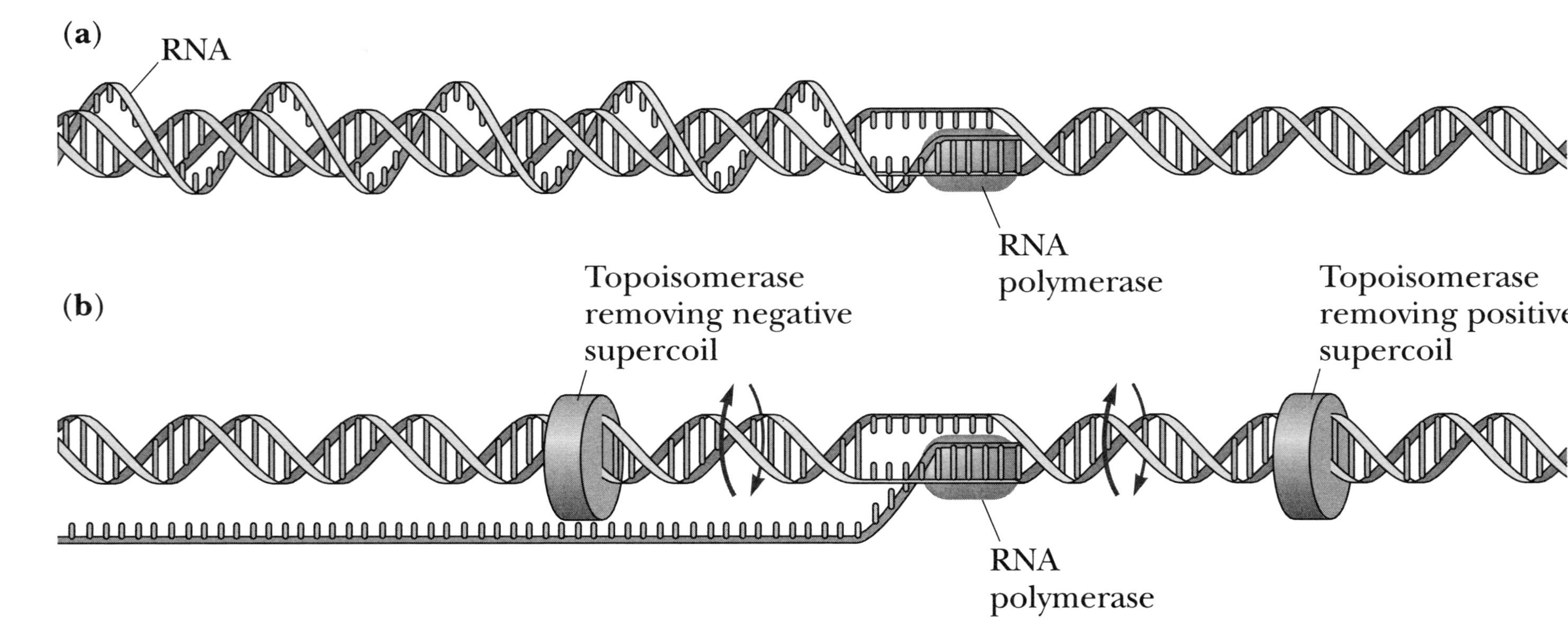

Figure 24.15   Control circuits governing the expression of genes.

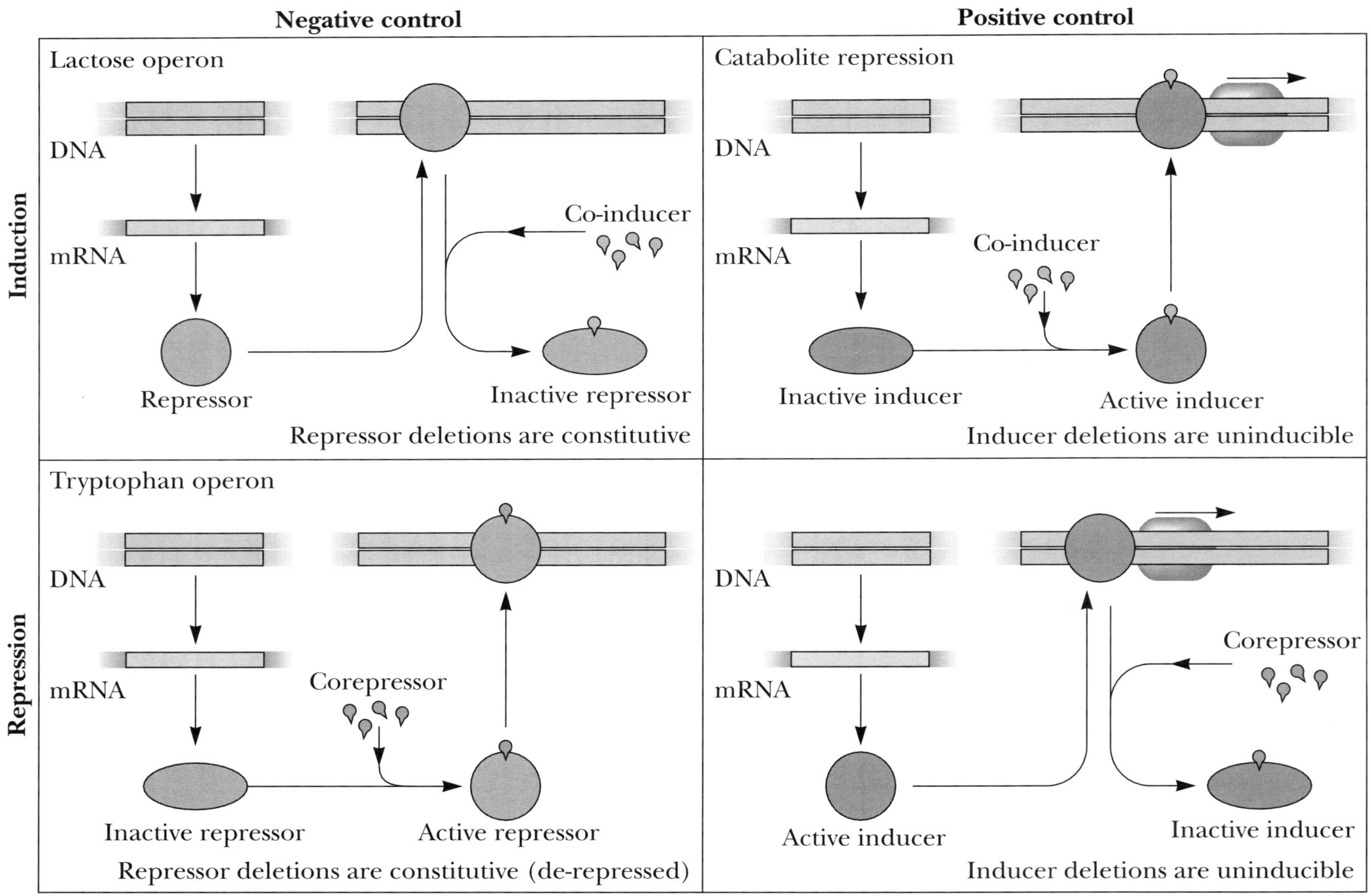

© Harcourt, Inc.

Table 25.1 The genetic code.

Table 25.1　The Genetic Code

First Position (5′-end)	Second Position				Third Position (3′-end)
	U	C	A	G	
U	UUU Phe	UCU Ser	UAU Tyr	UGU Cys	U
	UUC Phe	UCC Ser	UAC Tyr	UGC Cys	C
	UUA Leu	UCA Ser	UAA Stop	UGA Stop	A
	UUG Leu	UCG Ser	UAG Stop	UGG Trp	G
C	CUU Leu	CCU Pro	CAU His	CGU Arg	U
	CUC Leu	CCC Pro	CAC His	CGC Arg	C
	CUA Leu	CCA Pro	CAA Gln	CGA Arg	A
	CUG Leu	CCG Pro	CAG Gln	CGG Arg	G
A	AUU Ile	ACU Thr	AAU Asn	AGU Ser	U
	AUC Ile	ACC Thr	AAC Asn	AGC Ser	C
	AUA Ile	ACA Thr	AAA Lys	AGA Arg	A
	AUG Met*	ACG Thr	AAG Lys	AGG Arg	G
G	GUU Val	GCU Ala	GAU Asp	GGU Gly	U
	GUC Val	GCC Ala	GAC Asp	GGC Gly	C
	GUA Val	GCA Ala	GAA Glu	GGA Gly	A
	GUG Val	GCG Ala	GAG Glu	GGG Gly	G

* AUG signals translation initiation as well as coding for Met residues.

Third-Base Degeneracy Is Color-Coded

Third-Base Relationship	Third Bases with Same Meaning	Number of Codons
Third base irrelevant	U, C, A, G	32 (8 families)
Purines	A or G	12 (6 pairs)
Pyrimidines	U or C	14 (7 pairs)
Three out of four	U, C, A	3 (AUX = Ile)
Unique definitions	G only	2 (AUG = Met) (UGG = Trp)
Unique definition	A only	1 (UGA = Stop)

Figure 25.3 The aminoacyl-tRNA synthetase reaction.

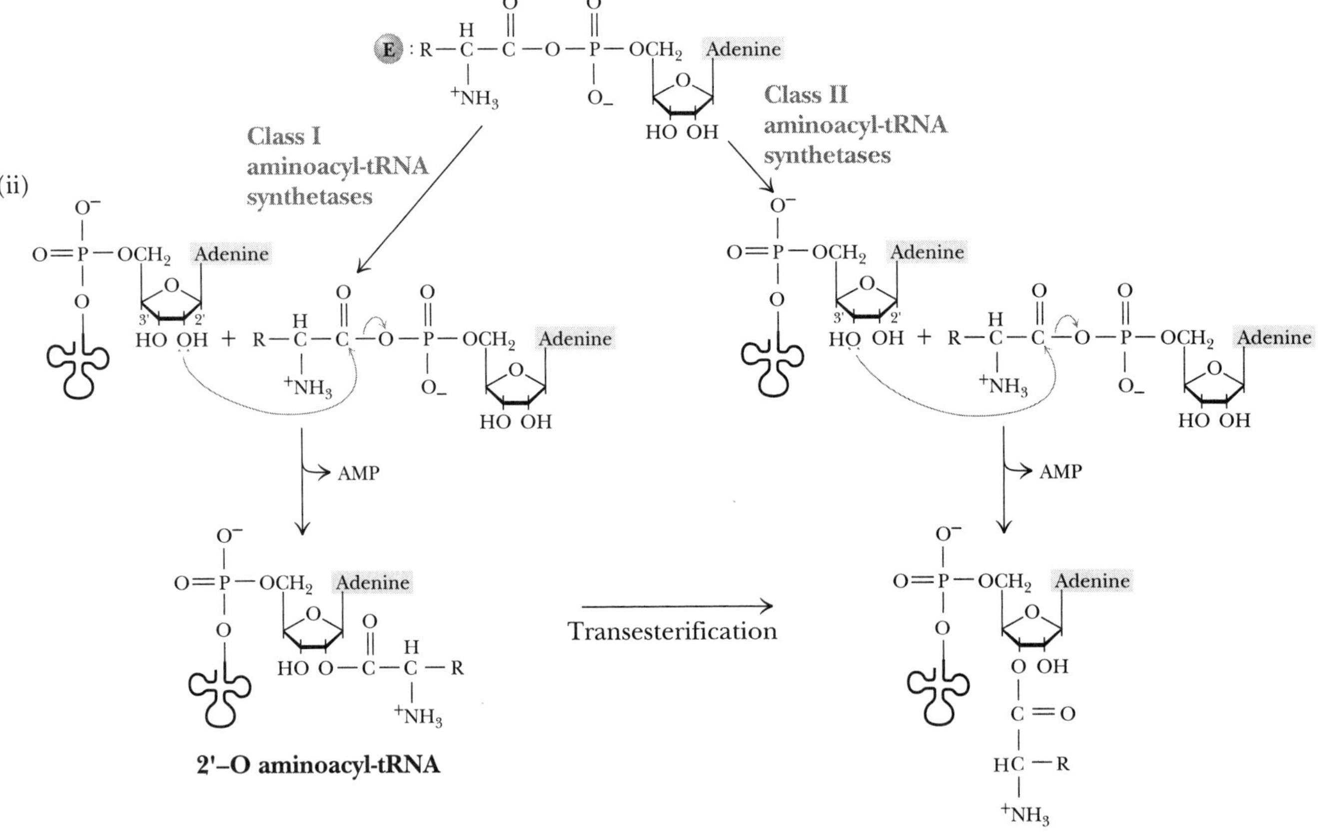

© Harcourt, Inc.

Figure 25.12   The basic steps in protein synthesis.

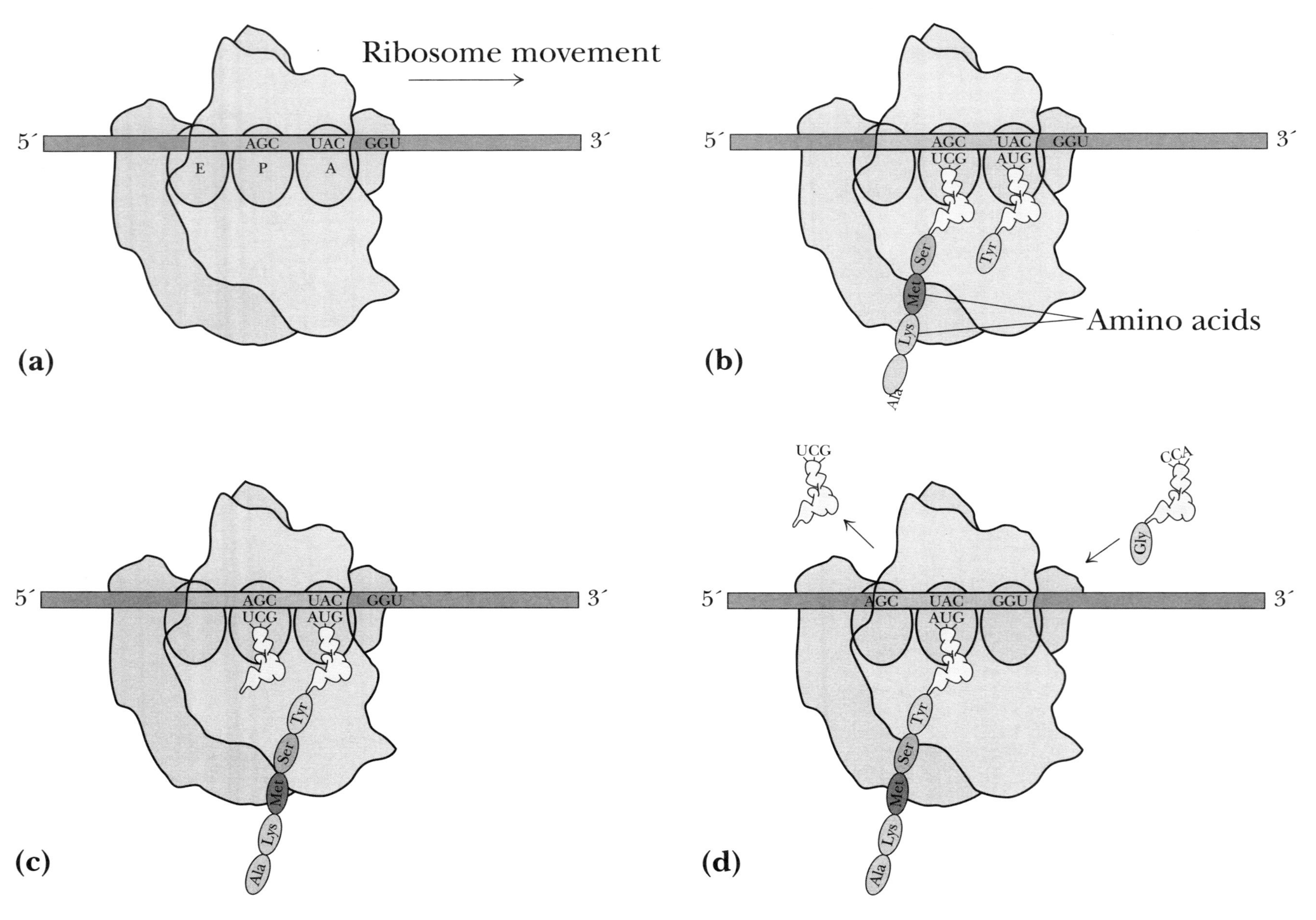

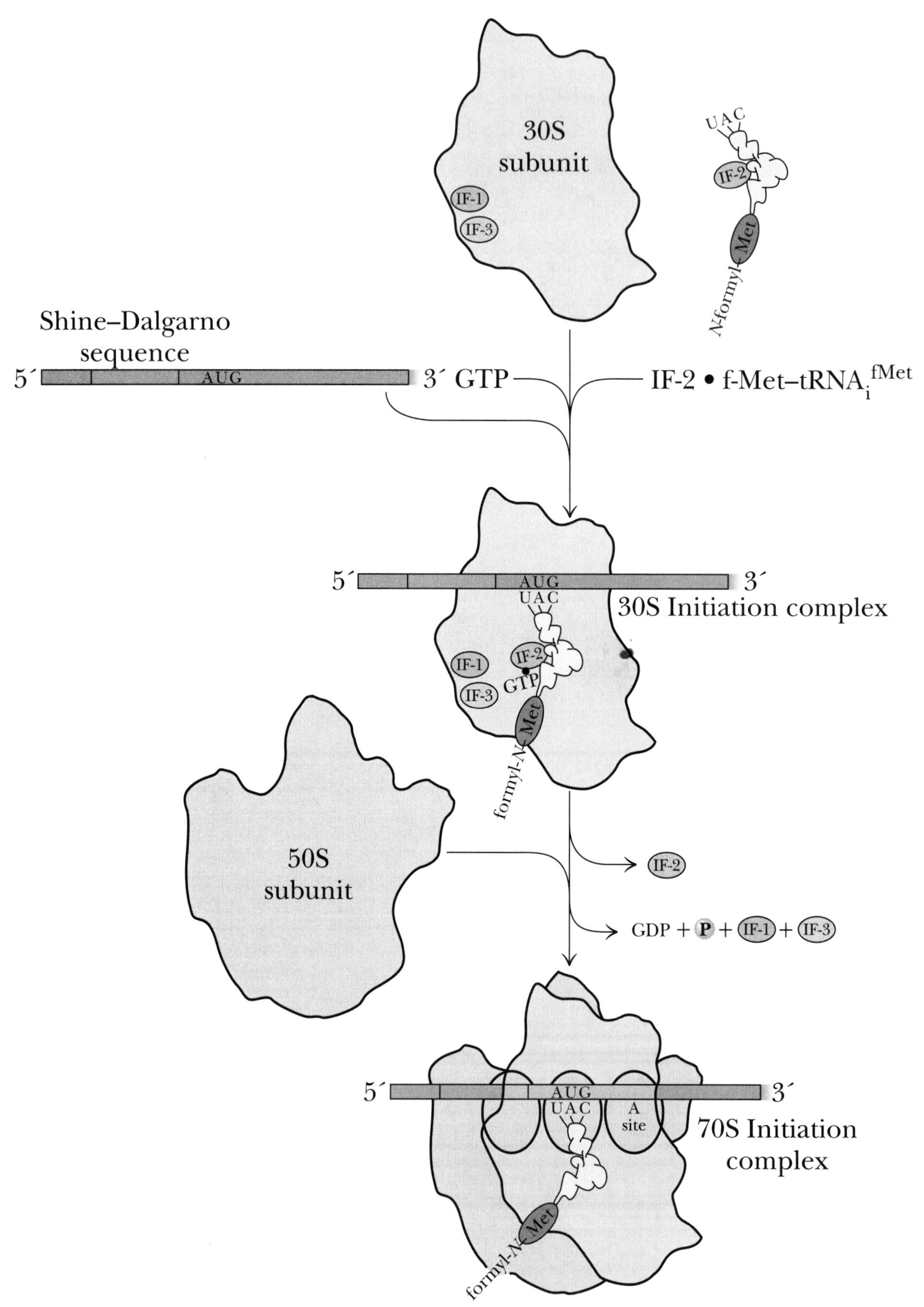

30S
subunit
IF-1
IF-3
UAC
IF-2
N-formyl-Met
Shine–Dalgarno
sequence
5′
AUG
3′ GTP
IF-2 • f-Met–tRNA$_i^{fMet}$
5′
AUG
UAC
3′
30S Initiation complex
IF-1
IF-2
IF-3
GTP
formyl-N-Met
50S
subunit
IF-2
GDP + P + IF-1 + IF-3
5′
AUG
UAC
A
site
3′
70S Initiation
complex
formyl-N-Met

Figure 25.18 The cycle of events in peptide chain elongation on *E. coli* ribosomes.

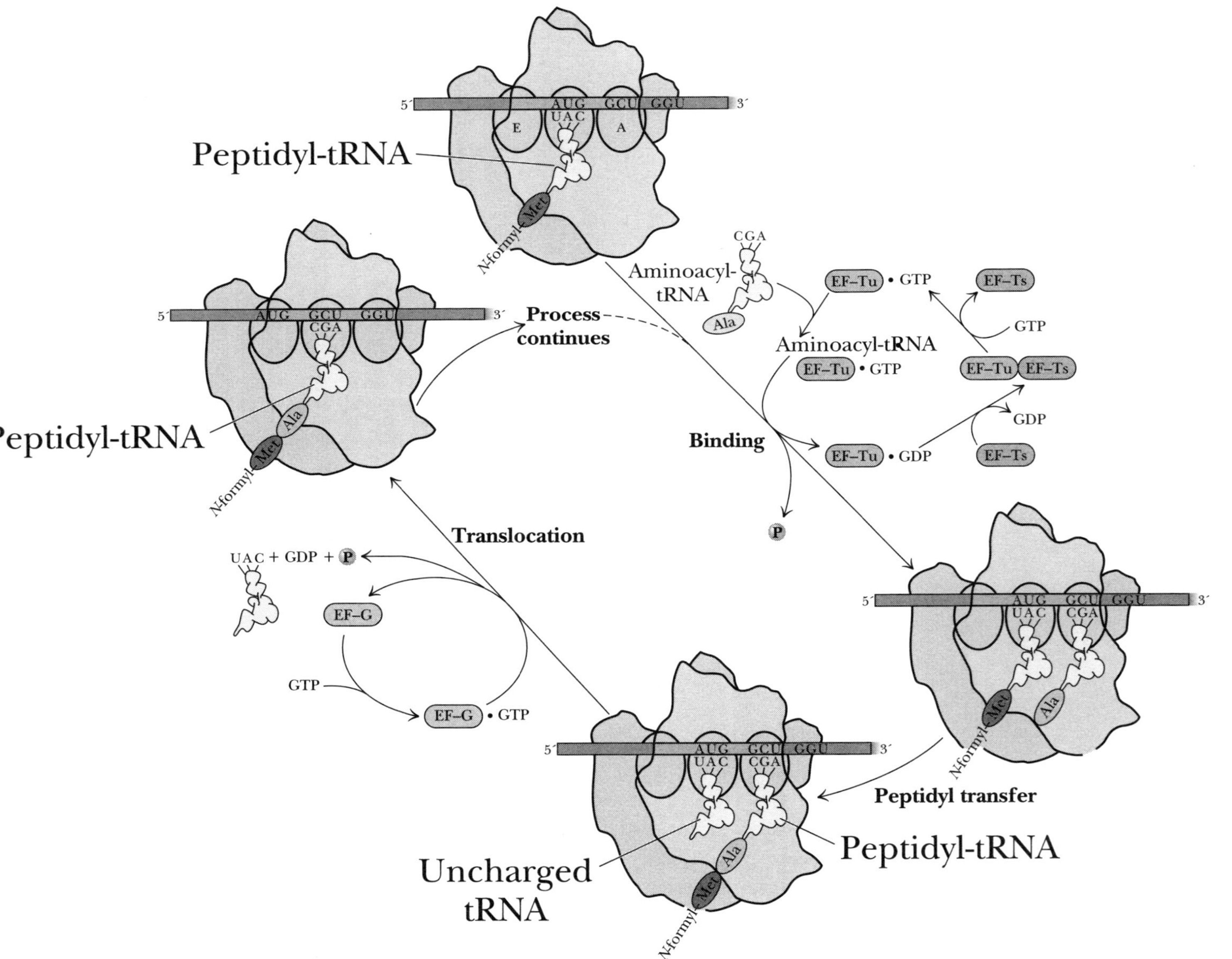

Figure 25.22 The events in peptide chain termination.

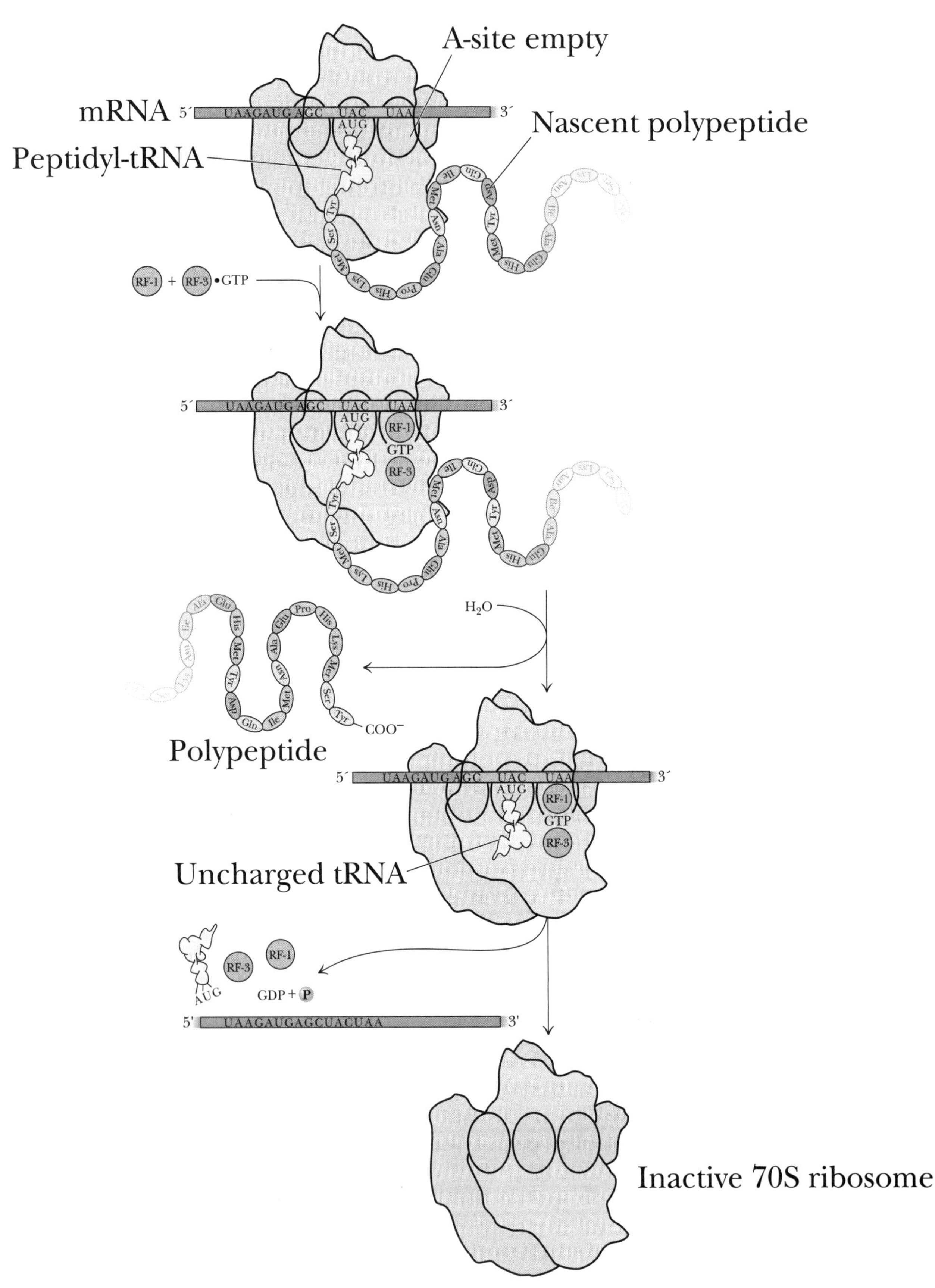

Figure 26.4 A complete signal transduction pathway that connects a hormone receptor with transcription events in the nucleus.

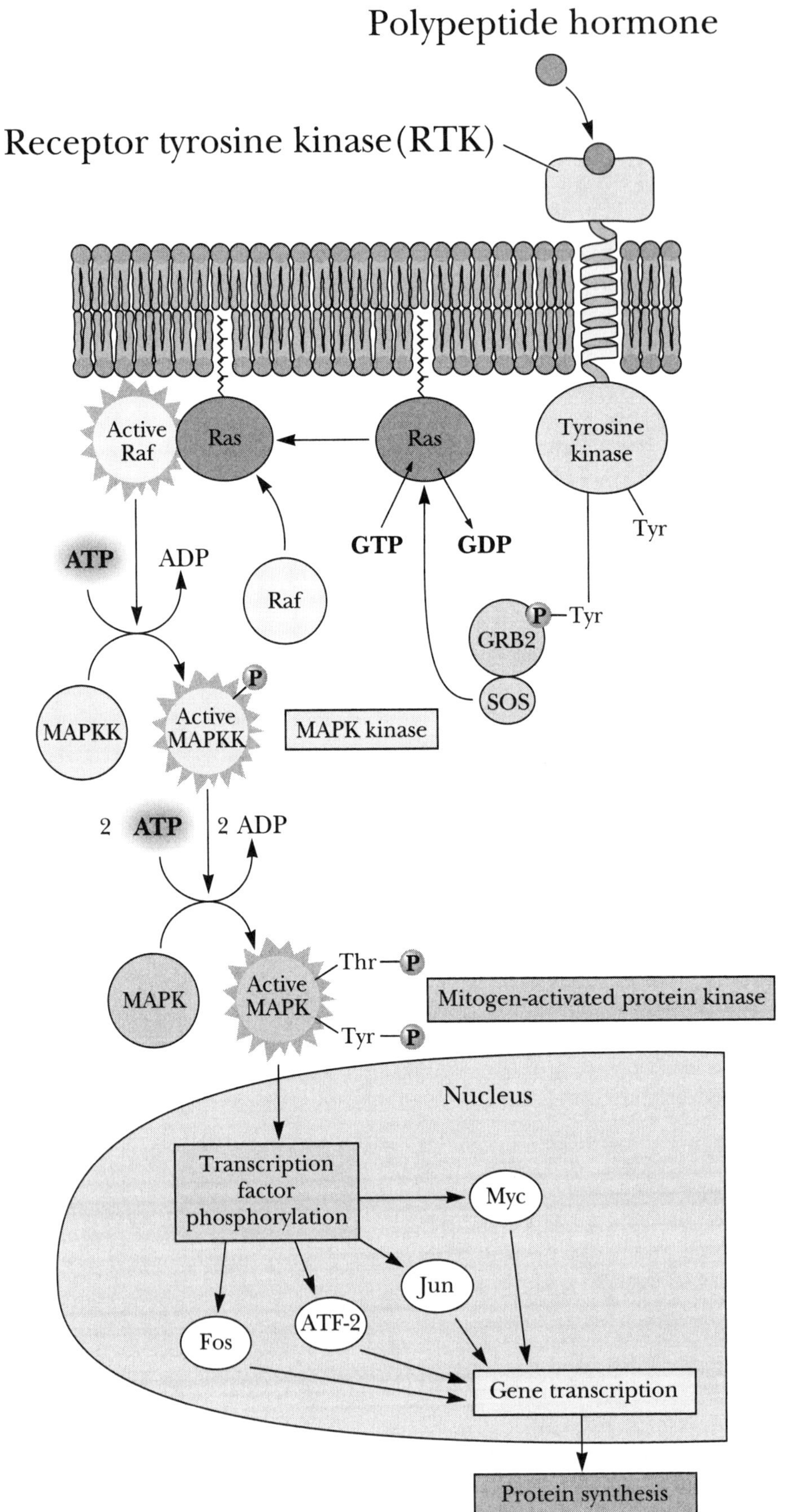

Figure 26.22   IP$_3$-mediated signal transduction pathways.

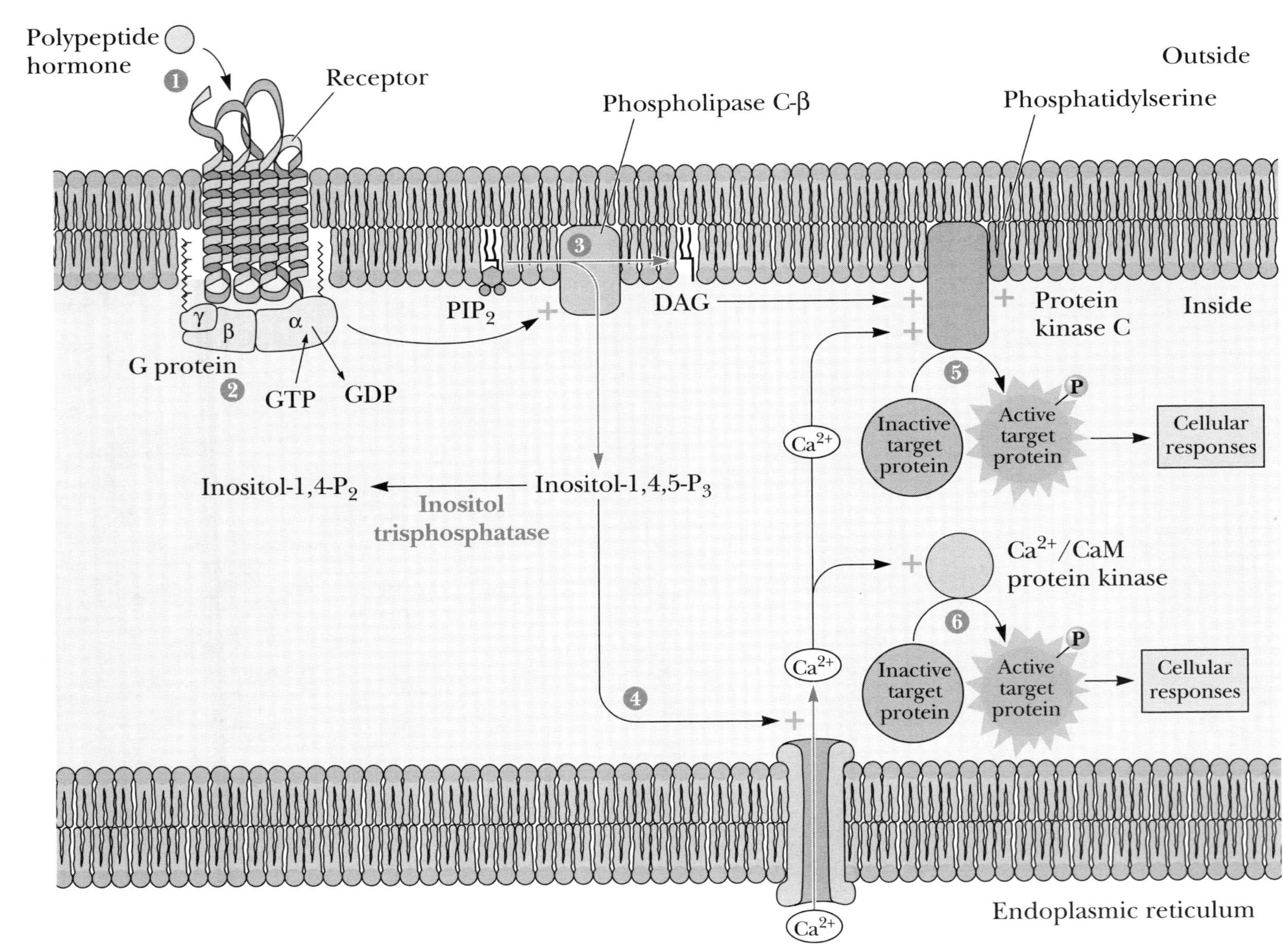

© Harcourt, Inc.

Figure 26.39   Cell-to-cell communication at the synapse.

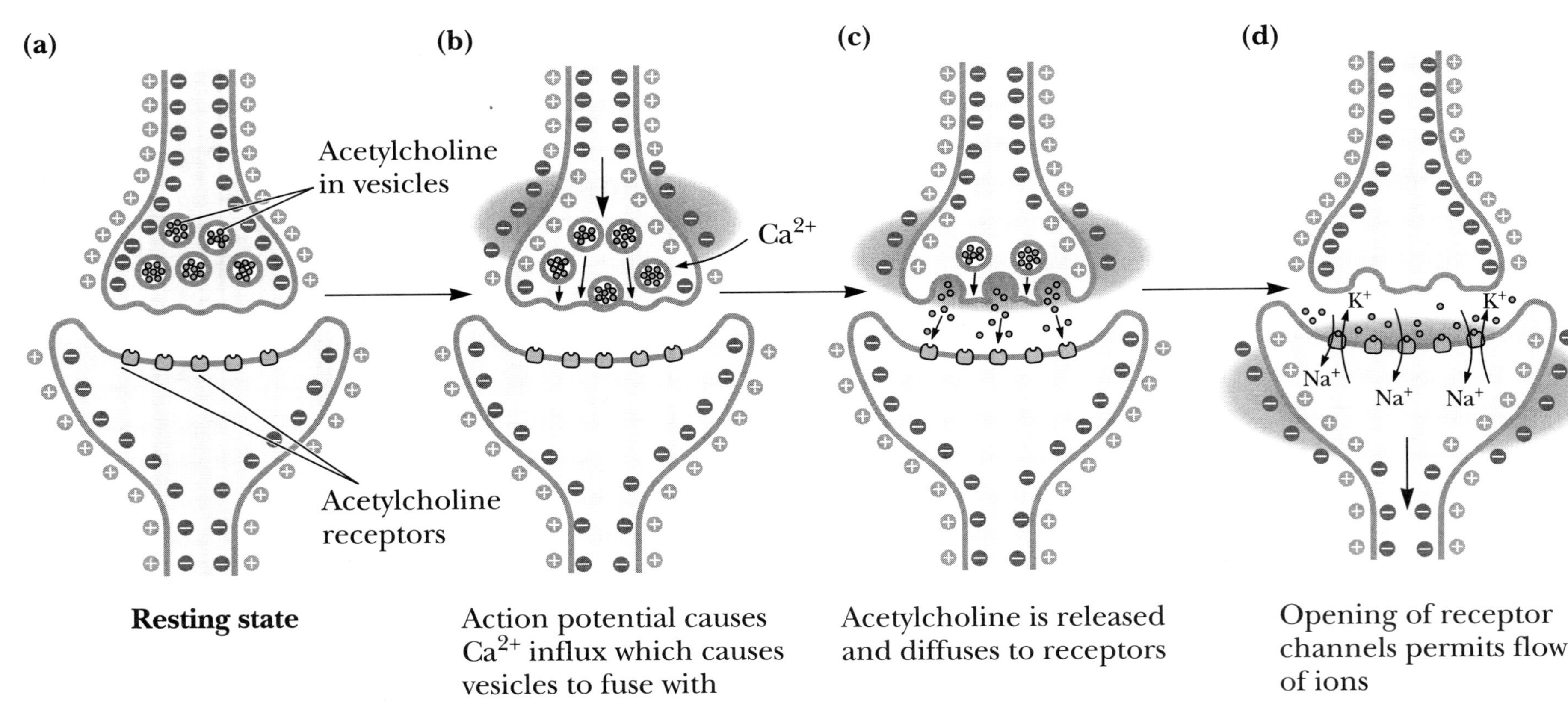